GRAPH THEORY AND MOBILE COMMUNICATIONS

ADVANCED SERIES IN ELECTRICAL AND COMPUTER ENGINEERING

The purpose of this series is to publish work of high quality by authors who are experts in their respective areas of electrical and computer engineering. Each volume contains the state-of-the-art coverage of a particular area, with emphasis throughout on practical applications. Sufficient introductory materials will ensure that a graduate and a professional engineer with some basic knowledge can benefit from it.

Published:

Vol. 23: *Graph Theory and Mobile Communications*
by Masakazu Sengoku

Vol. 22: *Silicon Systems for Wireless LAN*
edited by Zoran Stamenković, Gildas Leger and Alberto Bosio

Vol. 21: *Application of Omics, AI and Blockchain in Bioinformatics Research*
edited by Jeffrey J. P. Tsai and Ka-Lok Ng

Vol. 20: *Computational Methods with Applications in Bioinformatics Analysis*
edited by Jeffrey J. P. Tsai and Ka-Lok Ng

Vol. 18: *Broadband Matching: Theory and Implementations (Third Edition)*
by Wai-Kai Chen

Vol. 17: *Practical Signal Processing and Its Applications: With Solved Homework Problems*
by Sharad R. Laxpati and Vladimir Goncharoff

Vol. 16: *Design Techniques for Integrated CMOS Class-D Audio Amplifiers*
by Adrian I. Colli-Menchi, Miguel A. Rojas-Gonzalez and Edgar Sanchez-Sinencio

Vol. 15: *Active Network Analysis: Feedback Amplifier Theory (Second Edition)*
by Wai-Kai Chen

Vol. 14: *Linear Parameter-Varying System Identification: New Developments and Trends*
by Paulo Lopes dos Santos, Teresa Paula Azevedo Perdicoúlis, Carlo Novara, Jose A. Ramos and Daniel E. Rivera

Vol. 13: *Semiconductor Manufacturing Technology*
by C. S. Yoo

For the complete list of titles in this series, please visit
http://www.worldscientific.com/series/asece

Advanced Series in Electrical and Computer Engineering – Vol. 23

GRAPH THEORY AND MOBILE COMMUNICATIONS

Masakazu Sengoku
Graduate Institute for Entrepreneurial Studies, Japan

World Scientific

NEW JERSEY • LONDON • SINGAPORE • BEIJING • SHANGHAI • HONG KONG • TAIPEI • CHENNAI • TOKYO

Published by

World Scientific Publishing Co. Pte. Ltd.
5 Toh Tuck Link, Singapore 596224
USA office: 27 Warren Street, Suite 401-402, Hackensack, NJ 07601
UK office: 57 Shelton Street, Covent Garden, London WC2H 9HE

Library of Congress Cataloging-in-Publication Data
Names: Sengoku, Masakazu (Professor of engineering), author.
Title: Graph theory and mobile communications / Masakazu Sengoku, Graduate Institute for Entrepreneurial Studies, Japan.
Description: New Jersey : World Scientific, [2023] | Series: Advanced series in electrical and computer engineering ; vol. 23 | Includes bibliographical references and index.
Identifiers: LCCN 2022025581 | ISBN 9789811255298 (hardcover) | ISBN 9789811255304 (ebook for institutions) | ISBN 9789811255311 (ebook for individuals)
Subjects: LCSH: Mobile communication systems. | Graph theory.
Classification: LCC TK5103.2 .S463 2023 | DDC 621.384--dc23/eng/20220829
LC record available at https://lccn.loc.gov/2022025581

British Library Cataloguing-in-Publication Data
A catalogue record for this book is available from the British Library.

For any available supplementary material, please visit
https://www.worldscientific.com/worldscibooks/10.1142/12803#t=suppl

Desk Editors: Balasubramanian Shanmugam/Steven Patt

Typeset by Stallion Press
Email: enquiries@stallionpress.com

Advanced Series in Electrical and Computer Engineering – Vol. 23

GRAPH THEORY AND MOBILE COMMUNICATIONS

Masakazu Sengoku
Graduate Institute for Entrepreneurial Studies, Japan

World Scientific

NEW JERSEY • LONDON • SINGAPORE • BEIJING • SHANGHAI • HONG KONG • TAIPEI • CHENNAI • TOKYO

Published by

World Scientific Publishing Co. Pte. Ltd.
5 Toh Tuck Link, Singapore 596224
USA office: 27 Warren Street, Suite 401-402, Hackensack, NJ 07601
UK office: 57 Shelton Street, Covent Garden, London WC2H 9HE

Library of Congress Cataloging-in-Publication Data
Names: Sengoku, Masakazu (Professor of engineering), author.
Title: Graph theory and mobile communications / Masakazu Sengoku, Graduate Institute for Entrepreneurial Studies, Japan.
Description: New Jersey : World Scientific, [2023] | Series: Advanced series in electrical and computer engineering ; vol. 23 | Includes bibliographical references and index.
Identifiers: LCCN 2022025581 | ISBN 9789811255298 (hardcover) | ISBN 9789811255304 (ebook for institutions) | ISBN 9789811255311 (ebook for individuals)
Subjects: LCSH: Mobile communication systems. | Graph theory.
Classification: LCC TK5103.2 .S463 2023 | DDC 621.384--dc23/eng/20220829
LC record available at https://lccn.loc.gov/2022025581

British Library Cataloguing-in-Publication Data
A catalogue record for this book is available from the British Library.

For any available supplementary material, please visit
https://www.worldscientific.com/worldscibooks/10.1142/12803#t=suppl

Desk Editors: Balasubramanian Shanmugam/Steven Patt

Typeset by Stallion Press
Email: enquiries@stallionpress.com

Preface

There are various systems having a network structure. For example, the Internet, a communication network, an electrical network, a road network, a traffic network, a human network, and so forth. The field of studying this network structure is called graph theory. In this book, we will discuss from the basics of graph theory to its application and in particular, we will focus on the application of graph theory to mobile communications. A mobile communication connects a mobile terminal and a base station wirelessly, and the base station enables communication all over the world via a wired and satellite communication system. This means that the mobile communication system includes wire and wireless technologies, and also hardware part such as analog electric circuits, digital circuits and a software part such as computer algorithms. The systems of *1G* to *4G* of mobile communications in which G means generation have been developed as a tool for improving communication speed and mainly for communication between people. In *5G*, various things are connected to the network and the change of the way of communication are expected to develop new businesses. The characteristics of mobile communication systems are cellular systems. A cellular system is a system that can continuously communicate no matter how the mobile terminal moves. The shape of the cell has evolved from one-dimensional line, a two-dimensional plane to a three-dimensional space and, in $5G, 6G, \ldots$ and kG, the shape changes with time to a four-dimensional shape. We also discuss the application of graph theory to such multidimensional mobile communications.

In Chapters 1 and 2, we deal with introduction and the basic concept of graphs, respectively. In Chapters 3 and 4, we deal with subgraphs trees, cotrees and hybrid trees, and matrix and application of trees which are useful subgraphs and tools for main application area of graph theory, respectively. In Chapter 5, we deal with graphical views of electrical and electronic networks which are related to hardware part such as analog electric circuits, digital circuits in mobile communication systems. In Chapters 6 and 7, we deal with graphs and mobile communication systems in which Chapter 6 is concerned with location problems and Chapter 7 is concerned with cellular systems in mobile communication systems.

Although this book is a kind of monograph, it can also be used as a textbook for graduate students who are interested in applied graph theory to communication engineering, and it will be useful for finding future research themes. At the end of each chapter, a summary of the chapter and interesting challenges are given. Most of the existing books with titles such as network engineering deal with how-to materials such as the mechanism and usage of the Internet and communication networks. On the other hand, this book is based on the viewpoint of deeply studying how the network structure influences the performance of the corresponding system. In general, the performance of a system is determined by the network structure and the physical values associated with the network. Various network performances are determined only by physical values, determined only by the network structure, or determined by both. A deep exploration of which part of the network each performance is gained from is an essential process for understanding the system and creating new innovations. In that sense, clarifying this process in terms of the structure of the network is the important from an academic point of view and for creative innovation. This book uses many examples and figures for ease of understanding. The purpose of these examples and figures is to make it easier to understand and to have an image of responding to new research themes.

A graph is a kind of figure, and figures are closely related to geometry, so graph theory can be considered to belong to geometry. The symmetry of figures is beautiful, and the fluctuations and irregularities of geometric patterns are also attractive. Just as geometry has played a major role in the development of both the natural sciences and the humanities and social sciences, graph theory has applications in both the natural sciences and the humanities and social sciences including arts and business. The beauty of figures may be a major motivation for scientists, engineers and so on to study graph theory. It would be a great pleasure for me, if this book could serve as a strong motivation for such scientists, engineers and so on.

Masakazu Sengoku
Niigata, Japan

About the Author

Dr. Masakazu SENGOKU received the B.E. degree from Niigata University, Niigata Japan in 1967, and Ph.D. degree from Hokkaido University in 1972. In 1972, he joined as a Research Associate, Hokkaido University. In 1978, he was an Associate Professor, Niigata University and he was promoted to a Professor at Niigata University in 1989. He was Dean of Faculty of Engineering, Niigata University during 2003–2007. He was a Trustee and an Executive Vice President of Niigata University during 2008–2014, and since 2014 he has been a Professor emeritus of Niigata University. He was President of Graduate Institute for Entrepreneurial Studies during 2014–2022, and since 2022 he has been President emeritus and a Professor emeritus of Graduate Institute for Entrepreneurial Studies.

His research interests include Information and Communication Technology (ICT): especially mobile communications, and Network Engineering, and Applied Graph Theory. He received Best Paper Awards from IEICE (IEICE: The Institute of Electronics, Information and Communication Engineers) four times and IEEE conference Best Paper Award. He received the Achievement Award of IEICE in 2005 and Distinguished Achievement and Contributions Award of IEICE in 2009. He was a member of Editorial Board, ACM, URSI, *Wireless Networks*, Kluwer Academic Publishers, an editor of the *Journal of Communications and Networks* and a regional editor of *Journal of Circuits, Systems, and Computers*, World Scientific Publishing. He was the President of the Engineering Science Society, IEICE in 2003–2004. He is a Life Fellow member of IEEE, a Fellow/Honorary member of IEICE, and a Fellow/Honorary member of JSST.

Acknowledgments

First of all, I would like to thank Professor Wai-Kai Chen of the University of Illinois at Chicago for inviting me to write a book in World Sciences and for encouraging me. He was my academic advisor during my stay as a visiting scholar at the University of Illinois at Chicago. Since then, I am deeply grateful to him for all the valuable advice I have received.

Throughout my research life, I received various useful advice from the following people. I would like to express my sincere gratitude to these people.

Professor L. O. Chua (University of California at Berkeley), Dr. C. Y. Lee (Airtouch Communications, Inc.), Professor K. Thulasiraman (University of Oklahoma), Professor R. R. Chen (San Jose State University), Professor G. Setti (Politecnico di Torino), Professor K. Gong (Tsinghua University), Professor J. Yang (Tsinghua University), Professor D. Singh (IIT), Professor T. W. Rhee (Korea University), Professor S. S. Im (Hanyang University), Professor Y. J. Park (Hanyang University), Professor H. S. Hong (Inha University), Professor K. S. Kwak (Inha University), Professor D. K. Sung (KAIST), Dr. A. Motai (MIC), Dr. K. Okada (NICT), Dr. T. Aoki (NTT), Dr. M. Kuramoto(NTT), Dr. Y. Aoyama (NTT), Dr. H. Yasuda (NTT), Dr. K.. Mase (NTT), Dr. H. Watanabe (NEC), Dr. T. Ohtsuki (NEC), Dr. T. Tsuda, (Fujitsu), Dr. M. Aoyama (Fujitsu), President H. Ikeda (NSG), The late Professor M. Iri (The University of Tokyo), Professor M. Hatori (The University of Tokyo), Professor H.. Morikawa (The University of Tokyo), Professor H. Imai (The University of Tokyo), Professor S. Tsujii (Tokyo Institute of Technology), Professor Y. Kajitani (Tokyo Institute of Technology), Professor S. Shinoda (Chuo University), Professor S. Ohishi

(Waseda University), Professor T. Sato (Waseda University), Professor M. Nakagawa (Keio University), Professor I. Sasase (Keio University), Professor S. Yoshida (Kyoto University), Professor H. Harada (Kyoto University), Professor I. Shirakawa (Osaka University), Professor S. Sampei (Osaka University), Professor M. Tanimoto (Nagoya University), Professor K. Fujimoto (Tsukuba University), Professor K. Onaga (Ryukyu University), Professor T. Koga (Kyushu University), Professor T. Nishi (Kyushu University), Professor F. Adachi (Tohoku University), The late Professor T. Kurobe (Hokkaido University), The late Professor T. Matsumoto (Hokkaido University), The late Professor T. Abe (Niigata University), Professor Y. Saitoh (Niigata University) and Professor Y. Yamaguchi (Niigata University).

I am particularly indebted to collaborators Professor H. Tamura (Chuo University), Professor K. Nakano (Niigata University) and our students at Niigata University, and to my family.

Contents

Chapter 1

Introduction

There are various systems having a network structure. For example, the Internet, a communication network, an electrical network, a road network, a traffic network, a human network, and so forth. The field of studying this network structure is called graph theory and forms the foundation of network engineering. We will discuss how the network structure determines the performance of the systems, focusing on application to electrical network, location network systems and mobile communication systems.

As an example of a net, imagine a fishing net. The net is knitted with threads or wires. The net is structurally composed of the thread or wire part and the knot part. If it is represented graphically, it is a figure consisting of the lines of the thread and wire parts and the points of the knot parts. Figure 1.1(a) shows an example of a figure showing lines and points are connected. The theory of investigating the properties of such a figure with lines and points is called *graph theory*, and the figure is called a *graph*.

In graph theory, a line on a figure is called an *edge*, and a point is called a *vertex*. The weighted edges and/or points of the graph are called a *network*. For example, Fig. 1.1(b) shows a network with edge weights such as 1, 2, and 5. These weights mean, for example, the length and thickness of a net thread or wire.

As an example, consider an information communication network in which vertices are communication stations and edges are communication lines connecting between stations. If the weight of an edge means the length (or distance) of the communication line, it can be considered to represent

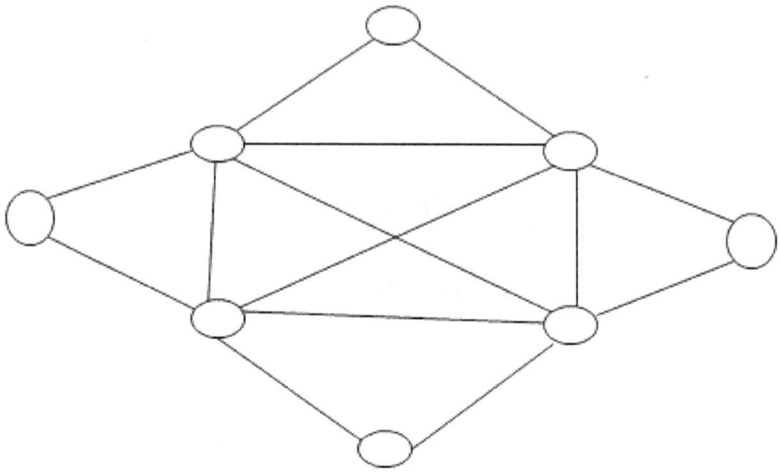

(a)

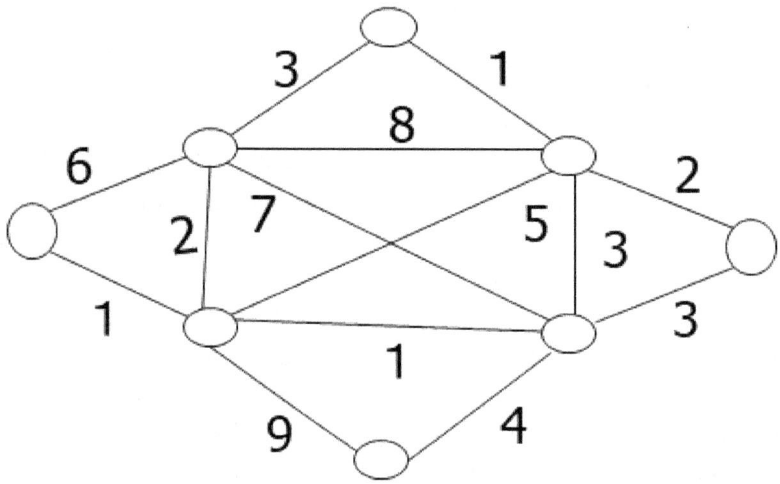

(b)

Fig. 1.1. (a) A figure showing lines and points are connected (a graph). (b) A network with edge weights.

the delay time. If the weight of an edge means the thickness of the communication line, it can be considered to represent the capacity through the communication line. In the network model, the shortest delay time between two stations and/or the maximum capacity between two stations are important factors for the overall performance evaluation analysis. The shortest delay and the maximum capacity are known as the basic problems as the *shortest path problem* and the *maximum flow problem* in the network, respectively. In a *mobile communication network*, *communication stations* (*terminals*) move, so *radio* (*wireless*) is used and there are many mobile stations (terminals) in small areas. Thus, the effective use of frequencies is the most important issue in mobile communication networks. The technology of the effective use of *frequencies* in mobile communication networks includes application of graph theory.

Taking a road network as an example, vertices represent cities, edges represent roads between cities, edge weights represent the length (distance) of a road or the width of a road, and the weight of a vertex may represent the population of the city. In the road network, the shortest path problem finds the shortest distance on the road between two cities, and the maximum flow problem finds the maximum traffic volume between two cities.

Mobile communications belong to the *electrical, electronics and computer science* (EECS) area. In EECS, the electrical network is most important basic subject and includes most useful application of graph theory. For example, Fig. 1.2(a) shows an electrical network in which L, C, R and E are *inductor*, *capacitor*, *resistor* and *voltage source*, respectively. Figure 1.2(b) shows the graph of the electrical network Fig. 1.2(a).

I would like to remark here that some networks showed above: communication networks, road networks and electrical networks have network structure really. However, there are many systems where the model such as their function has a network structure even if the system does not have a network structure. Then the graph theory is also applicable to the system.

The characteristics of the network system may be determined only by the graph, by the *weights* of the vertices and edges of the network, or by both the graph and the weights.

The structure of the network is interesting and profound with its beauty of *graphical structure* and the knowledge gained from it. We are confident that by having everyone enjoy such intelligent and human-friendly fields, new inventions and discoveries will be led and will lead to contributions that will be useful to future society.

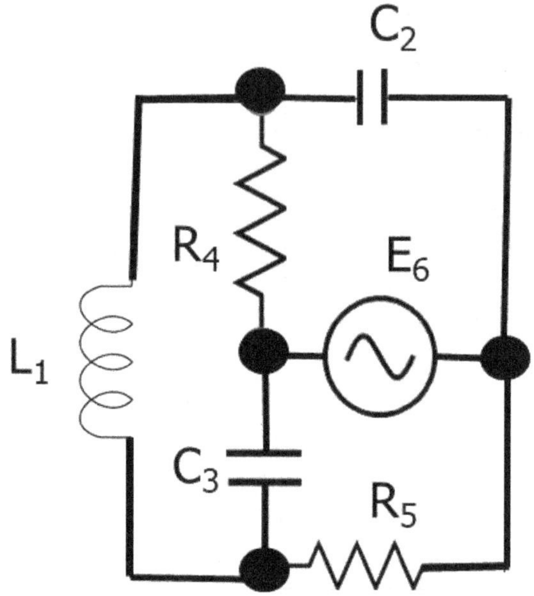

(a)

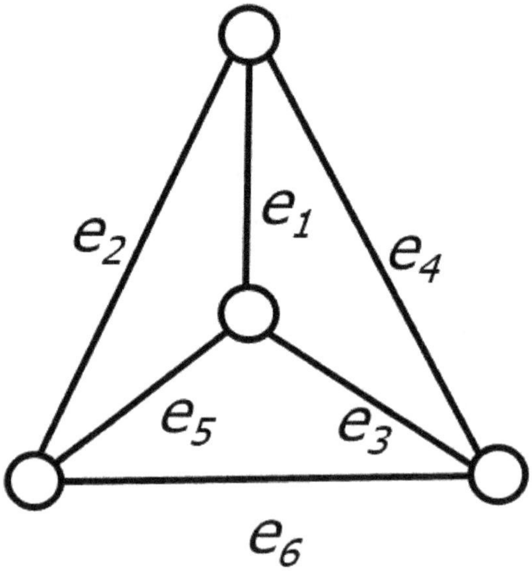

(b)

Fig. 1.2. (a) An electrical network. (b) The graph of figure (a).

- **Summary:** Introductions to graph theory and mobile communications are described.
- **Interesting challenges:** The beauty of the graph structure gives a new image to various issues. Consider why and how to make your image more exciting.

Bibliography

[1] C. Berge, *Graphs*, North-Holland, 1985.
[2] B. Bollobás, *Modern Graph Theory*, Springer-Verlag, 2002.
[3] W. K. Chen, *Applied Graph Theory*, North-Holland, Amsterdam, The Netherlands, 1976.
[4] N. Deo, *Graph Theory with Applications to Engineering and Computer Science*, Prentice-Hall, 1974.
[5] F. Harary, *Graph Theory*, Addison-Wesley, Reading, MA, 1969.
[6] O. Ore, *Theory of Graphs*, American Mathematical Society Colloquium Publications XXXVIII, 1962.
[7] K. Ruohonen, *Graph Theory* (Translation by Janne Tamminen, Kung-Chung Lee and Robert Piché), 2013.
[8] S. Seshu and M. B. Reed, *Linear Graphs and Electrical Networks*, Addison-Wesley, 1961.
[9] M. N. S. Swamy and K. Thulasiraman, *Graphs: Theory and Algorithms,* Wiley, 1992.
[10] M. Iri, I. Shirakawa, Y. Kajitani, S. Shinoda, H. Ariyoshi, K. Onaga, T. Ozawa, T. Ohtsuki, T. Nishi, N. Tomizawa, M. Sengoku, R. Ishii, T. Kashiwabara, T. Nishizeki, S. Fujishige and T. Asano, *Graph Theory with Exercise*, Corona Pub. 1983.
[11] G. Chartrand, L. Lesniak and P. Zhang, *Graphs & Digraphs*, Sixth Edition, Chapman and Hall/CRC, 2015.
[12] W. C. Y. Lee, *Mobile Communications Engineering*, McGraw-Hill, 1998.
[13] H. Tamura, K. Nakano and M. Sengoku, *Network Engineering*, IEICE, Corona Pub., 2020.

Chapter 2

Basic Concept of Graphs

2.1. Definitions

A *graph* G is formed by vertices and edges connecting the vertices. Formally, a graph is a pair of sets $G = (V, E)$, where V is the set of vertices and E is the set of edges, formed by pairs of vertices. If an edge e connects vertices u and v, vertices u and v are called *end vertices* of the edge $e = (u, v)$. Edges that have the same end vertices are called *parallel*. An edge of the form $e = (v, v)$ is a *loop* (or *self-loop*). A graph is *simple* if it has no parallel edges or loops. A graph with no edges (i.e., E is empty) is *empty*. A graph with no vertices (i.e. V and E are empty) is a *null graph*. A graph with only one vertex is *trivial*. Edges are *adjacent* if they share a common end vertex. Two vertices u and v are *adjacent* if they are connected by an edge. The *degree* of the vertex v, written as $d(v)$, is the number of edges with v as an end vertex. By convention, we count a loop twice and parallel edges contribute separately. A *pendant vertex* is a vertex whose degree is 1. An *isolated vertex* is a vertex whose degree is 0.

Example. In Fig. 2.1.1, $G = (V, E)$, where $V = \{v_1, v_2, v_3, v_4, v_5, v_6, v_7, v_8, v_9\}$, $E = \{e_1, e_2, e_3, e_4, e_5, e_6, e_7, e_8, e_9, e_{10}\}$ and $e_1 = (v_1, v_2)$, $e_2 = (v_2, v_3)$, $e_3 = (v_3, v_4)$, $e_4 = (v_1, v_4)$, $e_5 = (v_1, v_4)$, $e_6 = (v_3, v_3)$, $e_7 = (v_6, v_7)$, $e_8 = (v_7, v_8)$, $e_9 = (v_8, v_9)$, $e_{10} = (v_7, v_9)$, v_3 and v_4 are *end vertices* of e_3, e_4 and e_5 are *parallel*, and e_6 is a *loop* in G. Since G has *parallel* edges and *loop*, G is not *simple*. e_1 and e_2 are *adjacent*, v_1 and v_2 are *adjacent*, the *degree of* v_3 *is* 4, the *degree of* v_4 *is* 3, and the *degree of* v_6 is 1 so it is a *pendant vertex*. The *degree* of v_5 *is* 0 so it is an *isolated vertex*. **(End of Example)**

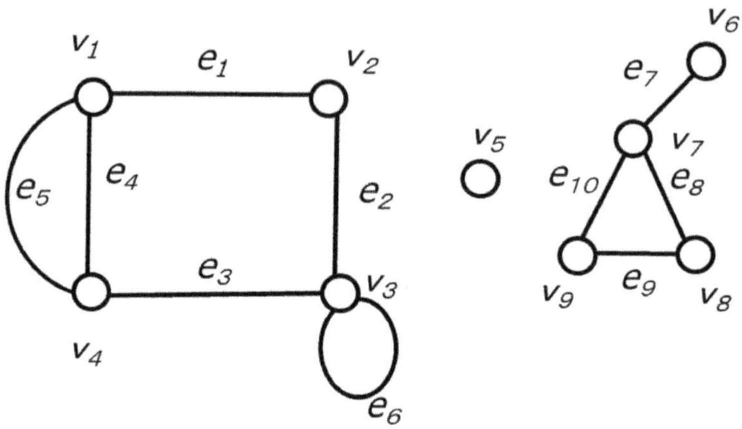

Fig. 2.1.1. A graph.

In $G = (V, E)$, if V and E are *finite sets*, G is called a *finite graph*. In this book, we consider only *finite graphs*.

Since every edge has two end vertices, we get the following theorem.

Theorem 2.1.1. *The graph* $G = (V, E)$, *where* $V = \{v_1, v_2, \ldots, v_n\}$ *and* $E = \{e_1, e_2, \ldots, e_m\}$, *satisfies*

$$\sum_{i=1}^{n} d(v_i) = 2m. \tag{2.1.1}$$

This means that the total number of degrees in a graph G is even.

Corollary 2.1.1. *Every graph has an even number of vertices of odd degree.*

Proof. Since the total number of degrees in a graph G is even, number of vertices of odd degree must be even. □

Example. In Fig. 2.1.1, $|V| = 9$, $|E| = 10$, $d(v_1)+d(v_2)+d(v_3)+d(v_4)+d(v_5)+d(v_6)+d(v_7)+d(v_8)+d(v_9) = 3+2+4+3+0+1+3+2+2 = 2 \cdot 10 = 20$. There are four vertices of odd degree, namely v_1, v_4, v_6 and v_7.

(End of Example)

A *directed graph* or *digraph* G is formed by vertices connected by *directed edges* or *oriented edges*. Formally, a digraph is a pair of sets $G = (V, E)$, where V is the set of vertices and E is the set of edges

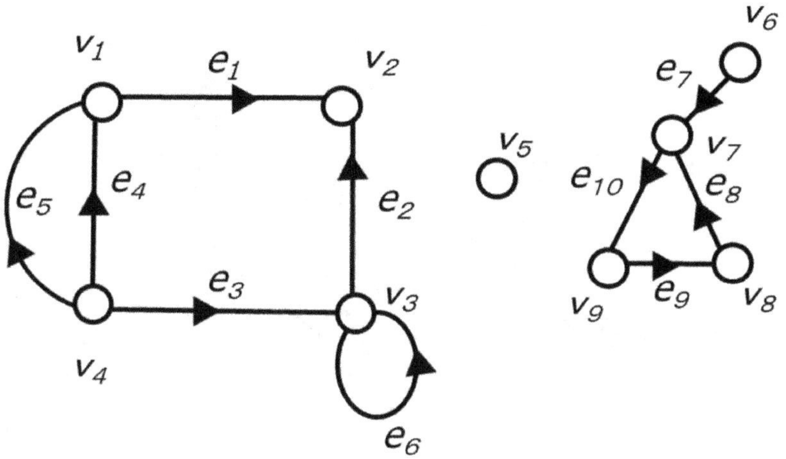

Fig. 2.1.2. A digraph.

with orientation. The difference is that the elements of E are ordered pairs: The edge e from vertex u to vertex v is written as $e = (u, v)$ and the other pair $e = (v, u)$ is the opposite direction edge. We also have to keep track of the multiplicity of the edge (direction of a loop is irrelevant). We can pretty much use the same notions and results for digraphs. In digraphs, vertex u is the *initial vertex* and vertex v is the *terminal vertex* of $e = (u, v)$. We also say that the edge is *incident out* of u and *incident into* v. The *out-degree* of the vertex v is the number of arcs out of it (denoted $d^+(v)$) and the *in-degree* of v is the number of arcs going into it (denoted $d^-(v)$). In a diagram, the orientation is shown by an arrowhead on the edge pointing toward the second vertex of the ordered pair.

Example. In Fig. 2.1.2, $G = (V, E)$, where $V = \{v_1, v_2, v_3, v_4, v_5, v_6, v_7, v_8, v_9\}$, the *directed edge set* $E = \{e_1, e_2, e_3, e_4, e_5, e_6, e_7, e_8, e_9, e_{10}\}$ and $e_1 = (v_1, v_2)$, $e_2 = (v_3, v_2)$, $e_3 = (v_4, v_3)$, $e_4 = (v_4\, v_1)$, $e_5 = (v_4\, v_1)$, $e_6 = (v_3, v_3)$, $e_7 = (v_6, v_7)$, $e_8 = (v_8, v_7)$, $e_9 = (v_9, v_8)$, $e_{10} = (v_9, v_7)$. The vertex v_4 is the *initial vertex* and vertex v_3 is the *terminal vertex* of $e_3 = (v_4, v_3)$. e_4 and e_5 are *parallel*, and e_6 is a *loop* in G. The *out-degree* of v_3 *is* 2, the *in-degree* of v_3 *is* 2, and the *out-degree of* v_6 is 1. The *out-degree* and *in-degree* of v_5 *is* 0 so it is an *isolated vertex*.

(End of Example)

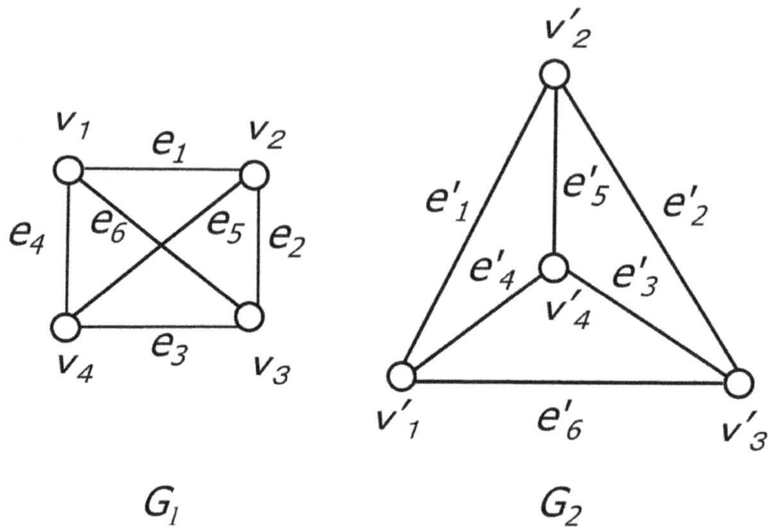

Fig. 2.1.3. *Isomorphic* graphs G_1 and G_2.

Since every edge has the *initial vertex* and the *terminal vertex*, we get

Theorem 2.1.2. *The graph $G = (V, E)$, where $V = \{v_1, v_2, \ldots, v_n\}$ and $\{e_1, e_2, \ldots, e_m\}$, satisfies*

$$\sum_{i=1}^{n} d^+(v_i) = \sum_{i=1}^{n} d^-(v_i) = m. \tag{2.1.2}$$

Example. In Fig. 2.1.2, $|V| = 9$, $|E| = 10$, $d^+(v_1) + d^+(v_2) + d^+(v_3) + d^+(v_4) + d^+(v_5) + d^+(v_6) + d^+(v_7) + d^+(v_8) + d^+(v_9) = 1+0+2+3+0+1+1+1+1 = 10$, and $d^-(v_1) + d^-(v_2) + d^-(v_3) + d^-(v_4) + d^-(v_5) + d^-(v_6) + d^-(v_7) + d^-(v_8) + d^-(v_9) = 2+2+2+0+0+0+2+1+1 = 10$. **(End of Example)**

Let $G_1 = (V_1, E_1)$ and $G_2 = (V_2, E_2)$ be graphs. An *isomorphism* φ: $G_1 \to G_2$ is a bijection (a one-to-one correspondence) from V_1 to V_2 such that $e = (u, v) \in E_1$ if and only if $e' = (\varphi(u), \varphi(v)) \in E_2$. We say G_1 is *isomorphic* to G_2 if there is an isomorphism between them.

Example. In Fig. 2.1.3, $G_1 = (V_1, E_1)$ and $G_2 = (V_2, E_2)$, where $V = \{v_1, v_2, v_3, v_4\}$, $E = \{e_1, e_2, e_3, e_4, e_5, e_6\}$, $V' = \{v'_1, v'_2, v'_3, v'_4\}$ and

$E = \{e'_1, e'_2, e'_3, e'_4, e'_5, e'_6\}$. φ: $G_1 \to G_2$ given by $\varphi(v_i) = v'_i$ $(i = 1, 2, 3, 4)$. G_1 is *isomorphic* to G_2. **(End of Example)**

2.2. Special graphs

A simple graph that contains every possible edge between all the vertices is called a *complete graph.*

A complete graph with n vertices is denoted as K_n. A simple graph $G = (V, E)$ is *bipartite* if there is a partition $V = V_1 \cup V_2$ into two disjoint sets such that each $e \in E$ intersects both V_1 and V_2. A simple graph $G = (V, E)$ is the *complete bipartite* graph if there is a partition $V = V_1 \cup V_2$ into two disjoint sets such that E consists of every possible edge intersecting both V_1 and V_2 where $|V| = n+m$, $|V_1| = n$ and $|V_2| = m$. If there are n vertices in V_1 and m vertices in V_2, we denote it as $K_{n,m}$.

A graph G is said to be *planar* if there exists some geometric representation of G which can be drawn on a plane such that no two of its edges intersect. A graph that cannot be drawn on a plane without a crossover between its edges is called *nonplanar*.

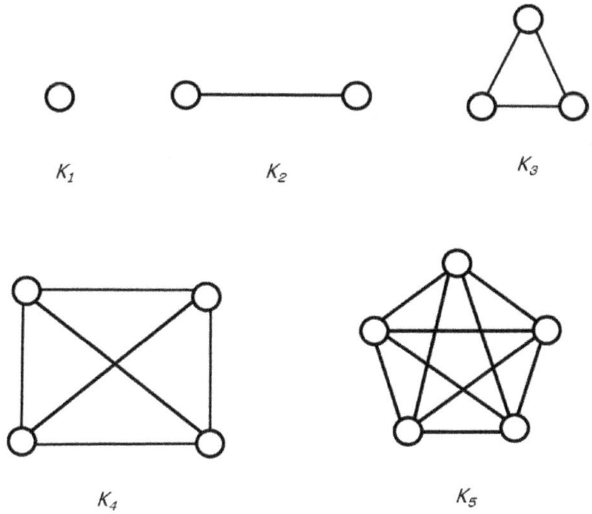

Fig. 2.2.1. The *complete* graphs K_1, K_2, K_3, K_4 and K_5.

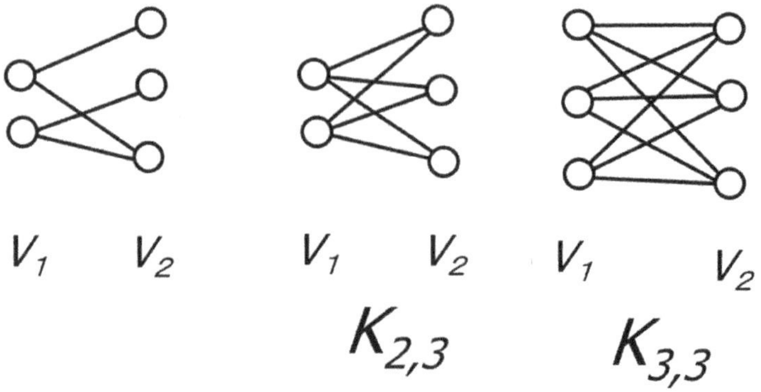

Fig. 2.2.2. Two *bipartite* graphs (the right side graphs are $K_{2,3}$ and $K_{3,3}$).

Example. In Fig. 2.2.1, the *complete* graphs K_1, K_2, K_3, K_4 and K_5 are shown. In Fig. 2.2.2, two *bipartite* graphs are shown where $V = V_1 \cup V_2$ and $|V_1| = 2$ and $|V_2| = 3$. The right side graphs are a *complete bipartite* graph $K_{2,3}$ and $K_{3,3}$. K_1, K_2 and K_3 are *planar.* K_4 seems nonplanar in Fig. 2.2.1, but it can be drawn as a G_2 in Fig. 2.1.3, thus K_4 is planar. $K_{2,3}$ is planar. K_5 and $K_{3,3}$ are *nonplanar*, called *Kuratowski's two graphs.*

(End of Example)

2.3. Subgraphs

The graph $G_1 = (V_1, E_1)$ is a *subgraph* of $G = (V, E)$ if $V_1 \subseteq V, E_1 \subseteq E$ and every edge of G_1 is also an edge of G. If $V_1 = V$ then G_1 is called *spanning.*

The *subgraph* of $G = (V, E)$ *induced by the edge set* $E_1 \subseteq E$ is $G_1 = (V_1, E_1)$, where V_1 consists of every end vertex of the edges in E_1. The *subgraph* of $G = (V, E)$ *induced by the vertex set* $V_1 \subseteq V$ is $G_1 = (V_1, E_1)$, where E_1 consists of every edge between the vertices in V_1.

Example. In Fig. 2.3.1, $G_1 = (V_1, E_1)$, where $V_1 = \{v_1, v_2, v_3, v_5, v_7, v_8\}$, $E_1 = \{e_1, e_2, e_6, e_8\}$, is a *subgraph* of $G = (V, E)$ in Fig. 2.1.1. In Fig. 2.3.2,

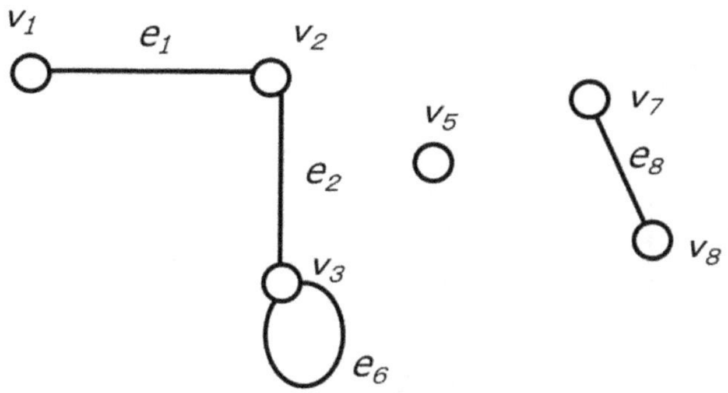

Fig. 2.3.1. A *subgraph* G_1 of G in Fig. 2.1.1.

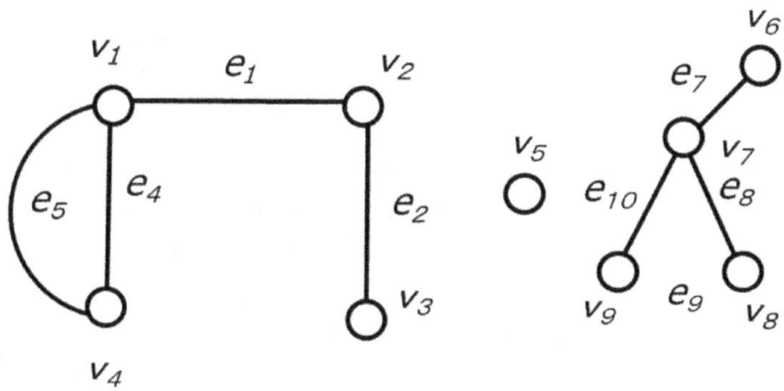

Fig. 2.3.2. A *spanning subgraph* G_2 of G in Fig. 2.1.1.

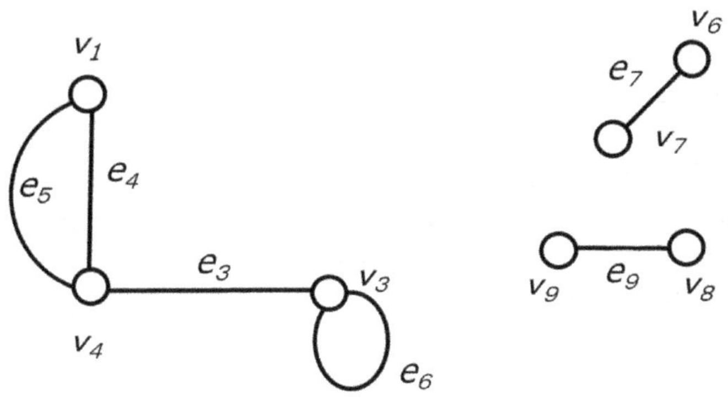

Fig. 2.3.3. The subgraph G_3 of G *induced* by the edge set $E_3 = \{e_3, e_4, e_5, e_6, e_7, e_9\}$.

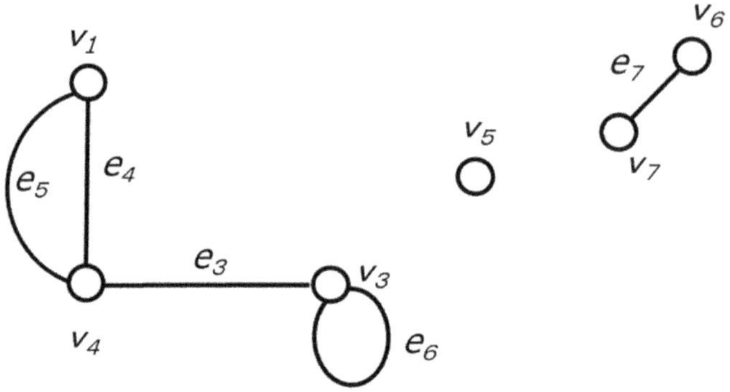

Fig. 2.3.4. The subgraph G_4 of G *induced* by the vertex set $V_4 = \{v_1, v_3, v_4, v_5, v_6, v_7\}$.

$G_2 = (V_2, E_2)$, where $V_2 = \{v_1, v_2, v_3, v_4, v_5, v_6, v_7, v_8, v_9\}$, $E_1 = \{e_1, e_2, e_4, e_5, e_7, e_8, e_{10}\}$, is a *spanning subgraph* of $G = (V, E)$ in Fig. 2.1.1, since $V_2 = V$. In Fig. 2.3.3, $G_3 = (V_3, E_3)$ is the subgraph of G *induced* by the edge set $E_3 = \{e_3, e_4, e_5, e_6, e_7, e_9\}$, where $E_3 \subseteq E$ and $V_3 = \{v_1, v_3, v_4, v_6, v_7, v_8, v_9\}$, which consists of every end vertex of the edges in E_3 of $G = (V, E)$. In Fig. 2.3.4, $G_4 = (V_4, E_4)$ is the subgraph of G *induced* by the vertex set $V_4 = \{v_1, v_3, v_4, v_5, v_6, v_7\}$, where $V_4 \subseteq V$ and $E_4 = \{e_3, e_4, e_5, e_6, e_7\}$, which consists of every edge between the vertices in V_4 of $G = (V, E)$. **(End of Example)**

2.4. Paths, Ties and Components

A *walk* in the graph $G = (V, E)$ is a finite sequence of the form: $v_{a0}, e_{b1}, v_{a1}, e_{b2}, \ldots, e_{bk}, v_{ak}$ which consists of alternating vertices and edges of G. Vertices v_{ax-1} and v_{ax} are end vertices of e_{bx} $(x = 1, \ldots, k)$. v_{a0} is the *initial vertex* and v_{ak} is the *terminal vertex*. k is the *length* of the walk. A zero length walk is just a single vertex v_{a0}. It is allowed to visit a vertex or go through an edge more than once. A walk is *open* if $v_{a0} \neq v_{ak}$, otherwise it is *closed*. A walk is a *path* if any vertex is visited at most once except possibly the initial and terminal vertices when they are the same. The *length* of a path is the number of edges in the path. A closed path is a *tie*. Since any vertex is visited at most once except possibly the initial and

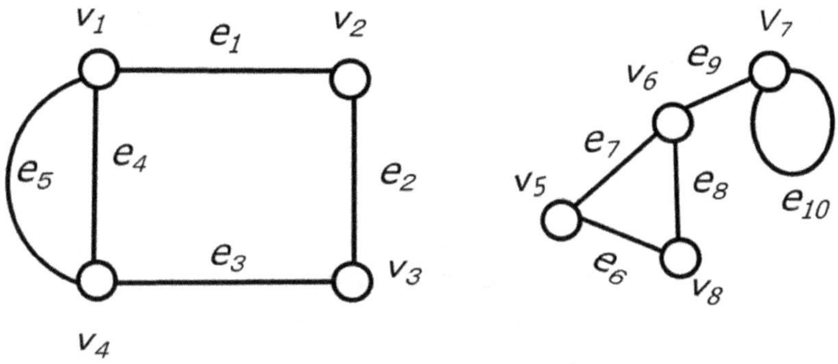

Fig. 2.4.1. A graph.

terminal vertices in a path and a tie, a path and a tie can be represented by edge set in them. A path $v_{a0}, e_{b1}, v_{a1}, e_{b2}, \ldots, e_{bk}, v_{ak}$ is considered as a subgraph $G' = (V', E')$ of G where $V' = \{v_{a0}, v_{a1}, \ldots, v_{ak}\}$ and $E' = \{e_{b1}, e_{b2}, \ldots, e_{bk},\}$ and the subgraph is *induced by the edge set* $E' \subseteq E$ where V' consists of every end vertex of the edges in E'. Then we call E' a *path-set* or a *pathset.* A tie $v_{a0}, e_{b1}, v_{a1}, e_{b2}, \ldots, e_{bk}, v_{ak}, (v_{a0} = v_{ak}$ is considered as a subgraph $G'' = (V'', E'')$ of G where $V'' = \{v_{a0}, v_{a1}, \ldots, v_{ak}\}$ and $E'' = \{e_{b1}, e_{b2}, \ldots, e_{bk},\}$ and the subgraph is *induced by the edge set* $E'' \subseteq E$ where V'' consists of every end vertex of the edges in E''. Then we call E'' a *tie-set* or a *tieset.*

Example. In Fig. 2.4.1, $G = (V, E)$, where $V = \{v_1, v_2, v_3, v_4, v_5, v_6, v_7, v_8\}$, $E = \{e_1, e_2, e_3, e_4, e_5, e_6, e_7, e_8, e_9, e_{10}\}$ and $e_1 = (v_1, v_2)$, $e_2 = (v_2, v_3)$, $e_3 = (v_3, v_4)$, $e_4 = (v_1, v_4)$, $e_5 = (v_1, v_4)$, $e_6 = (v_5, v_6)$, $e_7 = (v_5, v_8)$, $e_8 = (v_6, v_8)$, $e_9 = (v_6, v_7)$, $e_{10} = (v_7, v_7)$. $v_5, e_7, v_6, e_9, v_7, e_{10}, v_7, e_9, v_6, e_8, v_8$ is a *walk* and *open* where v_5 is the *initial vertex* and v_8 is the *terminal vertex.* The length of the walk is 5. $v_5, e_7, v_6, e_9, v_7, e_{10}, v_7, e_9, v_6, e_8, v_8, e_6, v_5$ is a walk and *closed,* where v_5 is the *initial vertex* and the *terminal vertex.* $v_1, e_1, v_2, e_2, v_3, e_3, v_4$ is a *walk* and *open.* It is a *path* since any vertex is visited at most once in the walk. The length of the path is 3. $v_1, e_1, v_2, e_2, v_3, e_3, v_4\ e_5, v_1$ is a *tie* since it is a *closed path.* The subgraph $G' = (V', E')$ is a path, where $V' = \{v_1, v_2, v_3, v_4\}$ and $E'' = \{e_1, e_2, e_3\}$ and the subgraph is *induced by the edge set* $E' \subseteq E$ where V' consists of every end vertex of the edges in E'. $E' = \{e_1, e_2, e_3\}$ is the *path-set*. The subgraph $G'' = (V'', E'')$ is a tie, where $V'' = \{v_1, v_2, v_3, v_4\}$ and $E'' = \{e_1, e_2, e_3, e_5\}$ and the subgraph is *induced by the edge set* $E'' \subseteq E$ where V'' consists of every end vertex of the edges in E''. $E'' = \{e_1, e_2, e_3, e_5\}$ is the *tieset*. **(End of Example)**

v_i and v_j are *connected* if there is a path from v_i to v_j. If v_i and v_j are connected and v_j and v_k are connected, then v_i and v_k are also connected, that is, if there is a path from v_i to v_j and a path from v_j to v_k, then there is also a path from v_i to v_k. A graph is *connected* if all the vertices are connected to each other. The subgraph G_1 of the graph G is a *component* of G if G_1 is connected and G_1 is the subgraph induced by those edges of G that have one end vertex in G_1. Every isolated vertex forms its own component. A *connected graph* has only one component, namely, itself. A graph G with $|V|$ vertices, $|E|$ edges and p components has the *rank*

$$\rho(G) = |V| - p. \tag{2.4.1}$$

The *nullity* of the graph is

$$\mu(G) = |E| - |V| + p. \tag{2.4.2}$$

Example. In Fig. 2.4.1, $G = (V, E)$, the subgraph $G_1 = (V_1, E_1)$ where $V_1 = \{v_1, v_2, v_3, v_4\}$, $E_1 = \{e_1, e_2, e_3, e_4, e_5\}$, is a *component* of G. The subgraph $G_2 = (V_2, E_2)$ where $V_2 = \{v_5, v_6, v_7, v_8\}$, $E_2 = \{e_6, e_7, e_8, e_9, e_{10}\}$, is also a *component* of G. Thus, the number of *component* of G: $p = 2$. The *rank* of G: $\rho(G) = |V| - p = 8 - 2 = 6$. The *nullity* of G: $\mu(G) = |E| - |V| + p = 10 - 8 + 2 = 4$.

(End of Example)

2.5. Graph Operations

If e is an edge of the graph $G = (V, E)$, then $G - e$ is graph $G' = (V, E')$, where E' is obtained by removing e from E. This operation is known as *removal of an edge*. The operation does not include the removal of the end vertices of the edge e. If $v_i, v_j \in V$ and $e = (v_i, v_j) \notin E$ in the graph

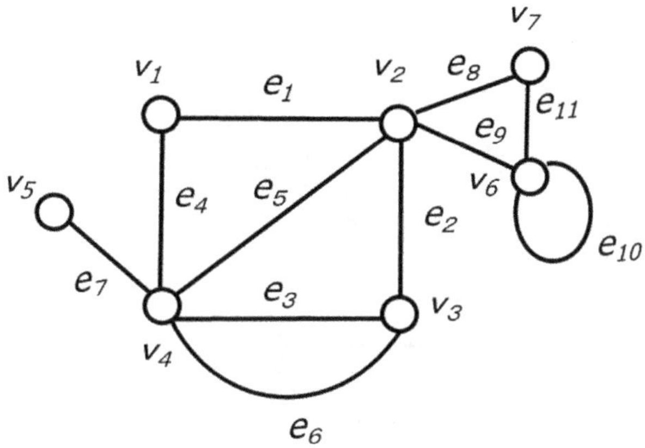

Fig. 2.5.1. A graph G.

$G = (V, E)$, then $G + e$ is graph $G' = (V, E')$, where E' is obtained by adding e to E. This operation is known as *addition of an edge*.

If v is a vertex of the graph $G = (V, E)$, then $G - v$ is the subgraph $G' = (V', E)$ induced by the vertex set $V' = V - \{v\}$. We call this operation the *removal of a vertex*. It is clear that the operation includes the removal of a vertex and the removal of the adjacent edges to the vertex. If $v \notin V$ in the graph $G = (V, E)$, then $G + v$ is graph $G' = (V', E)$, where V' is obtained by adding v to V. This operation is known as *addition of a vertex*.

If $v_i, v_j \in V$ in the graph $G = (V, E)$, then the graph $G' = (V', E')$, where V' is obtained by *coinciding* v_i and v_j, and the new vertex v_k by coinciding v_i and v_j is added to V. That is $V' = V - \{v_i, v_j\} + \{v_k\}$. We call this operation the *coincident of two vertices*.

If $e = (v_i, v_j) \in E$ in the graph $G = (V, E)$, then the graph $G' = (V', E')$, where V' is obtained by *coinciding* v_i and v_j, and E' is obtained by *removing e* from E. We call this operation the *short-circuiting of an edge*. When focusing on the operations of an edge, we can call the operation of *removal of an edge* the *open-circuiting of an edge*. If $E_1 \subset E$, $E_2 \subset E$, $E_1 \bigcap E_2 = \emptyset$ in the graph $G = (V, E)$ then the graph $G' = (V', E')$, where V' is obtained by *open-circuiting of edges* of E_1 and *short-circuiting of edges* of E_2. Since the graph $G' = (V', E')$ is obtained by *open-circuiting of edges* of E_1 and *short-circuiting of edges* of E_2 and

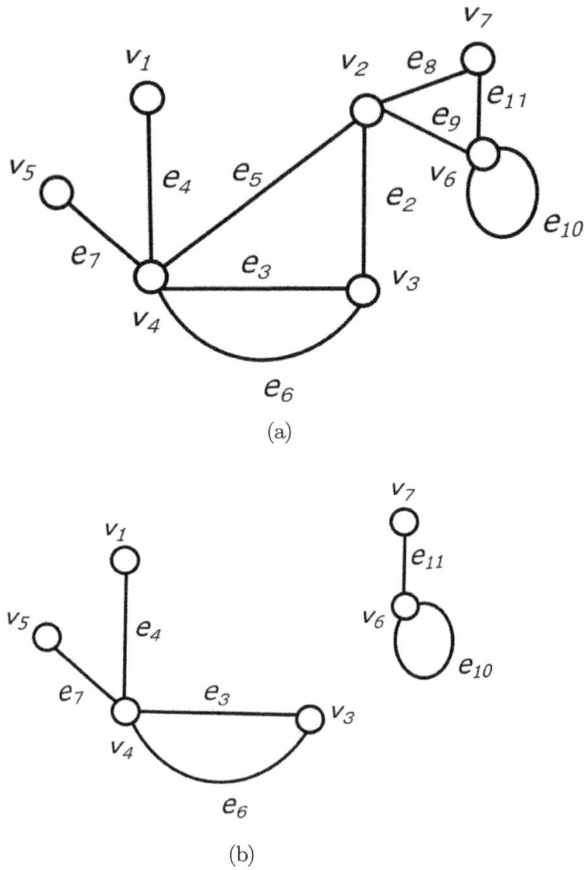

Fig. 2.5.2. Graphs $G-e_1$ (*removal of an edge*) and $G-v_2$ (*removal of a vertex*) in a graph of Fig. 2.5.1. (a) A graph $G-e_1$. (b) A graph $G-v_2$.

the graphs obtained from G are isomorphic regardless of the different operation sequence of open-circuiting of edges and short-circuiting of edges, we denote by $G_{os}[E_1; E_2]$ the graph $G' = (V', E')$.

Example. In Fig. 2.5.1, $G = (V, E)$, where $V = \{v_1, v_2, v_3, v_4, v_5, v_6, v_7\}$, $E = \{e_1, e_2, e_3, e_4, e_5, e_6, e_7, e_8, e_9, e_{10}, e_{11}\}$. In Fig. 2.5.2(a), graphs $G-e_1$ (*removal of an edge*) and $G-v_2$ (*removal of a vertex*) of G are shown.

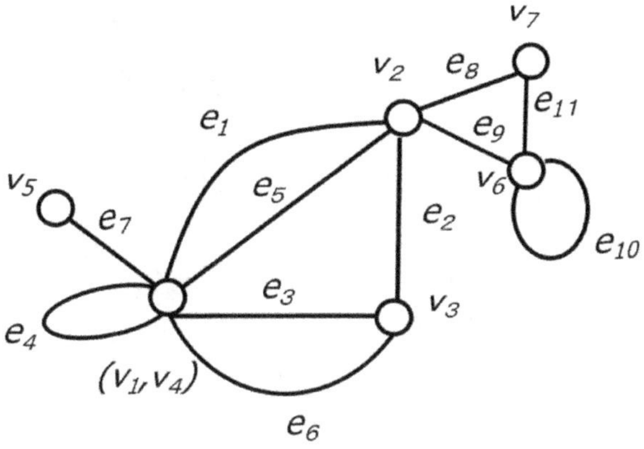

(a)

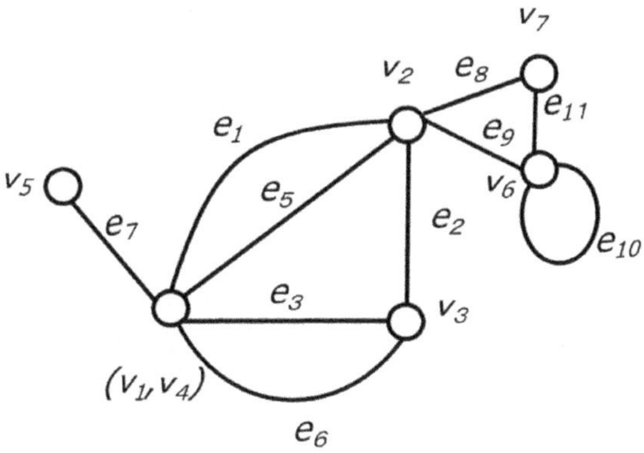

(b)

Fig. 2.5.3. Graphs obtained by *coinciding* v_1 and v_4, and obtained by *short-circuiting of an edge* e_4 from G in Fig. 2.5.1. (a) A graph G′ obtained by *coinciding* v_1 and v_4. (b) A graph G′ obtained by *short-circuiting of an edge* e_4.

Fig. 2.5.3(a) shows the graph obtained by *coinciding* v_1 and v_4, and Fig. 2.5.3(b) shows obtained by *short-circuiting of an edge* e_4 from G in Fig. 2.5.1. **(End of Example)**

Let $E_1 = \{e_1, e_3, e_{10}\}$ and $E_2 = \{e_4, e_8\}$ be subsets of E, where $E_1 \subset E$, $E_2 \subset E$ $E_1 \bigcap E_2 = \emptyset$ in the graph $G = (V, E)$ in Fig. 2.5.1. Figure 2.5.4 shows the graph $G_{os}[E_1; E_2]$ obtained by *open-circuiting of edges* of E_1 and *short-circuiting of edges* of E_2 from G in Fig. 2.5.1.

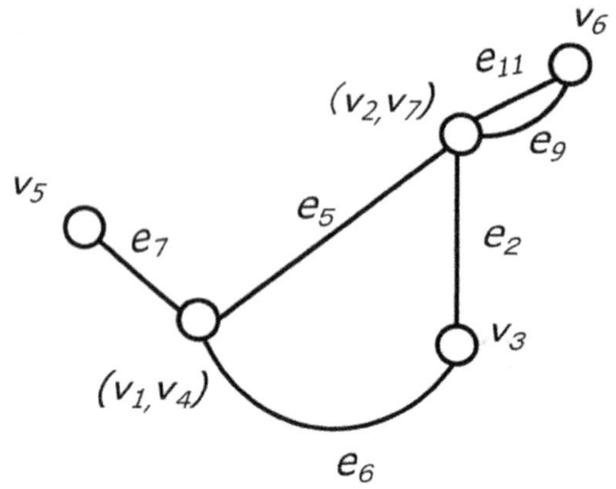

Fig. 2.5.4. The Graph $G_{os}[E_1; E_2]$ obtained by *open-circuiting of edges* of E_1 and *short-circuiting of edges* of E_2 from G in Fig. 2.5.1, where $E_1 = \{e_1, e_3, e_{10}\}$ and $E_2 = \{e_4, e_8\}$.

Conversely, a graph of Fig. 2.5.1 is obtained from the graphs of Figs. 2.5.2–2.5.4 by some operations *addition of a vertex* and/or *addition of an edge*.

2.6. Cuts

A vertex v of a graph G is a *cut vertex* of G if the graph $G - v$ consists of a greater number of components than G. In Fig. 2.5.1, the number of components of G is 1 and in Fig. 2.5.2(b), $G - v_2$ (*removal of a vertex*) is 2. Then v_2 is a cut vertex.

Let us consider *removal of edges* which makes an increase the number of components. A *cutset* of a graph $G = (V, E)$ is an edge set $E_1 \subseteq E$ whose *removal* makes increase of 1 in the number of components of G and no proper subset of E_1 makes increase the number of components. That is, a *cutset* is a *minimal edge set* whose *removal* makes increase of 1 in the number of components of G. The subgraph *induced by the cutset* $E' \subseteq E$ in G is called a *cut*.

Let $E\langle V_1, V_2\rangle$ be an edge set of a partition $V = V_1 \cup V_2$, $V_1 \cap V_2 = \emptyset$, $V_1 \neq \emptyset$, $V_2 \neq \emptyset$ such that one end vertex in every edge of $E\langle V_1, V_2\rangle$ is in V_1 and the other end vertex is in V_2. We denote the partition of vertices $\langle V_1, V_2\rangle$. It is clear that $E\langle V_1, V_2\rangle$ includes cutsets and consists of disjoint cutsets. Since a cutset E' is a *minimal edge set* whose *removal* makes increase of 1 in the number of components of G, a component of G is divided into two components by the removal. Let the set of vertices in one component of two components be V_1, and V_2 be $V_2 = V - V_1$. Then $\langle V_1, V_2\rangle$ is a partition of V and $E\langle V_1, V_2\rangle = E'$. If G is a connected graph, the subgraphs induced by V_1 and V_2 are connected. That is, the graph G' obtained by *removal of a cutset* of a connected graph G has two components which are the subgraphs induced by V_1 and V_2 of G.

Example. In Fig. 2.5.1, $G = (V, E)$ where $V = \{v_1, v_2, v_3, v_4, v_5, v_6, v_7\}$, $E = \{e_1, e_2, e_3, e_4, e_5, e_6, e_7, e_8, e_9, e_{10}, e_{11}\}$, let us consider *removal of*

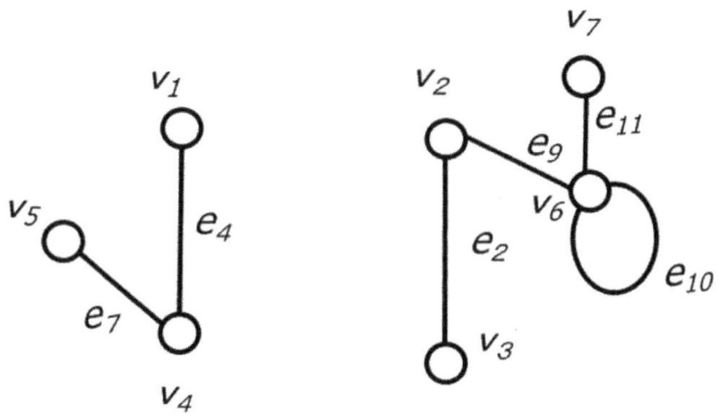

(a)

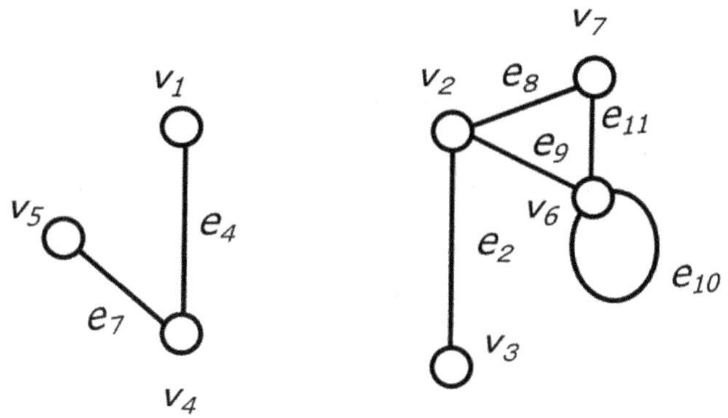

(b)

Fig. 2.6.1. The graphs obtained by removal of edge set. (a) A graph G_1 obtained by removal of an edge set $E_1 = \{e_1, e_3, e_5, e_6, e_8\}$. (b) A graph G_2 obtained by removal of an edge set $E_2 = \{e_1, e_3, e_5, e_6\}$.

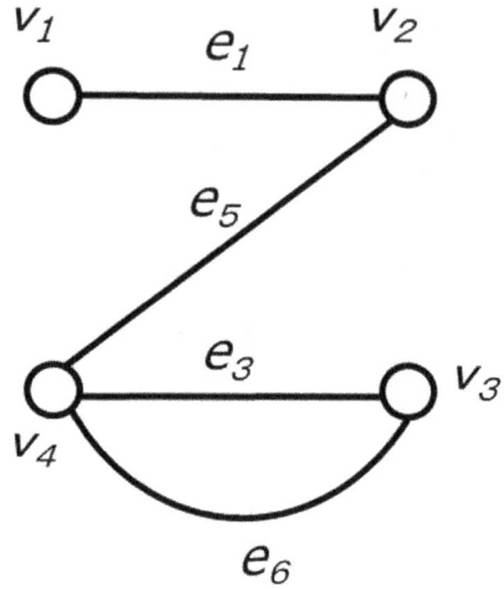

Fig. 2.6.2. The *cut* of a cutset $E_2 = \{e_1, e_3, e_5, e_6\}$.

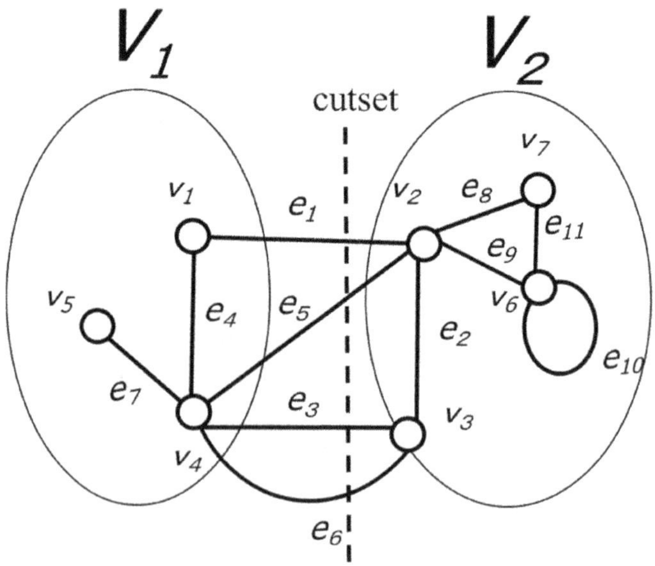

Fig. 2.6.3. The partition $\langle V_1, V_2\rangle$ of V by a cutset $E_2 = \{e_1, e_3, e_5, e_6\}$.

edge sets from G. Figure 2.6.1(a) shows a graph G_1 obtained by removal of an edge set $E_1 = \{e_1, e_3, e_5, e_6, e_8\}$. Figure 2.6.1(b) shows a graph G_2 obtained by removal of an edge set $E_2 = \{e_1, e_3, e_5, e_6\}$. The number of components of G_1 or G_2 increase from 1 to 2 by removal of edge set E_1 or E_2 of G. However, an edge set $E_1 = \{e_1, e_3, e_5, e_6, e_8\}$ includes every edge of E_2. No proper subset of E_2 makes increase the number of components. That is, E_1 is not a *cutset* and E_2 is a *cutset*. Figure 2.6.2 shows the subgraph *cut*. Figure 2.6.3 shows the partition $\langle V_1, V_2\rangle$ of V by a cutset $E_2 = \{e_1, e_3, e_5, e_6\}$, where $V_1 = \{v_1, v_4, v_5\}$ and $V_2 = \{v_2, v_3, v_6, v_7\}$.

(End of Example)

2.7. Tiesets and Cutsets

In a graph G, ties and cuts are subgraphs induced by tiesets and cutsets, respectively, and since they have one-to-one correspondence, they are sometimes used without distinction. There are some valuable relationships between tiesets and cutsets.

Theorem 2.7.1. *A tieset and a cutset have an even number (including 0) of common edges.*

Proof. Let F and H be a tieset and a cutset of a graph $G = (V, E)$, respectively. Since a cutset H is a *minimal edge set* whose *removal* makes increase of 1 in the number of components of G, a component of G is divided into two components by the removal. Let the set of vertices in one component of two components be V_1, and V_2 be $V_2 = V - V_1$. Then $\langle V_1, V_2\rangle$ is a partition of V and $E\langle V_1, V_2\rangle = H$. Let G_1 and G_2 be induced graphs by V_1 and V_2, respectively. If F is a subgraph of G_1 or G_2, then the theorem is obvious because they have no common edges. Let us assume that F and H have common edges. In Fig. 2.7.1, we traverse around a tie (tieset F: bold lines) by starting at some vertex v of G_2. Since we come back to v, there has to be an even number of edges of the cutset $E\langle V_1, V_2\rangle(= H)$ in F. □

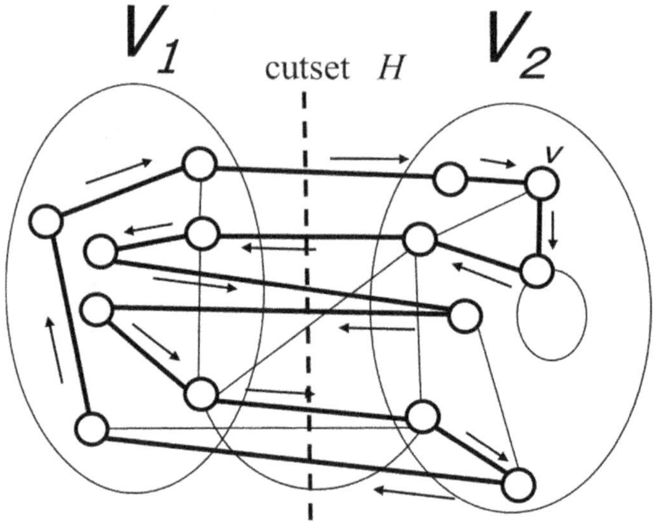

Fig. 2.7.1. A tieset (F: bold lines) and a cutset H.

If the number of edges of a tieset and a cutset is one, we call them a self tieset and a self cutset, respectively. For example, $\{e_{10}\}$ is a self tieset and $\{e_7\}$ is a self cutset of a graph in Fig. 2.4.1.

Theorem 2.7.2. *Let an edge subset be $E_1 \subset E$, $E_1 \neq \emptyset$ in a graph $G = (V, E)$.*

(1) *E_1 is union of tiesets if and only if $\left|E_1 \cap H\right| \neq 1$ for any cutset H in G.*
(2) *E_1 is union of cutsets if and only if $\left|E_1 \cap F\right| \neq 1$ for any tieset F in G.*

Proof. (1) We assume $|E_1 \cap H| \neq 1$ for any cutset H in G. Let $e = (v_i, v_j) \in E_1$ and $V(v_i)$ be the vertex set such that $V(v_i)$ consists of vertices included in the paths that pass through $E_1 - \{e\}$ from v_i and also $V(v_j)$ be the vertex set such that $V(v_j)$ consists of vertices included in the paths that pass through $E_1 - \{e\}$ from v_j. If $V(v_i) \cap V(v_j) \neq \emptyset$ (including the case of $v_i = v_j$), there a path-set from v_i *to* v_j in $E_1 - \{e\}$, and the path-set and $\{e\}$ constitute a tieset (a self tieset in the case of $v_i = v_j$,). If $V(v_i) \cap V(v_j) = \emptyset$ (including e: a self cutset), there is a partition $V = V_1 \cup V_2$, $V_1 \cap V_2 = \emptyset$, $V_1 \neq \emptyset$, $V_2 \neq \emptyset$ such that $V(v_i) \subset V_1, V(v_j) \subset V_2$ and $E\langle V_1, V_2\rangle = H$ is a cutset. In the cutset, only one edge e of E_1 is included, that is $|E_1 \cap H| = 1$ which is a *contradiction* from $|E_1 \cap H| \neq 1$. □

Conversely, let E_1 be union of tiesets. Then, for any cutset H in G, $|E_1 \cap H| = 0$ or $|E_1 \cap H| \geq 2$ from Theorem 2.7.1.

(2) We assume $|E_1 \cap F| \neq 1$ for any tietset F in G. Let $e = (v_i, v_j) \in E_1$ and $V'(v_i)$ be the vertex set such that $V'(v_i)$ consists of vertices included in the paths that pass through $E - E_1$ from v_i and also $V'(v_j)$ be the vertex set such that $V'(v_j)$ consists of vertices included in the paths that pass through $E - E_1$ from v_j. If $V'(v_i) \cap V'(v_j) = \emptyset$ (including e: a self cutset), there is a partition $V = V_1 \cup V_2$, $V_1 \cap V_2 = \emptyset$, $V_1 \neq \emptyset$, $V_2 \neq \emptyset$ such that $V'(v_i), \subset V_1$, $V'(v_j) \subset V_2$ and $E\langle V_1, V_2\rangle \subseteq E_1$ is a cutset. If $V'(v_i) \cap V'(v_j) \neq \emptyset$ (including the case of $v_i = v_j$), there a path-set from v_i *to* v_j in $E - E_1$, and the path-set and $\{e\}$ constitute a tieset F (a self tieset in the case of $v_i = v_j$). In the tieset, only one edge e of E_1 is included, that is $|E_1 \cap F| = 1$ which is a contradiction from $|E_1 \cap F| \neq 1$.

Conversely, let E_1 be union of cutsets. Then, for any tieset F in G, $|E_1 \cap F| = 0$ or $|E_1 \cap F| \geq 2$ from Theorem 2.7.1.

Corollary 2.7.2. *For an edge* e $\in E$ *in a graph* $G = (V, E)$,

(1) *e is a self tieset if and only if there is no cutsets including e.*
(2) *e is a self cutset if and only if there is no tiesets including e.*

Proof. (1) If there is no cutsets including e, $|\{e\} \cap H| = 0$ for any cutset H in G. Form Theorem 2.7.2(1), $\{e\}$ includes tiesets, that is a self tieset. Conversely, let $\{e\}$ be a self tieset. Since $\{e\}$ is tieset, $|\{e\} \cap H| \neq 1$ for any cutset H in G form Theorem 2.7.2(1). Since $\{e\}$ has one element, $|\{e\} \cap H| = 0$, that is, $\{e\}$ is not included in any cutsets.

(2) If there is no tiesets including e, $|\{e\} \cap F| = 0$ for any tietset F in G. Form Theorem 2.7.2(2), $\{e\}$ includes cutsets, that is a self cutset. Conversely, let $\{e\}$ be a self cutset. Since $\{e\}$ is cutset, $|\{e\} \cap F| \neq 1$ for any tietset F in G form Theorem 2.7.2(2). Since $\{e\}$ has one element, $|\{e\} \cap F| = 0$, that is, $\{e\}$ is not included in any tiesets. □

Theorem 2.7.3. *Let e_i and e_j be two edges in a graph $G = (V, E)$, e_i and e_j are included in a cutset, if and only if e_i and e_j are included in a tieset.*

Proof. Let F be a tieset including $e_i = (v_{i1}, v_{i2})$ and $e_j = (v_{j1}, v_{j2})$. Let F consists of a path-set $\mathrm{P}(v_{i1}, v_{j1})$ from v_{i1} to v_{j1}, a path-set $\mathrm{P}(v_{i2}, v_{j2})$ from v_{i2} to v_{j2} and two edges e_i, e_j. Then there is a cutset $E\langle V_1, V_2\rangle = H$ which includes e_i and e_j, where V_1 consists of vertices on a path-set $\mathrm{P}(v_{i1}, v_{j1})$ and $V_2 = V - V_1$. Conversely, let $H = E\langle V_1, V_2\rangle$ be a cutset including $e_i = (v_{i1}, v_{i2})$ and $e_j = (v_{j1}, v_{j2})$, where $v_{i1}, v_{j1} \in V_1$, and $v_{i2}, v_{j2} \in V_2$. Then there is a path-set $\mathrm{P}(v_{i1}, v_{j1})$ from v_{i1} to v_{j1} in a subgraph induced by V_1 and a path-set $\mathrm{P}(v_{i2}, v_{j2})$ from v_{i2} to v_{j2} in a subgraph induced by V_2. These two path-sets and $\{e_i, e_j\}$ constitute a tieset. □

Let us consider the binary relation between edges in a graph $G = (V, E)$.

(1) $e_i \sim_t e_j : e_i = e_j$ or there is a tieset with e_i and e_j.
(2) $e_i \sim_c e_j : e_i = e_j$ or there is a cutset with e_i and e_j.

It is clear that the binary relations $\sim_t, \sim_c$ are reflexive, symmetric and transitive, that is, equivalence relations. Each equivalence relation provides a partition of the underlying set into disjoint equivalence classes. The binary relations $\sim_t, \sim_c$ give the same equivalence class from Theorem 2.7.3. The subgraphs of the disjoint equivalence classes are called *nonseparable components*.

Example. Figure 2.7.2 shows a partition of the underlying set into disjoint equivalence classes by the binary relations $\sim_t, \sim_c$. The subgraphs of the disjoint equivalence classes are shown by dotted-line circles called *nonseparable components*. **(End of Example)**

Theorem 2.7.4. *Let G be a graph where the degree of every vertex is even (not including 0). Then G is a graph induced by sum of disjoint tiesets.*

Proof. Let us start from v in G and go along a path where any vertex is visited at most once except possibly the initial and terminal vertices when they are the same. Since every vertex is even, there are pair of edges, that is, an edge to a vertex and an edge form the vertex. This means that we can return to v and the path is a tie. Let F_1 be the tieset of the tie. Let G_1 be a graph by open-circuiting of F_1, then the degree of every vertex in G_1 is even. By repeating this operation, we obtain $F_2, F_3, \ldots,$ and G is a graph induced by sum of disjoint tiesets. □

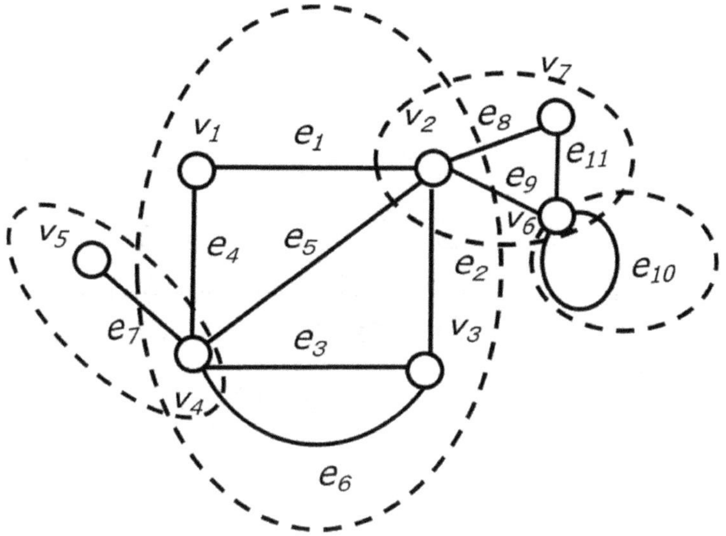

Fig. 2.7.2. Four nonseparable components which are the disjoint equivalence classes.

Let us define the set operation $\oplus$ as follows.

$$E_1 \oplus E_2 = (E_1 - E_2) \cup (E_2 - E_1) = (E_1 \cup E_2) - (E_1 \cap E_2). \tag{2.7.1}$$

The set operation $\oplus$ is called *the symmetric difference* or *the ring sum.*

Theorem 2.7.5. *In a graph $G = (V, E)$,*

(1) *for two tiesets F_1 and F_2, $F_1 \oplus F_2$ is a tieset or sum of disjoint tiesets.*
(2) *for two cutsets H_1 and H_2, $H_1 \oplus H_2$ is a cutset or sum of disjoint cutsets.*

Proof. (1) Let G_1 and G_2 be induced by tieset F_1 and F_2, respectively. The degree of every vertex of G_1 and G_2 is even. Let G_3 be induced by $F_1 \oplus F_2$. Since $F_1 \oplus F_2$ consists of the edges which are $(F_1 \cup F_2) - (F_1 \cap F_2)$, the degree of every vertex of G_3 is also even. From Theorem 2.7.4, $F_1 \oplus F_2$ is a tieset or sum of disjoint tiesets.

(2) Let H_1 and H_2 be cutsets where $H_1 = E\langle V_1, V_2\rangle$, $V = V_1 \cup V_2$, $V_1 \cap V_2 = \emptyset$ and $H_2 = E\langle V_3; V_4\rangle$, $V = V_3 \cup V_4, V_3 \cap V_4 = \emptyset$. Figure 2.7.3 shows H_1, H_2 and their partitions $\langle V_1, V_2\rangle$, $\langle V_3, V_4\rangle$ of V. In Fig. 2.7.3, $(V_1 \cap V_4) \cup (V_2 \cap V_3) = V_1 \oplus V_3$ and $(V_1 \cap V_3) \cup (V_2 \cap V_4) = V_2 \oplus V_3$.

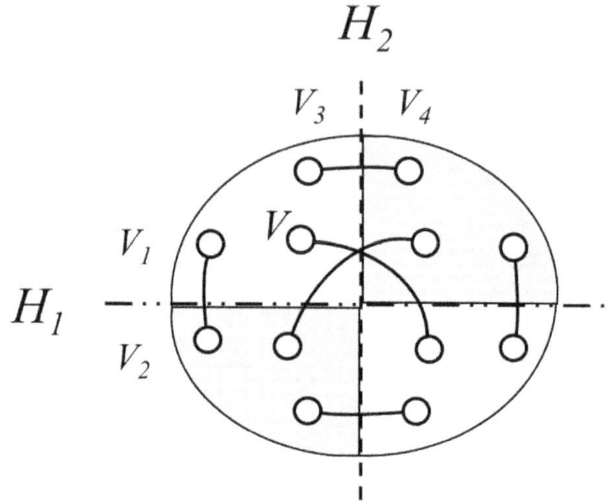

Fig. 2.7.3. Cutsets H_1, H_2, and partitions $\langle V_1, V_2\rangle$ and $\langle V_3, V_4\rangle$ of V.

Let $V_1 \oplus V_3 = V_5$ and $V_2 \oplus V_3 = V_6$, then every edge of $H_1 \oplus H_2$ connects between a vertex of V_5 and a vertex of V_6, and $H_1 \oplus H_2 = E\langle V_5, V_6\rangle$ by the partition $\langle V_5, V_6\rangle$ of V where $V = V_5 \cup V_6$, $V_5 \cap V_6 = \emptyset$. Let G_5 and G_6 be induced graphs by V_5 and V_6, respectively, then if G_5 and G_6 are connected graphs, $H_1 \oplus H_2$ is a cutset, and if G_5 and G_6 are not connected graphs, $H_1 \oplus H_2$ is sum of disjoint cutsets. □

A *directed graph* or *digraph* G is formed by vertices connected by *directed edges*. In digraphs, let us consider *a directed path, a directed tieset* and *a directed cutset*. If every edges in a path or a tieset has same direction, it is called *a directed path, or a directed tieset*, respectively. If every edges in a cutset has same direction, it is called *a directed cutset*. We obtain the following theorem.

Theorem 2.7.6. *A directed tieset and a directed cutset have no common edges. Every edge of a digraph is included only in directed tiesets or directed cutsets.*

Proof. If an edge is included in a directed tieset, a cutset with the edge has an edge of the directed tieset and the direction of the edge is opposite for the cutset. Similarly, if an edge is included in a directed cutset, a tieset

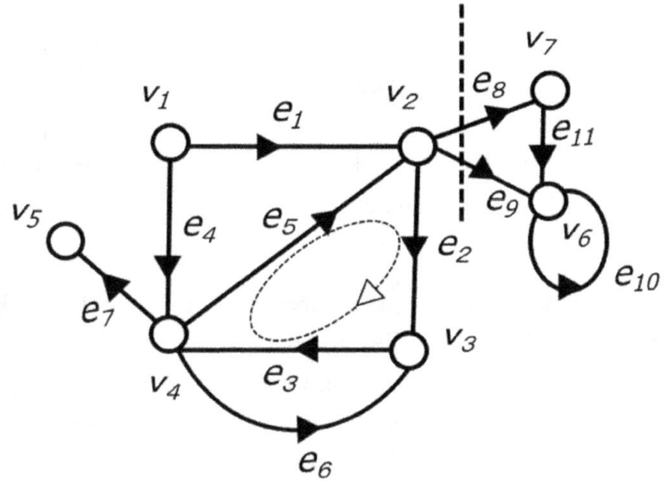

Fig. 2.7.4. A directed tieset and a directed cutset in a directed graph.

with the edge has an edge of the directed cutset and the direction of the edge is opposite for the tieset. Therefore, a directed tieset and a directed cutset have no common edges. Every edge of a digraph is included only in directed tiesets or directed cutsets. □

Example. Figure 2.7.4 shows a directed path, a directed tieset and a directed cutset in a digraph $G = (V, E)$, where $V = \{v_1, v_2, v_3, v_4, v_5, v_6, v_7\}$, the directed edge set $E = \{e_1, e_2, e_3, e_4, e_5, e_6, e_7, e_8, e_9, e_{10}, e_{11}\}$. $\{e_1, e_2\}$ is a directed path from v_1 to v_3, $\{e_2, e_3, e_5\}$ is *a directed tieset* and $\{e_8, e_9\}$ is *a directed cutset*. Every edge of $V_1 = \{e_2, e_3, e_5, e_6, e_{10}\}$ is included in some directed tiesets and every edge of $V_2 = \{e_1, e_4, e_7, e_8, e_9, e_{11}\}$ is included in some directed cutsets, where $V = V_1 \cup V_2$, $V_1 \cap V_2 = \emptyset$.

(End of Example)

A digraph is represented as the same notation $G = (V, E)$ as an undirected graph, where V is the set of vertices and E is the set of edges with orientation. In digraphs, we sometimes consider a tieset and a cutset in the graph obtained by ignoring the orientation of edges and the orientation of the edges are represented by + or – according to the orientation of a tieset and a cutset. For example, let us consider a tieset $\{e_1, e_2, e_3, e_4\}$ (not *a directed tieset*) in the graph obtained by ignoring the orientation of edges in a digraph of Fig. 2.7.4. Let the orientation of the tieset be $e_1 \to e_2 \to e_3 \to e_4$, then the tieset is represented as $\{+e_1, +e_2, +e_3, -e_4\}$ in a digraph G. Let us consider a cutset $\{e_1, e_3, e_5, e_6\}$ (not *a directed cutset*) in the graph obtained by ignoring the orientation of edges in a digraph of Fig. 2.7.4. Let the orientation of the cutset $E\langle V_1, V_2\rangle$ where $V_1 = \{v_1, v_4, v_5\}$ and $V_2 = \{v_2, v_3, v_6, v_7\}$ be $V_1 \to V_2$, then the cutset is represented as $\{+e_1, -e_3, +e_5, +e_6\}$ in a digraph G.

- **Summary:** Introductions of basic concept of graphs and the relationship between tieset and cutset are described.
- **Interesting challenges:** In a graph, there are some interesting relationships between the two concepts of tieset and cutset. Consider if there are any other interesting relationships among the multiple subgraphs of the graph.

Bibliography

[1] B. Bollobás, *Modern Graph Theory*, Springer–Verlag, 2002.

[2] W. K. Chen, *Applied Graph Theory*, North-Holland, Amsterdam, The Netherlands, 1976.

[3] N. Deo, *Graph Theory with Applications to Engineering and Computer Science*, Prentice-Hall, 1974.

[4] F. Harary, *Graph Theory*, Addison-Wesley, Reading, MA., 1969.

[5] O. Ore, *Theory of Graphs*, American Mathematical Society Colloquium Publications XXXVIII, 1962.

[6] K. Ruohonen, *Graph Theory* (Translation by Janne Tamminen, Kung-Chung Lee and Robert Piché), 2013.

[7] S. Seshu and M. B. Reed, *Linear Graphs and Electrical Networks*, Addison-Wesley, 1961.

[8] M. N. S. Swamy and K. Thulasiraman, *Graphs: Theory and Algorithms*, Wiley, 1992.

[9] M. Iri, I. Shirakawa, Y. Kajitani, S. Shinoda, H. Ariyoshi, K. Onaga, T. Ozawa, T. Ohtsuki, T. Nishi, N. Tomizawa, M. Sengoku, R. Ishii, T. Kashiwabara, T. Nishizeki, S. Fujishige and T. Asano, *Graph Theory with Exercise*, Corona Pub. 1983.

[10] G. Chartrand, L. Lesniak and P. Zhang, *Graphs & Digraphs*, Sixth Edition, Chapman and Hall/CRC, 2015.

Chapter 3

Trees, Cotrees and Hybrid Trees

3.1. Trees

In general, trees grow in the forest where plants grow. For example, three trees are shown in Fig. 3.1.1 in a forest. From the form of the forest and trees, the concept of the forest and trees are defined in the graph theory.

A *forest* is a graph that contains no ties, and a connected forest is a *tree*. A *subforest* is a subgraph of a forest. A connected subgraph of a tree is a *subtree*. Generally speaking, a subforest (respectively, subtree) of a graph is its subgraph, which is also a forest (respectively, tree). A tree is shown in Fig. 3.1.2(a), where the vertex at the bottom is *root* vertex. In applications of algorithm etc. the tree that is drawn upside has a hierarchical structure, with one vertex at the top (called the *root*), and the other vertices branching

Fig. 3.1.1. A forest.

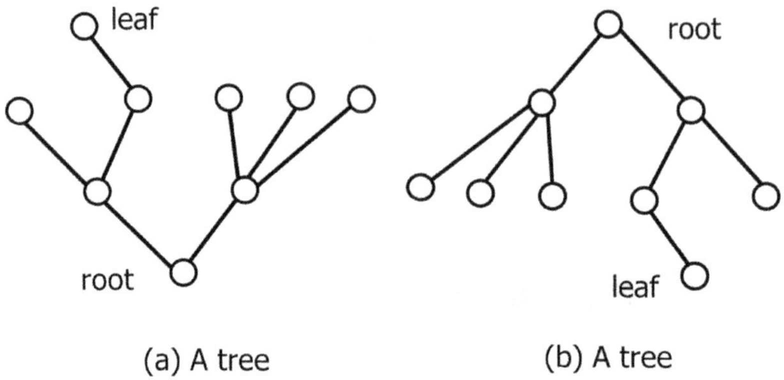

Fig. 3.1.2. A tree and the isomorphic tree.

down from it, as in Fig. 3.1.2(b) where the trees (a) and (b) are isomorphic. A *leaf* (or *pendant vertex*) is a vertex of degree 1.

In a connected graph G, a spanning connected subgraph that contains no ties is called a *tree* of G. That is, a tree of G is a connected subgraph that includes all the vertices of G and no ties. For example, Fig. 3.1.3 shows two trees (a) and (b) of G in Fig. 2.5.1. Since a tree of a connected graph G is a subgraph *induced by the edge set* of the tree, the tree is often represented by the edge set. For example, two trees (a) and (b) of G in Fig. 2.5.1 shown in Fig. 3.1.3 can be represented as $\{e_2, e_3, e_4, e_7, e_8, e_9\}$ and $\{e_4, e_5, e_6, e_7, e_8, e_{11}\}$, respectively.

Lemma 3.1.1. *A tree with at least two vertices has at least one leaf.*

Proof. We assume that a tree with at least two vertices has no leaves. Let v_1 be a vertex of the tree, then v_1 is adjacent to v_2, v_2 is adjacent to v_3 and v_3 is adjacent to v_4 and $v_4, v_5, \ldots$, because the tree has no leaves and the degree of every vertex is larger than 2 in the tree. Since the number of vertex is finite, there is the same vertex in the sequence of vertices v_1, $v_2, \ldots, v_i, \ldots, v_k, v_{k+1}$, If $v_i = v_{k+1}$, there is a tie in a subgraph of induced by the vertex set in the sequence of vertices $v_i, \ldots, v_k, v_{k+1}$. That is, the tree includes a tie which is a contradiction. □

Theorem 3.1.1. *A tree in a connected graph $G = (V, E)$ has $|V| - 1(= \rho(G))$ edges.*

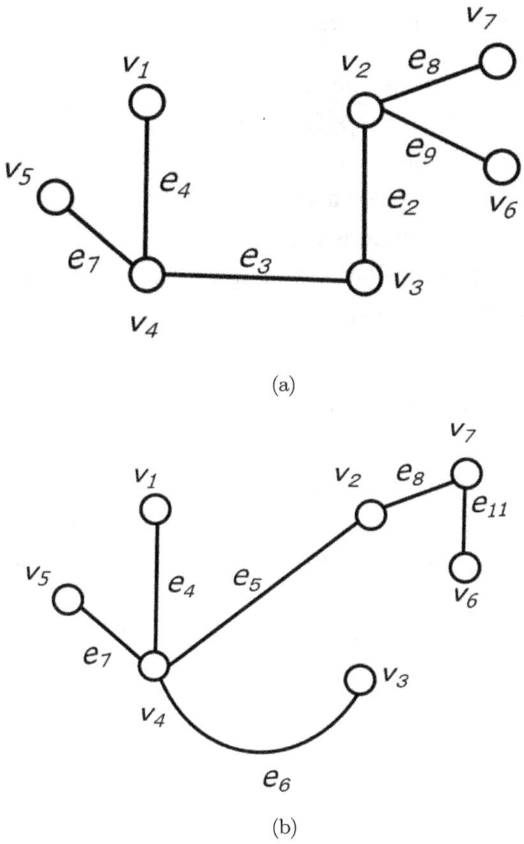

Fig. 3.1.3. Two trees of G in Fig. 2.5.1.

Proof. The proof is by *induction* on the number of edges of a tree. Suppose that a tree with k vertices has $k-1$ edges. Let us consider a tree T with $k+1$ vertices. The tree T with $k+1$ vertices has a leaf from Lemma 3.1. We remove a leaf from the tree. Since the removing a leaf from a tree maintains the connectedness and not having ties, the new graph is a tree T' in which the number of vertices and edges decrease by one. By the induction hypothesis, the tree T' with k vertices has $k-1$ edges. Since the tree T is reconstructed by adding one edge and one vertex, the tree T with $k+1$ vertices has k edges. That is, a tree in a connected graph $G = (V, E)$ has $|V| - 1(= \rho(G))$ edges. □

Theorem 3.1.2. *A subgraph in a connected graph $G = (V, E)$ is a tree if and only if it is connected and the number of edges is $|V| - 1(= \rho(G))$.*

Proof. If a subgraph in a connected graph $G = (V, E)$ is a tree, the number of edges is $|V| - 1(= \rho(G))$ from Theorem 3.1.1. We assume that a connected subgraph with $|V| - 1(= \rho(G))$ edges is not a tree. Since the connected subgraph is not a tree, it has ties. Since the removing an edge in one of the ties maintains the connectedness of the subgraph. If we repeat this removing operation, we obtain a tree which has not ties. The number of edges of this tree is smaller than $|V| - 1(= \rho(G))$ which is a contradiction. □

As I pointed out a moment ago, a tree is often represented by the edge set since a tree of a connected graph G is a subgraph *induced by the edge set* of the tree. Similarly, we often use tiesets and cutsets for ties and cuts, and we often define a tree using the concepts of *tiesets* and *cutsets*. We also use the terms *tiesetfree* and *cutsetfree* to mean that tieset and cutset are not included, respectively.

Theorem 3.1.3. *For a connected graph $G = (V, E)$, the following definitions of a tree are equivalent (and characterize the tree).*

(1) *A tree is connected and tiesetfree.*
(2) *A tree is tiesetfree and has $|V| - 1(= \rho(G))$ edges.*
(3) *A tree shares edges with every cutset and minimal.*
(4) *A tree is maximal tiesetfree.*

Proof. (1) This is the definition of a tree.

(2) If a subgraph S is a tree, S is tiesetfree from the definition and has $|V| - 1(= \rho(G))$ edges from Theorem 3.1.1. Conversely, let us prove that a subgraph S is a tree if S is tiesetfree and has $|V| - 1(= \rho(G))$ edges. The proof is by induction on the number of vertices. If $|V| = 1$, S is one vertex with no edges which is a tree. Let $|V| \geq 2$ and suppose that S is a tree for $|V| = k$. Let us consider S with $k + 1$ vertices. Since S has no tiesets, S has a leaf from the similar way in the proof of Lemma 3.1.1. Let us remove the leaf from S, then the new subgraph S' in which the number of vertices and edges decrease by one has no tiesets. By the induction hypothesis, S' is a

tree. Since S is reconstructed by adding one edge and one vertex to a tree S', S is connected and tiesetfree.

(3) If a subgraph S is a tree, S shares edges with every cutset and minimal. Because a tree is connected and tiesetfree, a connected subgraph in a connected graph has common edges with every cutset and if the subgraph is not minimal, then it has tiesets. Conversely, let us prove that a subgraph S is a tree if S shares edges with every cutset and minimal. Let a subgraph S share edges with every cutset, then it is connected. Because, if a subgraph S dose not share edges of a cutset H_1, there is a partition of vertices $\langle V_1, V_2\rangle$ such that $H_1 = E\langle V_1, V_2\rangle$, $V = V_1 \cup V_2$, $V_1 \cap V_2 = \emptyset$, where H_1 dose not include edges of S. This means that $S$is not connected. That is, if a subgraph S shares edges with every cutset, then it is connected. Let a subgraph S share edges with every cutset and minimal. Then the subgraph S dose not include a self tieset, since a self tieset is not included in any cutset from Corollary 2.7.2(1) and this means a self tieset does not contribute to any cutset. We assume that the subgraph S which dose not include self tiesets, has a tieset. Let e_i be an edge of the tieset of S. Since a tieset and a cutset have an even number of common edges from Theorem 2.7.1 and the tieset of S has an even number of common edges with every cutset, $S - \{e_i\}$ shares edges with every cutset and S is not minimal. This means that S has no tiesets which is a contradiction. Thus a subgraph S is a tree if S shares edges with every cutset and minimal.

(4) If a subgraph S is a tree, S is maximal tiesetfree. Because if an edge e_i other than S is added to S, $S + \{e_i\}$ has a tieset since a tree is connected and tiesetfree. This means that if a subgraph S is a tree, S is a maximal tiesetfree. Conversely, let us prove that a subgraph S is a tree if S is a maximal tiesetfree. Let a subgraph S be a maximal tiesetfree. Then the subgraph S includes all self cutsets in G, since a self cutset is not included in any tieset from Corollary 2.7.2(2). We assume that the subgraph S which includes all self cutsets is a maximal tiesetfree and not connected. If S is not connected, there is a partition of vertices $\langle V_1, V_2\rangle$, $V = V_1 \cup V_2$, $V_1 \cap V_2 = \emptyset$, where V_1 is included in a connected component of S, such that one end vertex in every edge of $E\langle V_1, V_2\rangle$ is in V_1 and the other end vertex is in $V_2 E\langle V_1, V_2\rangle$

includes cutsets which dose not include edges of S. Let e_i be an edge of one of the cutsets. If $S + \{e_i\}$ has a tieset, the common edge of the tieset and the cutset of $E\langle V_1, V_2\rangle$ is only e_i which is impossible from Theorem 2.7.1. That is $S + \{e_i\}$ is tiesetfree which means that S is not maximal tiesetfree. This is a contradiction. That is, S is connected and tiesetfree. □

3.2. Cotrees

The *complement* of the subgraph $G_1 = (V_1, E_1)$ of a graph $G = (V, E)$ is the subgraph $G_2 = (V_2, E_2)$, where $V = V_1 = V_2$ and $E_2 = E - E_1$. A *cotree* is defined as the *complement* of a tree.

For example, Fig. 3.2.1 shows two cotrees (a) and (b) of G in Fig. 2.5.1 which are the *complement* of trees in Fig. 3.1.3. Since a cotree of a connected graph G is a subgraph *induced by the edge set* of the cotree, the cotree is often represented by the edge set. For example, two cotrees of G in Fig. 2.5.1 shown in Fig. 3.2.1 can be represented as $\{e_1, e_5, e_6, e_{10}, e_{11}\}$ and $\{e_1, e_2, e_3, e_9, e_{10}\}$.

Theorem 3.2.1. *For a connected graph $G = (V, E)$, the following definitions of a cotree are equivalent (and characterize the cotree).*

(1) *A cotree is a complement of a tree.*
(2) *A cotree is cutsetfree and has $|E| - |V| + 1(= \mu(G))$ edges.*
(3) *A cotree shares edges with every tietset and minimal.*
(4) *A cotree is maximal cutsetfree.*

Proof. (1) This is the definition of a cotree.

(2) First, if a subgraph S is a cotree, S is cutsetfree and has $|E| - |V| + 1(= \mu(G))$ edges since S is a complement of a tree. Next, let us prove that a subgraph S is a cotree if S is cutsetfree and has $|E| - |V| + 1(= \mu(G))$ edges. Let a subgraph S be cutsetfree, then the complement of S is connected. If the number of edges of S is $|E| - |V| + 1(= \mu(G))$, the complement of S has $|V| - 1$ edges since the complement of S is connected. That is, the complement of S is a tree.

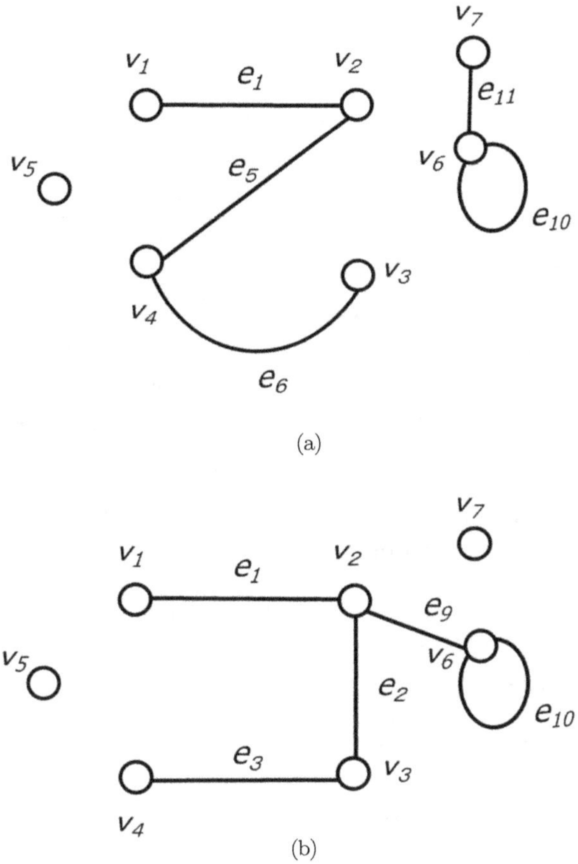

Fig. 3.2.1. Two cotrees of G in Fig. 2.5.1. (a) A cotree of a tree in Fig. 3.1.3(a). (b) A cotree of a tree in Fig. 3.1.3(b)

(3) First let us prove that if a subgraph S is a cotree, S shares edges with every tieset and minimal. Suppose that a tieset in G does not share with S, then the tieset is included in the complement of the cotree which is a tree with a tieset. This is a contradiction. That is, S shares edges with every tieset. Since a tree is maximal tiesetfree (Theorem 3.1.3(4)), S shares edges with every tieset and minimal. Next let us prove that if S shares edges with every tietset and minimal, a subgraph S is a cotree. Since S shares edges with every tietset, the complement of S is tiesetfree. If the complement of

S is tiesetfree and maximal, it is a tree. That is, if S shares edges with every tietset and minimal, a subgraph S is a cotree.

(4) First let us prove that if a subgraph S is a cotree, S is maximal cutsetfree. Suppose that a cutset in G is included in S, then the complement of the cotree is not connected. Since the complement of the cotree is a tree which is connected. This is a contradiction. That is, S is cutsetfree. Suppose that S is not maximal, then there is an edge e_i of S such that $S - \{e_i\}$ is cutsetfree. The complement of $S - \{e_i\}$ has tieset which is a contradiction. That is, if a subgraph S is a cotree, S is maximal cutsetfree. Conversely, let us prove that if S is maximal cutsetfree, S is a cotree. Since a subgraph S which is maximal cutsetfree means that the complement of S shares edges with every cutset and minimal (Theorem 1.3,(3)), S is the complement of a tree. That is, if S is maximal cutsetfree, S is a cotree. □

In Theorems 3.1.3 and 3.2.1, a tree and a cotree of a connected graph G are defined. There is another definition of a tree and a cotree.

Theorem 3.2.2. *E_1 and E_2 are a tree and a cotree respectively in a connected graph $G = (V, E)$ where $E_1 \cup E_2 = E$, $E_1 \cap E_2 = \emptyset$ if and only if E_1 and E_2 are tiesetfree and cutsetfree, respectively, in G.*

Proof. Let E_1 and E_2 are a tree and a cotree respectively in a connected graph $G = (V, E)$ where $E_1 \cup E_2 = E$, $E_1 \cap E_2 = \emptyset$, then E_1 and E_2 are tiesetfree and cutsetfree respectively in G from Theorems 3.1.3(2) and 3.2.1(2). Conversely, if E_2 is cutsetfree where $E_1 \cup E_2 = E$, $E_1 \cap E_2 = \emptyset$, it is clear that E_1 is connected. If E_1 is tiesetfree and connected, E_1 is a tree. That is, if E_1 and E_2 are tiesetfree and cutsetfree, respectively, in G, E_1 and E_2 are a tree and a cotree respectively in G. □

3.3. Fundamental Cutsets and Fundamental Tiesets

Let t and c be a tree and the cotree of a connected graph $G = (V, E)$.

$$t = \{e_{t1}, e_{t2}, \ldots, e_{t\rho(G)}\}, \tag{3.3.1}$$

$$c = \{e_{c1}, e_{c2}, \ldots, e_{c\mu(G)}\}, \tag{3.3.2}$$

$$E = t + c, \quad t \cap c = \emptyset. \tag{3.3.3}$$

A cotree is maximal cutsetfree from Theorem 3.2.1(4). This means that if an edge e_{ti} of t is added to a cotree c, there exists one and only one cutset h_i in $c + \{e_{ti}\}$ which is the subgraph of G induced by the edges of c and e_{ti}. We call this cutset a *fundamental cutset.* Every cotree defines $\rho(G)$ fundamental cutsets $h_1, h_2, \ldots, h_{\rho(G)}$, which together form a fundamental set of cutsets. Every fundamental cutset includes exactly one edge of t and every edge of t belongs to exactly one fundamental cutset. Therefore, every tree defines a unique fundamental set of cutsets for G.

A tree is maximal tiesetfree from Theorem 3.1.3(4). This means that if an edge e_{cj} of c is added to a tree t, there exists one and only one tieset f_j in $t + \{e_{cj}\}$ which is the subgraph of G induced by the edges of t and e_{cj} We call this tieset a *fundamental tieset.* Every tree defines $\mu(G)$ fundamental tiesets $f_1, f_2, \ldots, f_{\mu(G)}$, which together form a fundamental set of tiesets. Every fundamental tieset includes exactly one edge of c and every edge of c belongs to exactly one fundamental tieset. Therefore, every cotree defines a unique fundamental set of tiesets for G.

Theorem 3.3.1. *Let t and c be a tree and the cotree of a connected graph $G = (V, E)$, where $t = \{e_{t1}, e_{t2}, \ldots, e_{t\rho(G)}\}$, $c = \{e_{c1}, e_{c2}, \ldots, e_{c\mu(G)}\}$ and $E = t + c, t \cap c = \emptyset$. For $e_{ti} \in t$ and $e_{cj} \in c$, let h_i and f_j be the fundamental cutset including in $c + \{e_{ti}\}$ and the fundamental tieset including in $t + \{e_{cj}\}$, respectively. If $e_{cj} \in h_i$, then $e_{ti} \in f_j$ and if $e_{ti} \in f_j$, then $e_{cj} \in h_i$.*

Proof. Since $h_i \subset c + \{e_{ti}\}$ and $f_j \subset t + \{e_{cj}\}$, $h_i \cap f_j \subset \{e_{ti}, e_{cj}\}$. From Theorem 2.7.1, h_i and f_j have an even number (including 0) of common edges. That is, $h_i \cap f_j = \emptyset$ or $h_i \cap f_j = \{e_{ti}, e_{cj}\}$. From a given selection of e_{ti} and e_{cj}, $h_i \cap f_j = \{e_{ti}, e_{cj}\}$. Thus, if $e_{cj} \in h_i$, then $e_{ti} \in f_j$ and if $e_{ti} \in f_j$, then $e_{cj} \in h_i$. □

Example. Figure. 3.3.1 is a connected graph G of Fig. 2.5.1. $G = (V, E)$, where $V = \{v_1, v_2, v_3, v_4, v_5, v_6, v_7\}$, $E = \{e_1, e_2, e_3, e_4, e_5, e_6, e_7, e_8, e_9, e_{10}, e_{11}\}$. $\rho(G) = |V| - 1 = 7 - 1 = 6, \mu(G) = |E| - |V| + 1 = 11 - 7 + 1 = 5$.

Let a tree and the cotree be t and c, where $t = \{e_2, e_3, e_4, e_7, e_8, e_9\}$, $c = \{e_1, e_5, e_6, e_{10}, e_{11}\}$ and $E = t + c$. In Fig. 3.3.1, bold lines show the

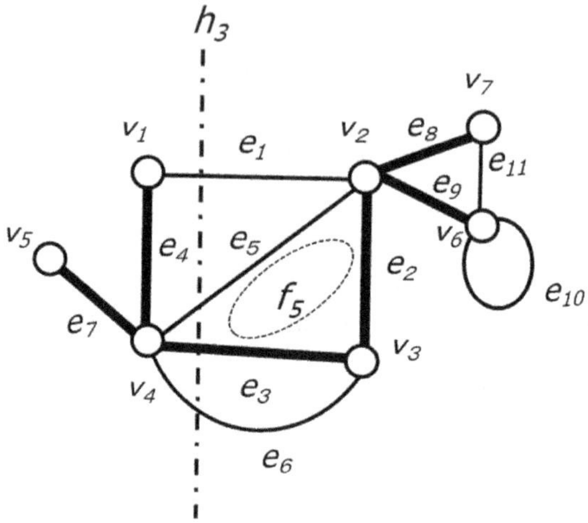

Fig. 3.3.1. Fundamental cutset $h_3$and fundamental tieset f_5 (Bold lines show a tree).

tree. There exists one and only one cutset h_3 in $c+\{e_3\}$ where e_3 is an edge of t and $h_3 = \{e_1, e_3, e_5, e_6\}$ shown in Fig. 3.3.1. h_3 is a fundamental cutset and there are $\rho(G)$ fundamental cutsets for a tree t. There exists one and only one tieset f_5 in $t+\{e_5\}$ where e_5 is an edge of c and $f_5 = \{e_2, e_3, e_5\}$ shown in Fig. 3.3.1. f_5 is a fundamental tietset and there are $\mu(G)$ fundamental tiesets for a cotree c. In h_3 and f_5, $e_5 \in h_3$ and $e_3 \in f_5$.

(End of Example)

Every fundamental cutset and tieset has exactly one edge which is not in any other fundamental cutset and tieset in the fundamental set of cutsets and tiesets, respectively. Therefore, we cannot write any fundamental cutset(tieset) as a ring sum of other fundamental cutsets(tiesets) in the same set. In other words, the fundamental set of cutsets(tiesets) is linearly independent under the ring sum operation.

Thus, using Theorem 2.7.5, we obtain the following theorem.

Theorem 3.3.2. *In a connected graph G,*

(1) *the fundamental set of cutsets is linearly independent under the ring sum operation.*

(2) *the fundamental set of tiesets is linearly independent under the ring sum operation.*

3.4. Short-Circuiting and Open-Circuiting

In Chapter 2 (Section 2.5), we introduce *short-circuiting of edges* and *open-circuiting of edges* in a graph $G = (V, E)$. If $E_1 \subset E$, $E_2 \subset E$, $E_1 \cap E_2 = \emptyset$ in the graph $G = (V, E)$, then the graph $G' = (V', E')$, where G' is obtained by *open-circuiting of edges* of E_1 and *short-circuiting of edges* of E_2. Let us denote by $G_{os}[E_1; E_2]$ the graph $G' = (V', E')$. We will examine the relations among trees, cotrees and short-circuiting, open-circuiting of edges in G.

Lemma 3.4.1. *In a connected graph $G = (V, E)$, we have the following conditions*:

(1) *Let $E_1 \subset E$ be a cutsetfree. The cutsets of Graph $G' = G_{os}[E_1; \emptyset]$ obtained by open-circuiting of E_1 are cutsets of G or cutsets deleting edges of E_1 in G.*
(2) *Let $E_2 \subset E$ be a tiesetfree. The tiesets of graph $G' = G_{os}[\emptyset; E_2]$ obtained by short-circuiting of E_2 are tiesets of G or tiesets deleting edges of E_2 in G.*

Proof. (1) Since $E_1 \subset E$ is a cutsetfree, the subgraph induced by $E - E_1$ is connected and maintains every cutset of G except edges of E_1. That is, the cutsets of Graph G' obtained by open-circuiting of E_1 are cutsets of G or cutsets deleting edges of E_1 in G.

(2) Since $E_2 \subset E$ be a tiesetfree, The graph G' obtained by short-circuiting of E_2 maintains every tiesets of G except edges of E_2. That is, the tiesets of graph G' obtained by short-circuiting of E_2 are tiesets of G or tiesets deleting edges of E_2 in G. □

Theorem 3.4.1. *In a connected graph $G = (V, E)$,*

(1) *$E_c \subset E$ is a cotree if and only if E_1 is a cotree in the graph $G_{os}[E_2; \emptyset]$ where $E_c = E_1 \cup E_2$, $E_1 \cap E_2 = \emptyset$ and E_2 is cutsetfree in G;*

(2) *$E_t \subset E$ is a tree if and only if E_1 is a tree in the graph $G_{os}[\emptyset; E_2]$ where $E_t = E_1 \cup E_2$, $E_1 \cap E_2 = \emptyset$ and E_2 is tiesetfree in G.*

Proof. (1) Let $E_c \subset E$ be a cotree, then E_c is cutsetfree and $|E_c| = |E| - |V| + 1$. Let $E_c = E_1 \cup E_2$, $E_1 \cap E_2 = \emptyset$, then E_2 is cutsetfree and the numbers of vertices and edges are $|V|$ and $|E| - |E_2|$ in $G_{os}[E_2; \emptyset]$. The number of edges of a cotree of $G_{os}[E_2; \emptyset]$ is $(|E| - |E_2|) - |V| + 1$ which means $|E_c| - |E_2| = |E_1|$. Since E_1 is cutsetfree and the number edges is nullity of $G_{os}[E_2; \emptyset]$, E_1 is a cotree from Theorem 3.2.1(2). Conversely, since $E_2 \subset E$ is cutsetfree in G, the cutsets of graph $G_{os}[E_2; \emptyset]$ are cutsets of G or cutsets deleting edges of E_1 in G from Lemma 3.4.1(1). This means that $Ec = E_1 \cup E_2$ is cutsetfree in G. The number of edges of a cotree in $G_{os}[E_2; \emptyset]$ is $|E_1| = |E| - |E_2| - (|V| - 1)$.$|Ec| = |E_1| + |E_2| = (|E| - |E_2| - (|V| - 1)) + |E_2| = |E| - |V| + 1 (= \mu(G))$. Hence, $E_c \subset E$ is a cotree from Theorem 3.2.1(2).

(2) Let $E_c \subset E$ be a tree, then E_c is tiesetfree and $|E_c| = |V| - 1$. Let $Ec = E_1 \cup E_2$, $E_1 \cap E_2 = \emptyset$, then E_2 is tiesetfree and the numbers of vertices and edges are $|V| - |E_2|$ and $|E| - |E_2|$ in $G_{os}[\emptyset; E_2]$. The number of edges of a tree of $G_{os}[E_2; \emptyset]$ is $(|V| - |E_2|) - 1$ which means $|E_c| - |E_2| = |E_1|$. Since E_1 is tiesetfree and the number edges is rank of $G_{os}[E_2; \emptyset]$, E_1 is a tree from Theorem 3.1.3(2). Conversely, since $E_2 \subset E$ is tiesetfree in G, the tiesets of graph $G_{os}[\emptyset; E_2]$ are tiesets of G or tietsets deleting edges of E_1 in G from Lemma 3.4.1(2). This means that $Et = E_1 \cup E_2$ is tiesetfree in G. The number of edges of a tree in $G_{os}[\emptyset; E_2]$ is $|E_1| = \rho(G_{os}[\emptyset; E_2]) = \rho(G) - |E_2|$. $|Et| = |E_1| + |E_2| = (\rho(G) - |E_2|) + |E_2| = |V| - 1 (= \rho(G))$. Hence, $E_t \subset E$ is a tree from Theorem 3.1.3(2). □

Corollary 3.4.1. *In a connected graph $G = (V, E)$, let e_k be an edge which is neither a self cutset nor a self tieset.*

Let the sets of trees of $G_{os}[\emptyset; \{e_k\}]$ and $G_{os}[\{e_k\}; \emptyset]$ be $T\begin{pmatrix} e_k \\ 1 \end{pmatrix}$ and $T\begin{pmatrix} e_k \\ 0 \end{pmatrix}$, respectively. Then the set of trees of G is as follows:

$$T(G) = \{e_k\} \times T\begin{pmatrix} e_k \\ 1 \end{pmatrix} + T\begin{pmatrix} e_k \\ 0 \end{pmatrix}. \tag{3.4.1}$$

where $T(G)$ is the set of trees of G and the set operation $\times$ is define as follows.

$$A \times B = \{a_i \cup b_j | a_i \in A,\ b_j \in B\}. \tag{3.4.2}$$

Proof. Since e_k is an edge which is neither a self cutset nor a self tieset, $\{e_k\}$ can be an element of $G_{os}[\emptyset; \{e_k\}]$ and $G_{os}[\{e_k\}; \emptyset]$. It is obvious that adding $\{e_k\}$ to a tree of $G_{os}[\emptyset; \{e_k\}]$ makes a tree of G and a tree of $G_{os}[\{e_k\}; \emptyset]$ represents a tree not containing e_k. □

Theorem 3.4.2. *In a connected graph $G = (V, E)$, $E_t \subset E$ without loops is a tree if and only if there is one path between any two given vertices.*

Proof. If E_t is a tree, there is one path between any two given vertices, since a tree is connected and tiesetfree. Suppose that there is at least one path between any two given vertices. If there is at least one path between any two given vertices, E_t is connected. If there is two paths between some two vertices, the two paths include tiesets which is a contradiction. That is, E_t is a tree. □

Let $T(G)$ and $C(G)$ be the set of trees and cotrees of a connected graph $G = (V, E)$.

$$T(G) = \{t_1, t_2, \ldots, t_m\}, \tag{3.4.3}$$

$$C(G) = \{c_1, c_2, \ldots, c_m\}. \tag{3.4.4}$$

Theorem 3.4.3. *Let $G_1, G_2, \ldots$ be separable components in a connected graph $G = (V, E)$. Let $T_1, T_2, \ldots (C_1, C_2, \ldots)$ be the set of trees (cotree) of $G_1, G_2, \ldots$, respectively. Then,*

$$T(G) = T(G_1) \times T(G_2) \times \cdots, \tag{3.4.5}$$

$$C(G) = C(G_1) \times C(G_2) \times \cdots. \tag{3.4.6}$$

Proof. Since every two separable components has at most one common vertex in which the paths between two separable components always go through the vertex, the part of a tree consists of disjoint sets of trees of the two separable components, a tree of G consists of sum of trees of every separable component $G_1, G_2, \ldots$ Since a cotree is a complement of a tree, it is clear that cotrees are similar as trees. □

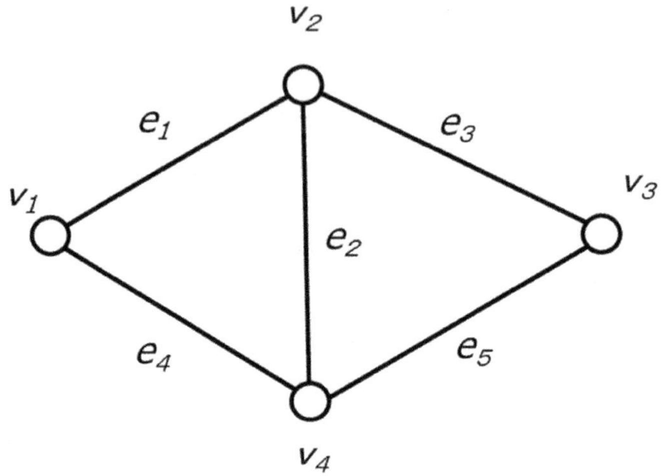

Fig. 3.4.1. A connected graph $G = (V, E)$.

In order to generate tree set it is easier for a small graph than a big graph. Using Theorem 3.4.1–3.4.3, we will show a method to generate tree set of a connected graph G.

Let us consider all paths between two vertices. Since a path is tiesetfree and a tree includes always paths, we get the following conditions.

$$P(v_i, v_j) = \{p_1, p_2, \ldots, p_n\}, \tag{3.4.7}$$

$$T(G) = \bigcup_{k=1}^{n} (p_k \times (G_{as}[\emptyset; p_k])). \tag{3.4.8}$$

Example. (1) Let us generate the set of trees in a connected graph $G = (V, E)$ of Fig. 3.4.1, where $V = \{v_1, v_2, v_3, v_4\}$, $E = \{e_1, e_2, e_3, e_4, e_5\}$. Let $P(v_2, v_4) = \{p_1, p_2, p_3\}$, where $p_1 = \{e_1, e_4\}$, $p_2 = \{e_2\}$, $p_3 = \{e_3, e_5\}$.

In Fig. 3.4.2, we can find $T(G_{os}[\emptyset; p_1])$, $T(G_{os}[\emptyset; p_1])$ and $T(G_{os}[\emptyset; p_3])$ easily because they are small graph.

$$T(G_{os}[\emptyset; p_1]) = \{\{e_3\}, \{e_5\}\},$$

$$T(G_{os}[\emptyset; p_2]) = \{\{e_1e_3\}, \{e_1e_5\}, \{e_3e_4\}, \{e_4e_5\}\},$$

$$T(G_{os}[\emptyset; p_3]) = \{\{e_1\}, \{e_4\}\}.$$

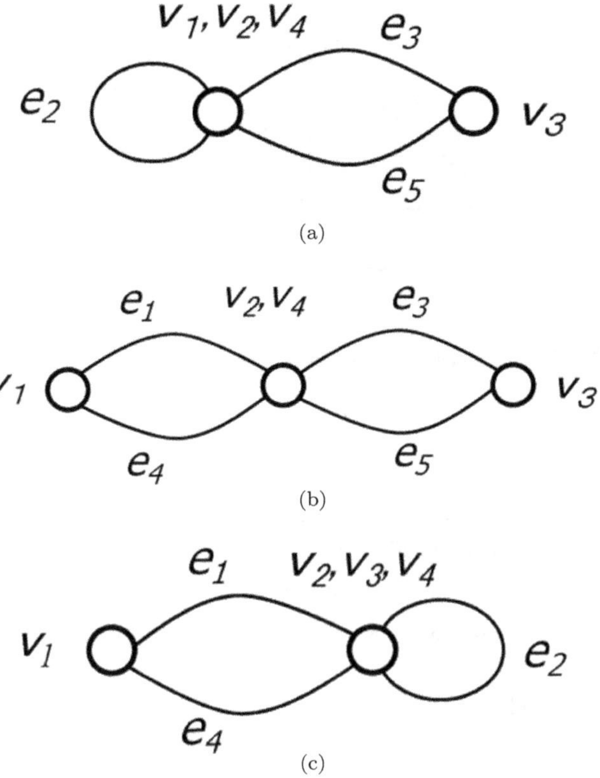

Fig. 3.4.2. $G_{os}[\emptyset; p_k]$ for $P(v_2, v_4)$ in a connected graph $G = (V, E)$. (a) A graph $G_{os}[\emptyset; p_1]$ where $p_1 = \{e_1, e_4\}$. (b) A graph $G_{os}[\emptyset; p_2]$ where $p_2 = \{e_2\}$. (c) A graph $G_{os}[\emptyset; p_3]$ where $p_3 = \{e_3, e_5\}$.

Thus,

$$T(G) = \bigcup_{k=1}^{3}(p_k \times T(G_{as}[\emptyset; p_k])) = \{e_1, e_4\} \times \{\{e_3\}, \{e_5\}\} \cup \{e_2\} \times \{\{e_1e_3\}, \{e_1e_5\}, \{e_3e_4\}, \{e_4e_5\}\} \cup \{e_3, e_5\} \times \{\{e_1\}, \{e_4\}\} = \{\{e_1, e_3, e_4\}, \{e_1, e_4, e_5\}, \{e_1, e_2, e_3\}, \{e_1, e_2, e_5\}, \{e_2, e_3, e_4\}, \{e_2, e_4, e_5\}, \{e_1, e_3, e_5\}, \{e_3, e_4, e_5\}\}, \tag{3.4.9}$$

$$C(G) = \{\{e_2, e_5\}, \{e_2, e_3\}, \{e_4, e_5\}, \{e_3, e_4\}, \{e_1, e_5\}, \{e_1, e_3\}, \{e_2, e_4\}, \{e_1, e_2\}\}. \tag{3.4.10}$$

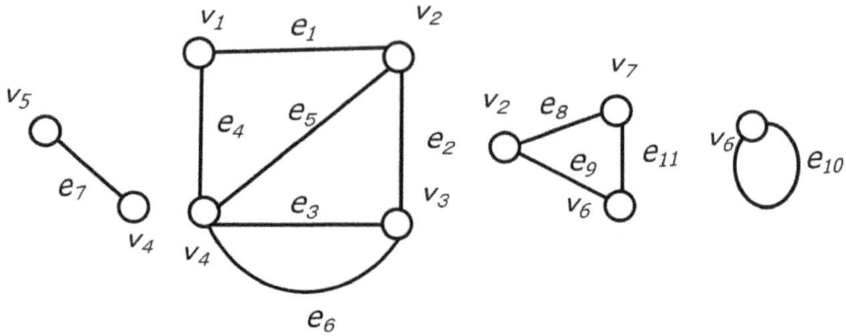

Fig. 3.4.3. The four nonseparable components G_1, G_2. G_3 and G_4 of a connected graph G in Fig. 2.5.1.

(2) Let us generate the set of trees in a connected graph $G = (V, E)$ of Fig. 2.5.1, where $V = \{v_1, v_2, v_3, v_4, v_5, v_6, v_7\}$, $E = \{e_1, e_2, e_3, e_4, e_5, e_6, e_7, e_8, e_9, e_{10}, e_{11}\}$. Figure 2.7.2 shows four nonseparable components and Fig. 3.4.3 shows the four nonseparable components G_1, G_2, G_3 and G_4 of a connected graph G in Fig. 2.5.1. From (3.4.5),

$$
\begin{aligned}
T(G) &= T(G_1) \times T(G_2) \times T(G_3) \times T(G_4),\\
T(G_1) &= \{\{e_7\}\},\\
T(G_2) &= \{\{e_1, e_2, e_4\}, \{e_1, e_3, e_4\}, \{e_1, e_4, e_6\}, \{e_1, e_2, e_5\},\\
&\qquad \{e_1, e_3, e_5\}, \{e_1, e_5, e_6\}, \{e_2, e_4, e_5\}, \{e_3, e_4, e_5\},\\
&\qquad \{e_4, e_5, e_6\}, \{e_1, e_2, e_3\}, \{e_2, e_3, e_4\}, \{e_1, e_2, e_6\}, \{e_2, e_4, e_6\}\},\\
T(G_3) &= \{\{e_8, e_9\}, \{e_8, e_{11}\}, \{e_9, e_{11}\}\},\\
T(G_4) &= \{\emptyset\}.
\end{aligned}
$$

Figure 3.4.4 shows a method to generate the tree set of G_2. Let $P(v_2, v_4) = \{p_1, p_2, p_3, p_4\}$, where $p_1 = \{e_1, e_4\}$, $p_2 = \{e_5\}$, $p_3 = \{e_2, e_3\}$, $p_4 = \{e_2, e_6\}$ in G_2.

$$
\begin{aligned}
T(G_{2os}[\emptyset; p_1]) &= \{\{e_2\}, \{e_3\}, \{e_6\}\},\\
T(G_{2os}[\emptyset; p_2]) &= \{\{e_1e_2\}, \{e_1e_3\}, \{e_1e_6\}, \{e_4e_5\}, \{e_2e_4\}, \{e_4e_6\}\},
\end{aligned}
$$

$$T(G_{2os}[\emptyset; p_3]) = \{\{e_1\}, \{e_4\}\},$$
$$T(G_{2os}[\emptyset; p_4]) = \{\{e_1\}, \{e_4\}\}.$$

Thus,

$$\begin{aligned}
T(G_2) = \bigcup_{k=1}^{4}(p_k \times T(G_{2as}[\emptyset; p_k])) &= \{e_1, e_4\} \times \{\{e_2\}, \{e_3\}, \{e_6\}\} \cup \{e_5\} \\
&\times \{\{e_1e_2\}, \{e_1e_3\}, \{e_1e_6\}, \{e_4e_5\}, \{e_2e_4\}, \{e_4e_6\}\} \cup \{e_2, e_3\} \\
&\times \{\{e_1\}, \{e_4\}\} \cup \{e_2, e_6\} \times \{\{e_1\}, \{e_4\}\} \\
&= \{\{e_1, e_2, e_4\}, \{e_1, e_3, e_4\}, \{e_1, e_4, e_6\}, \{e_1, e_2, e_5\}, \{e_1, e_3, e_5\}, \\
&\{e_1, e_5, e_6\}, \{e_2, e_4, e_5\}, \{e_3, e_4, e_5\}, \{e_4, e_5, e_6\}, \{e_1, e_2, e_3\}, \\
&\{e_2, e_3, e_4\}, \{e_1, e_2, e_6\}, \{e_2, e_4, e_6\}\}.
\end{aligned}$$

Therefore,

$$\begin{aligned}
T(G) &= T(G_1) \times T(G_2) \times T(G_3) \times T(G_4) \\
&= \{\{e_7\}\} \times \{\{e_1, e_2, e_4\}, \{e_1, e_3, e_4\}, \{e_1, e_4, e_6\}, \\
&\{e_1, e_2, e_5\}, \{e_1, e_3, e_5\}, \{e_1, e_5, e_6\}, \{e_2, e_4, e_5\}, \\
&\{e_3, e_4, e_5\}, \{e_4, e_5, e_6\}, \{e_1, e_2, e_3\}, \{e_2, e_3, e_4\}, \\
&\{e_1, e_2, e_6\}, \{e_2, e_4, e_6\}\} \times \{\{e_8, e_9\}, \{e_8, e_{11}\}, \{e_9, e_{11}\}\} \times \{\emptyset\} \\
&= \{\{e_1, e_2, e_4, e_7, e_8, e_9\}, \{e_1, e_2, e_4, e_7, e_8, e_{11}\}, \\
&\{e_1, e_2, e_4, e_7, e_9, e_{11}\}, \{e_1, e_3, e_4, e_7, e_8, e_9\}, \\
&\{e_1, e_3, e_4, e_7, e_8, e_{11}\}, \{e_1, e_3, e_4, e_7, e_9, e_{11}\}, \\
&\{e_1, e_4, e_6, e_7, e_8, e_9\}, \{e_1, e_4, e_6, e_7, e_8, e_{11}\}, \\
&\{e_1, e_4, e_6, e_7, e_9, e_{11}\}, \{e_1, e_2, e_5, e_7, e_8, e_9\}, \\
&\{e_1, e_2, e_5, e_7, e_8, e_{11}\}, \{e_1, e_2, e_5, e_7, e_9, e_{11}\}, \\
&\{e_1, e_3, e_5, e_7, e_8, e_9\}, \{e_1, e_3, e_5, e_7, e_8, e_{11}\},
\end{aligned}$$

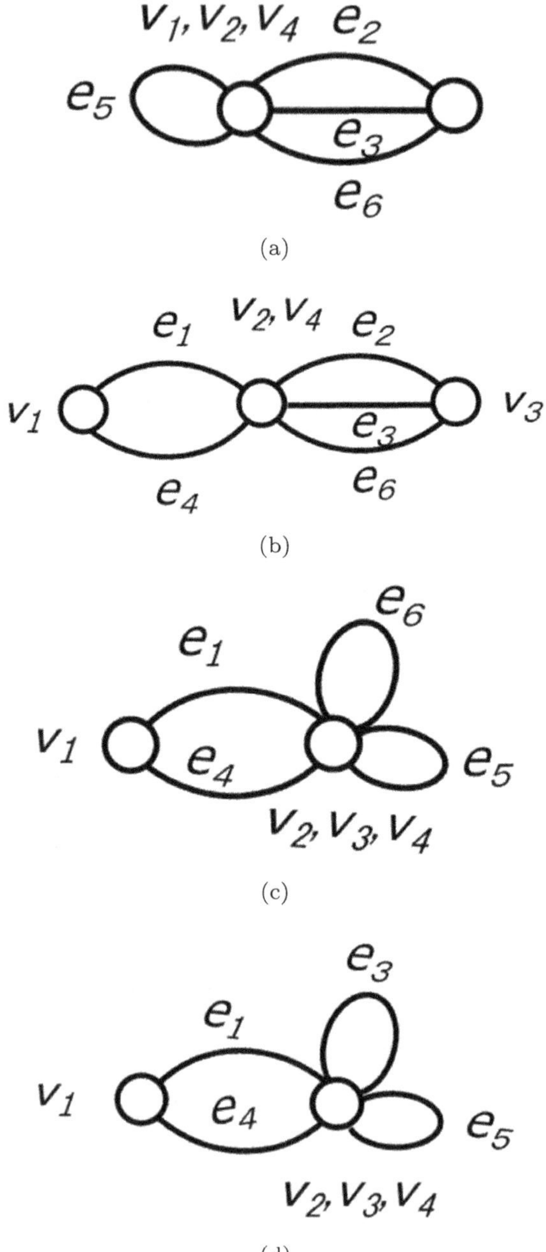

Fig. 3.4.4. $G_{2os}[\emptyset; p_k]$ for $P(v_2, v_4)$ in a nonseparable component G_2. (a) A graph $G_{2os}[\emptyset; p_1]$ where $p_1 = \{e_1, e_4\}$. (b) A graph $G_{2os}[\emptyset; p_2]$ where $p_2 = \{e_5\}$. (c) A graph $G_{2os}[\emptyset; p_3]$ where $p_3 = \{e_2, e_3\}$. (d) A graph $G_{2os}[\emptyset; p_4]$ where $p_4 = \{e_2, e_6\}$.

$$\{e_1, e_3, e_5, e_7, e_9, e_{11}\}, \{e_1, e_5, e_6, e_7, e_8, e_9\},$$
$$\{e_1, e_5, e_6, e_7, e_8, e_{11}\}, \{e_1, e_5, e_6, e_7, e_9, e_{11}\},$$
$$\{e_2, e_4, e_5, e_7, e_8, e_9\}, \{e_2, e_4, e_5, e_7, e_8, e_{11}\},$$
$$\{e_2, e_4, e_5, e_7, e_9, e_{11}\}, \{e_3, e_4, e_5, e_7, e_8, e_9\},$$
$$\{e_3, e_4, e_5, e_7, e_8, e_{11}\}, \{e_3, e_4, e_5, e_7, e_9, e_{11}\},$$
$$\{e_4, e_5, e_6, e_7, e_8, e_9\}, \{e_4, e_5, e_6, e_7, e_8, e_{11}\},$$
$$\{e_4, e_5, e_6, e_7, e_9, e_{11}\}, \{e_1, e_2, e_3, e_7, e_8, e_9\},$$
$$\{e_1, e_2, e_3, e_7, e_8, e_{11}\}, \{e_1, e_2, e_3, e_7, e_9, e_{11}\},$$
$$\{e_2, e_3, e_4, e_7, e_8, e_9\}, \{e_2, e_3, e_4, e_7, e_8, e_{11}\},$$
$$\{e_2, e_3, e_4, e_7, e_9, e_{11}\}, \{e_1, e_2, e_6, e_7, e_8, e_9\},$$
$$\{e_1, e_2, e_6, e_7, e_8, e_{11}\}, \{e_1, e_2, e_6, e_7, e_9, e_{11}\},$$
$$\{e_2, e_4, e_6, e_7, e_8, e_9\}, \{e_2, e_4, e_6, e_7, e_8, e_{11}\},$$
$$\{e_2, e_4, e_6, e_7, e_9, e_{11}\}\}, \tag{3.4.11}$$

$$\begin{aligned} C(G) &= C(G_1) \times C(G_2) \times C(G_3) \times C(G_4) \\ &= \{\emptyset\} \times \{\{e_3, e_5, e_6\}, \{e_2, e_5, e_6\}, \\ &\quad \{e_2, e_3, e_5\}, \{e_3, e_4, e_6\}, \{e_2, e_4, e_6\}, \{e_2, e_3, e_4\}, \\ &\quad \{e_1, e_3, e_6\}, \{e_1, e_2, e_6\}, \{e_1, e_2, e_3\}, \{e_4, e_5, e_6\}, \\ &\quad \{e_1, e_5, e_6\}, \{e_3, e_4, e_5\}, \{e_1, e_3, e_5\}\} \times \{\{e_{11}\}, \{e_9\}, \\ &\quad \{e_8\}\} \times \{\{e_{10}\}\} = \{\{e_3, e_5, e_6, e_{10}, e_{11}\}, \{e_3, e_5, e_6, e_9, e_{10}\}, \\ &\quad \{e_3, e_5, e_6, e_8, e_{10}\}, \{e_2, e_5, e_6, e_{10}, e_{11}\}, \{e_2, e_5, e_6, e_9, e_{10}\}, \\ &\quad \{e_2, e_5, e_6, e_8, e_{10}\}, \{e_2, e_3, e_5, e_{10}, e_{11}\}, \{e_2, e_3, e_5, e_9, e_{10}\}, \\ &\quad \{e_2, e_3, e_5, e_8, e_{10}\}, \{e_3, e_4, e_6, e_{10}, e_{11}\}, \{e_3, e_4, e_6, e_9, e_{10}\}, \\ &\quad \{e_3, e_4, e_6, e_8, e_{10}\}, \{e_2, e_4, e_6, e_{10}, e_{11}\}, \{e_2, e_4, e_6, e_9, e_{10}\}, \\ &\quad \{e_2, e_4, e_6, e_8, e_{10}\}, \{e_2, e_3, e_4, e_{10}, e_{11}\}, \{e_2, e_3, e_4, e_9, e_{10}\}, \end{aligned}$$

$$\{e_2, e_3, e_4, e_8, e_{10}\}, \{e_1, e_3, e_6, e_{10}, e_{11}\}, \{e_1, e_3, e_6, e_9, e_{10}\},$$
$$\{e_1, e_3, e_6, e_8, e_{10}\}, \{e_1, e_2, e_6, e_{10}, e_{11}\}, \{e_1, e_2, e_6, e_9, e_{10}\},$$
$$\{e_1, e_2, e_6, e_8, e_{10}\}, \{e_1, e_2, e_3, e_{10}, e_{11}\}, \{e_1, e_2, e_3, e_9, e_{10}\},$$
$$\{e_1, e_2, e_3, e_8, e_{10}\}, \{e_4, e_5, e_6, e_{10}, e_{11}\}, \{e_4, e_5, e_6, e_9, e_{10}\},$$
$$\{e_4, e_5, e_6, e_8, e_{10}\}, \{e_1, e_5, e_6, e_{10}, e_{11}\}, \{e_1, e_5, e_6, e_9, e_{10}\},$$
$$\{e_1, e_5, e_6, e_8, e_{10}\}, \{e_3, e_4, e_5, e_{10}, e_{11}\}, \{e_3, e_4, e_5, e_9, e_{10}\},$$
$$\{e_3, e_4, e_5, e_8, e_{10}\}, \{e_1, e_3, e_5, e_{10}, e_{11}\}, \{e_1, e_3, e_5, e_9, e_{10}\},$$
$$\{e_1, e_3, e_5, e_8, e_{10}\}\}. \tag{3.4.12}$$

(End of Example)

3.5. Hybrid Trees

Trees and cotrees in a connected graph G are fundamental and important concepts not only for the sake of their applications to many different fields, but also to graph theory itself. They especially play an important role in the electrical network theory (see next chapter). Thus the properties of trees or cotrees have been investigated. [2,3] etc.

There are many ways of defining trees and cotrees which was shown in Theorems 3.1.3 and 3.2.1. In the definitions, trees and cotrees are defined using the concepts tiesets and cutsets, respectively. In this section, *hybrid trees* are defined using both tiesets and cutsets. They are concepts which have the properties of both trees and cotrees. As shown in Theorems 3.1.3 and 3.2.1, the number of edges in a tree or a cotree is identical with *rank* or *nullity* of the graph G, respectively. Every hybrid tree in G, however, does not necessarily contain the same number of edges. Some fundamental properties of hybrid trees especially on the number of edges in a hybrid tree are examined.

3.5.1. *Definition of hybrid trees*

$G = (V, E)$ is a connected graph. Let E be the set of edges of G and let the partition of E be (E_y, E_z) into two subsets E_y and E_z such that $E_y \cup E_z = E$ and $E_y \cap E_z = \emptyset$.

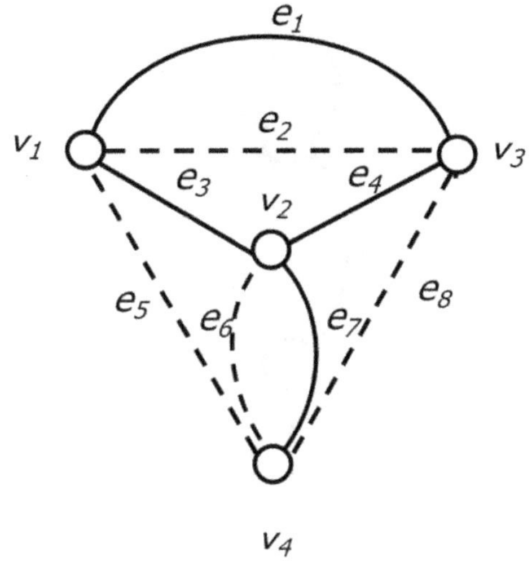

Fig. 3.5.1.1. A connected graph $G = (V, E)$, where $E_y = \{e_1, e_3, e_4, e_7\}$ and $E_z = \{e_2, e_5, e_6, e_8\}$.

Let ε_z and $\bar{\varepsilon_z}$ be cutsetfree and tiesetfree in G, respectively, where $\varepsilon_z \cup \bar{\varepsilon_z} = E_z$, $\varepsilon_z \cap \bar{\varepsilon_z} = \emptyset$. Let ε_y and $\bar{\varepsilon_y}$ where $\varepsilon_y \cup \bar{\varepsilon_y} = E_z$, $\varepsilon_y \cap \bar{\varepsilon_y} = \emptyset$ be tiesetfree and cutsetfree respectively in $G_y = G_{os}[\varepsilon_z;\ \bar{\varepsilon_z}]$ which is obtained by open-circuiting of ε_z and short-circuiting of $\bar{\varepsilon_z}$ in G. Then the subgraph $\varepsilon_y \cup \varepsilon_z$ is called a *hybrid tree* in G with respect to the partition (E_y, E_z) of E.

Example. Figure 3.5.1.1 is a connected graph $G = (V, E)$, where $V = \{v_1, v_2, v_3, v_4\}$, $E = \{e_1, e_2, e_3, e_4, e_5, e_6, e_7, e_8\}$.

Let E_y and E_z be $E_y = \{e_1, e_3, e_4, e_7\}$ and $E_z = \{e_2, e_5, e_6, e_8\}$. Edges of E_y and E_z are shown by solid lines and broken lines, respectively, in Fig. 3.5.1.1. Let $\varepsilon_z = \{e_5, e_6, e_8\}$ and $\bar{\varepsilon_z} = \{e_2\}$ where ε_z and $\bar{\varepsilon_z}$ are cutsetfree and tiesetfree in G, respectively. Fig. 3.5.1.2 shows $G_y = G_{os}[\varepsilon_z;\ \bar{\varepsilon_z}]$ which is obtained by open-circuiting of ε_z and short-circuiting of $\bar{\varepsilon_z}$ in G. In Fig. 3.5.1.2, $\varepsilon_y = \{e_3, e_7\}$ is tiesetfree and $\bar{\varepsilon_y} = \{e_1, e_4\}$ is cutsetfree. Then, $\varepsilon_y \cup \varepsilon_z = \{e_3, e_5, e_6, e_7, e_8\}$ is a *hybrid tree*. **(End of Example)**

Theorem 3.5.1.1. *$G = (V, E)$ is a connected graph. Let E be the set of edges of G and let the partition of E be (E_y, E_z) into two subsets E_y and*

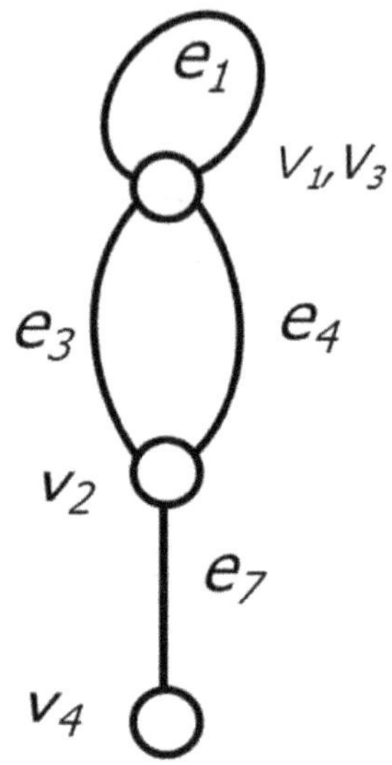

Fig. 3.5.1.2. $G_y = G_{os}[\varepsilon_z; \bar{\varepsilon_z}]$ which is obtained by open-circuiting of ε_z and short-circuiting of $\bar{\varepsilon_z}$ in G.

E_z such that $E_y \cup E_z = E$ and $E_y \cap E_z = \emptyset$. The following definitions of a hybrid tree are equivalent (and characterize the hybrid tree).

(1) *Let ε_z and $\bar{\varepsilon_z}$ be cutsetfree and tiesetfree in G respectively where $\varepsilon_z \cup \bar{\varepsilon_z} = E_z, \varepsilon_z \cap \bar{\varepsilon_z} = \emptyset$. Let ε_y and $\bar{\varepsilon_y}$ where $\varepsilon_y \cup \bar{\varepsilon_y} = E_y, \varepsilon_y \cap \bar{\varepsilon_y} = \emptyset$ be tiesetfree and cutsetfree respectively in $G_y = G_{os}[\varepsilon_z; \bar{\varepsilon_z}]$ which is obtained by open-circuiting of ε_z and short-circuiting of $\bar{\varepsilon_z}$ in G. Then the subgraph $\varepsilon_y \cup \varepsilon_z$ is a hybrid tree in G with respect to the partition (E_y, E_z) of E.*

(2) *The subgraph $\varepsilon_y \cup \varepsilon_z$ is a hybrid tree in G with respect to the partition (E_y, E_z) of E if and only if $\varepsilon_y \cup \bar{\varepsilon_z}$ is a tree where $\varepsilon_y \cup \bar{\varepsilon_y} = E_y, \varepsilon_y \cap \bar{\varepsilon_y} = \emptyset$ and $\varepsilon_z \cup \bar{\varepsilon_z} = E_z, \varepsilon_z \cap \bar{\varepsilon_z} = \emptyset$.*

(3) *The subgraph $\varepsilon_y \cup \varepsilon_z$ is a hybrid tree in G with respect to the partition (E_y, E_z) of E if and only if $\bar{\varepsilon_y} \cup \varepsilon_z$ is a cotree where $\varepsilon_y \cup \bar{\varepsilon_y} = E_y, \varepsilon_y \cap \bar{\varepsilon_y} = \emptyset$ and $\varepsilon_z \cup \bar{\varepsilon_z} = E_z, \varepsilon_z \cap \bar{\varepsilon_z} = \emptyset$.*

(4) *Let ε_y and $\bar{\varepsilon_y}$ be tiesetfree and cutsetfree in G respectively where $\varepsilon_y \cup \bar{\varepsilon_y} = E_y, \varepsilon_y \cap \bar{\varepsilon_y} = \emptyset$. Let ε_z and $\bar{\varepsilon_z}$ where $\varepsilon_z \cup \bar{\varepsilon_z} = E_z, \varepsilon_z \cap \bar{\varepsilon_z} = \emptyset$ be cutsetfree and tiesetfree respectively in $G_z = G_{os}[\bar{\varepsilon_y}; \varepsilon_y]$ which is*

obtained by open-circuiting of $\bar{\varepsilon_y}$ and short-circuiting of ε_y in G. Then the subgraph $\varepsilon_y \cup \varepsilon_z$ is a hybrid tree in G with respect to the partition (E_y, E_z) of E.

Proof. (1) This is the definition of a hybrid tree.

(2) Let $\varepsilon_y \cup \bar{\varepsilon_z}$ be a tree where $\varepsilon_y \cup \bar{\varepsilon_y} = E_y, \varepsilon_y \cap \bar{\varepsilon_y} = \emptyset$ and $\varepsilon_z \cup \bar{\varepsilon_z} = E_z, \varepsilon_z \cap \bar{\varepsilon_z} = \emptyset$, then $\bar{\varepsilon_y} \cup \varepsilon_z$ is a cotree. Since ε_z and $\bar{\varepsilon_z}$ are the parts of a cotree and a tree respectively, ε_z and $\bar{\varepsilon_z}$ are cutsetfree and tiesetfree in G respectively. Since ε_y and $\bar{\varepsilon_y}$ are the parts of a tree and a cotree respectively, ε_y and $\bar{\varepsilon_y}$ are tiesetfree and cutsetfree in G respectively. From Theorem 3.4.1(2), ε_y is a tree of $G_{os}[\emptyset;\ \bar{\varepsilon_z}]$. Since $G_y = G_{os}[\varepsilon_z;\ \bar{\varepsilon_z}]$ is obtained by open-circuiting of ε_z in $G_{os}[\emptyset;\ \bar{\varepsilon_z}]$ and ε_z is cutsetfree, ε_y is a tree and $\bar{\varepsilon_y}$ is a cotree of $G_y = G_{os}[\varepsilon_z;\ \bar{\varepsilon_z}]$. That is, ε_y and $\bar{\varepsilon_y}$ are tiesetfree and cutsetfree respectively in $G_y = G_{os}[\varepsilon_z;\ \bar{\varepsilon_z}]$ from Theorem 3.2.2. Conversely, let us prove that if $\varepsilon_y \cup \varepsilon_z$ is a hybrid tree in G, $\varepsilon_y \cup \bar{\varepsilon_z}$ is a tree in G. In the definition of hybrid tree, ε_y and $\bar{\varepsilon_y}$ are tiesetfree and cutsetfree respectively in $G_y = G_{os}[\varepsilon_z;\ \bar{\varepsilon_z}]$ which mean that ε_y and $\bar{\varepsilon_y}$ are a tree and a cotree respectively in G_y from Theorem 3.2.2. Therefore, from Theorem 3.4.1(2), $\varepsilon_y \cup \bar{\varepsilon_z}$ is a tree in G.

(3) This is proved by similar way as (2).

(4) In the partition (E_y, E_z) of E, if ε_y and $\bar{\varepsilon_y}$ are tiesetfree and cutsetfree in G respectively where $\varepsilon_y \cup \bar{\varepsilon_y} = E_y, \varepsilon_y \cap \bar{\varepsilon_y} = \emptyset$, and ε_z and $\bar{\varepsilon_z}$ where $\varepsilon_z \cup \bar{\varepsilon_z} = E_z, \varepsilon_z \cap \bar{\varepsilon_z} = \emptyset$ are cutsetfree and tiesetfree respectively in $G_z = G_{os}[\bar{\varepsilon_y};\ \varepsilon_y]$ which is obtained by open-circuiting of $\bar{\varepsilon_y}$ and short-circuiting of ε_y in G, then $\varepsilon_y \cup \bar{\varepsilon_z}$ is a tree of G. From Theorem 3.5.1.1(2), $\varepsilon_y \cup \varepsilon_z$ is a hybrid tree in G with respect to the partition (E_y, E_z) of E. □

3.5.2. *Number of edges of a hybrid tree*

In the definition of a hybrid tree, if $E_y \neq \emptyset$ and $E_z = \emptyset$, then since $G_y = G$ and ε_y and $\bar{\varepsilon_y}$ are tiesetfree and cutsetfree in G respectively, the hybrid tree ε_y is a tree of G. If $E_y = \emptyset$ and $E_z \neq \emptyset$, then since $G_y = \emptyset$ and, ε_z and $\bar{\varepsilon_z}$ are cutsetfree and tiesetfree in G respectively, the hybrid tree ε_z is a cotree of G. This means that a hybrid tree of G becomes a tree or a cotree of G

according to a partition of E. In a hybrid tree $\varepsilon_y \cup \varepsilon_z$, the subgraphs (induced by) ε_y and ε_z form a part of a tree and a cotree in G, respectively. So, it may be said that a hybrid tree has properties of both a tree and a cotree. Also, a hybrid tree is a generalization of a tree and a cotree.

Let $HT(G)$, $T(G)$, and $C(G)$ be the sets of hybrid trees, trees, and cotrees in G, respectively. From Theorem 3.11, $T(G)$ and $CT(G)$ are represented as follows:

$$T(G) = \{t | t = ht \oplus E_z, ht \in HT(G)\}, \tag{3.5.2.1}$$

$$C(G) = \{c | c = ht \oplus E_y, ht \in HT(G)\}. \tag{3.5.2.2}$$

Equations (3.5.2.1) and (3.5.2.2) show the relations among $HT(G)$, $T(G)$, and $C(G)$, and when $E_y \neq \emptyset$ and $E_z = \emptyset$, $T(G) = HT(G)$, and when $E_y = \emptyset$ and $E_z \neq \emptyset$, $C(G) = HT(G)$.

The number of edges in a tree and a cotree of G is identical with the rank $\rho(G)$ and the nullity $\mu(G)$ of G, respectively. The number of edges in a hybrid tree of G, however, is not always equal to $\rho(G)$ or $\mu(G)$ and every hybrid tree of G does not always contain the same number of edges. Let hT_k, be a subset of $HT(G)$ such that

$$hT_k = \{ht || ht| = k,\ ht \in HT(G)\}. \tag{3.5.2.3}$$

Theorem 3.5.2.1. *Let $HT(G)$ be the set of hybrid trees of G with respect to the partition (E_y, E_z) of E. Then $HT(G)$ is classified as follows:*

$$HT(G) = \{hT_m,\ hT_{m-2}, \ldots, hT_{l+2}, hT_l\} \tag{3.5.2.4}$$

where $hT_k \neq \emptyset$, $m \geq k \geq l \geq 0$ and, m, k, and l are nonnegative integers.

Proof. Let $ht \in hT_k$ be a hybrid tree in G with respect to the partition (E_y, E_z) of E such that $ht = \varepsilon_y \cup \varepsilon_z$, $\bar{\varepsilon_z} = E_z - \varepsilon_z$ and $\bar{\varepsilon_y} = E_y - \varepsilon_y$. If $|\varepsilon_z| = j$, then $|\varepsilon_y| = k - j$ since $|ht| = k$. The rank of the graph $G_y = G_{os}[\varepsilon_z;\ \bar{\varepsilon_z}]$ which is obtained by open-circuiting of ε_z and short-circuiting of $\bar{\varepsilon_z}$ in G is $k - j$, since ε_y and $\bar{\varepsilon_y}$ are tiesetfree and cutsetfree respectively in G_y, this means ε_y is a tree of G_y from Theorem 3.2.2 and the rank of G_y is equal to the number of edges of a tree ε_y. If E_z is partitioned into two subsets ε_z' and $\bar{\varepsilon_z}'$ such that $|\varepsilon_z'| = j - 1$ and ε_z' and $\bar{\varepsilon_z}'$ are cutsetfree and tiesetfree in G respectively, the rank of $G_y' = G_{os}[\varepsilon_z';\ \bar{\varepsilon_z}']$ derived from

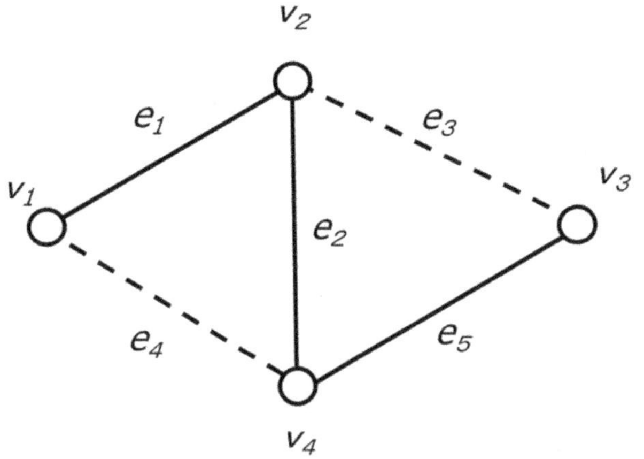

Fig. 3.5.2.1. A graph $G = (V, E)$ where $E = E_y \cup E_z$, $E_y \cap E_z = \emptyset$, $E_y = \{e_1, e_2, e_5\}$ and $E_z = \{e_3, e_4\}$.

G by open-circuiting all the edges of ε_z' and short-circuiting all the edges of $\bar{\varepsilon}_z'$, is $k - j - 1$ since $|\bar{\varepsilon}_z'| = |\bar{\varepsilon}_z| + 1$. Let ε_y' be a subset of E_y such that ε_y' and $\bar{\varepsilon}_y'(= E_y - \varepsilon_y')$ are tiesetfree and cutsetfree in G_y', respectively. Then $|\varepsilon_y'| = k - j - 1$ and the subgraph $\varepsilon_y' \cup \varepsilon_z'$ is a hybrid tree and $|\varepsilon_y' \cup \varepsilon_z'| = k - 2$. Similarly if $|\varepsilon_z'| = j + 1$, then $|\varepsilon_y' \cup \varepsilon_z'| = k + 2$. This means that in any partition of E_z, hybrid trees whose number of edges are $k + 1$ or $k - 1$ are not obtained.

In order to prove that $hT_k \neq \emptyset$ when $hT_{k-2} \neq \emptyset$ and $hT_{k+2} \neq \emptyset$, let us assume $hT_k = \emptyset$ when $hT_{k-2} \neq \emptyset$ and $hT_{k+2} \neq \emptyset$. Let $ht'hT_{k-2}$, $ht''hT_{k+2}$, $ht' = \varepsilon_y' \cup \varepsilon_z'$, $ht'' = \varepsilon_y' \cup \varepsilon_z''$, $\bar{\varepsilon}_z' = E_z\text{-}\varepsilon_z'$ and $\bar{\varepsilon}_z'' = E_z - \varepsilon_z''$. In $ht'' = \varepsilon_y'' \cup \varepsilon_z''$ where $|ht''| = k + 2$, ε_z'' and $\bar{\varepsilon}_z''$ are cutsetfree and tiesetfree respectively in G. For any $e_z \in \varepsilon_z''$, $\varepsilon_z'' - e_z$ is cutsetfree but $\bar{\varepsilon}_z'' + e_z$ is not tiesetfree, that is, $\bar{\varepsilon}_z'' + e_z$ includes a tieset, because $hT_k = \emptyset$ from the assumption. Let $\bar{\varepsilon}' = \bar{\varepsilon}_z'' \cap \bar{\varepsilon}_z'$ and $\bar{\varepsilon}'' = \bar{\varepsilon}_z'' - \bar{\varepsilon}'$, then $\bar{\varepsilon}_z'$ consists of $\bar{\varepsilon}'$ and $|\bar{\varepsilon}''| + 2$ edges of ε_z''. Since $\bar{\varepsilon}_z'$ is tiesetfree, $\bar{\varepsilon}_z' \cup \bar{\varepsilon}''$ contains at most $|\bar{\varepsilon}''|$ tiesets. However, $\bar{\varepsilon}_z' \cup \bar{\varepsilon}''$ contains $|\bar{\varepsilon}''| + 2$ tiesets, because $\bar{\varepsilon}_z' \cup \bar{\varepsilon}'$ consists of $\bar{\varepsilon}_z'$ and $|\bar{\varepsilon}''| + 2$ edges of $\bar{\varepsilon}_z''$. This contradicts the assumption. □

Example. As a simple example, let us consider the set of hybrid trees of the graph shown in Fig. 3.5.2.1. Let $V = \{v_1, v_2, v_3, v_4\}$, $E = \{e_1, e_2, e_3, e_4, e_5\}$, $E_y = \{e_1, e_2, e_5\}$ and $E_z = \{e_3, e_4\}$. The set of hybrid trees of G is given as follows:

$$\begin{aligned} HT(G) &= \{\{e_1, e_2, e_3, e_4, e_5\},\ \{e_1, e_2, e_4\}, \{e_1, e_4, e_5\},\ \{e_1, e_3, e_5\}, \\ &\quad \{e_2, e_3, e_5\},\ \{e_1\}, \{e_2\}, \{e_5\}\} \\ &= \{ht_1, ht_2, ht_3, ht_4, ht_5, ht_6, ht_7, ht_8\} = \{hT_5, hT_3, hT_1\}, \end{aligned} \tag{3.5.2.5}$$

where

$$\begin{aligned} hT_5, &= \{ht_1\} = \{\{e_1, e_2, e_3, e_4, e_5\}\},\ hT_3 = \{ht_2, ht_3, ht_4, ht_5\} \\ &= \{\{e_1, e_2, e_4\}, \{e_1, e_3, e_5\}, \{e_1, e_4, e_5\}, \\ &\quad \{e_2, e_3, e_5\}\} \quad \text{and} \quad hT_1 = \{\{e_1\}, \{e_2\}, \{e_5\}\}. \end{aligned}$$

(End of Example)

The values of m and l in the classification (3.5.2.4) of $HT(G)$ vary with the partition (E_y, E_z) of E. If the graph G is separable and is partitioned into two subsets E_y and E_z such that E_z consists of all the edges of some separable components and $E_y = E - E_z$, then $m = l$. If the graph G is nonseparable, $m = l$ if and only if $HT(G)$ coincides with $T(G)$ or $C(G)$ that is $m = l = \rho(G)$ or $\mu(G)$ respectively. Since

$$|E| \geq m \geq l \geq 0 \tag{3.5.2.6}$$

we obtain the following theorem.

Theorem 3.5.2.2. *In a connected graph $G = (V, E)$,*

(1) *if $E_z = c_i$ and $E_y = E - E_z$ where $c_i \in C(G)$, then m is maximum and $m = |E|$;*
(2) *if $E_z = t_i$ and $E_y = E - E_z$ where $t_i \in T(G)$, then l is minimum and $l = 0$.*

The proof of Theorem 3.5.2.2 is obvious and omitted. This theorem gives the partition of E for the maximum of m or the minimum of l. If $E_z \in C(G)$ and $E_z \in T(G)$, obviously $m = |E|$ and $l = 0$ respectively. However, the partitions of E giving the minimum of m or the maximum of l are not obtained so easy. Let m_p and l_p be the values of m and l corresponding to the partition (E_y^p, E_z^p) of E, respectively. Then from (3.5.2.1) and (3.5.2.2), we obtain

$$m_q = |E| - l_p, \tag{3.5.2.7}$$

$$l_q = |E| - m_p, \tag{3.5.2.8}$$

where m_q and l_q are the values of m and l corresponding to the partition (E_y^q, E_z^q) of E, respectively, and $E_y^q = E_z^p$ and $E_z^q = E_y^p$. These relations (3.5.2.7) and (3.5.2.8) mean that if the partition (E_y^p, E_z^p) gives the maximum of m, then under the partition (E_y^q, E_z^q) of E the value of l becomes minimum. Thus we shall only consider the partition of E giving the minimum of m. First of all, we investigate the properties of the elements of hT_m.

Lemma 3.5.2.3(1). *Let ht be a hybrid tree of a connected graph G with respect to the partition (E_y, E_z) of E. Then $ht = \varepsilon_y \cup \varepsilon_z$ is an element of hT_m, if and only if the graph $G' = G_{os}[\varepsilon_z;\ \varepsilon_y]$ derived from G by open-circuiting all the edges of ε_z, and short-circuiting all the edges of ε_y is connected and contains no ties (tiesets) but self-loops (self tiesets).*

Proof. In general, it is clear that the graph derived from a connected G by open-circuiting a cutsetfree set of edges and short-circuiting some of edges keeps connectivity.

Let $ht_m \in hT_m$, $ht_m = \varepsilon_y \cup \varepsilon_z$, $\bar{\varepsilon_y} = E_y - \varepsilon_y$ and $\bar{\varepsilon_z} = E_z - \varepsilon_z$. From the definition of hybrid trees, ε_z and $\bar{\varepsilon_z}$ are cutsetfree and tiesetfree in G respectively, and ε_y and $\bar{\varepsilon_y}$ are tiesetfree and cutsetfree respectively in $G_y = G_{os}[\varepsilon_z;\ \bar{\varepsilon_z}]$ which is obtained by open-circuiting of ε_z and short-circuiting of $\bar{\varepsilon_z}$ in G. Since ht_m is an element of hT_m, ε_z must be a maximal cutsetfree of E_z and $\bar{\varepsilon_z} \cup E_y$ contains no tiesets because if $\bar{\varepsilon_z} \cup E_y$ contains tiesets, the edge of $\bar{\varepsilon_z}$ including in the tieset can be transferred to ε_z which contradicts the maximality of ε_z. Then the graph G$' = G_{os}[\varepsilon_z;\ \varepsilon_y]$ is connected and contains ties(tiesets) except self-loops (self-tiesets).

Let $ht_k \in HT(G)$, $ht_k \notin hT_m$, $ht_k = \varepsilon_y' \cup \varepsilon_z'$, $\bar{\varepsilon_y}' = E_y - \varepsilon_y'$ and $\bar{\varepsilon_z}' = E_z - \varepsilon_z'$. Since ε_z' is not a maximal cutsetfree of E_z in G and $\bar{\varepsilon_z}'$ contains no tiesets of G, there exist some tiesets consisting of edges of $\bar{\varepsilon_z}'$ and E_y. Let one of them be L'. Since $\bar{\varepsilon_z}' \cup \varepsilon_y'$ is tiesetfree, L' contains edges of $\bar{\varepsilon_y}'$. Therefore, the graph $G' = G_{os}[\varepsilon_z';\ \varepsilon_y']$ derived from G by open-circuiting all the edges of ε'_z and short-circuiting all the edges of ε'_y contains some tiesets which are not self-loops (self tiesets).

Lemma 3.5.2.3(1) shows that only the element $ht_m = \varepsilon_y \cup \varepsilon_z$ of hT_m in $HT(G)$ has the properties such that the graph $G' = G_{os}[\varepsilon_z;\ \varepsilon_y]$ derived from G by short-circuiting all the edges of ε_y and open-circuiting all the edges of ε_z is connected and contains no ties (tiesets) but self-loops (self tiesets). This means that $G' = G_{os}[\varepsilon_z;\ \varepsilon_y]$ is connected and contains only self cutsets and self tiesets.

Let H be the set of subgraphs of satisfying the following condition:

The graph derived from G by short-circuiting some edges (denote h') of a subgraph $h \in H$ and open-circuiting all the other edges ($h'' = h - h'$) of h is connected and contains only self cutsets and self tiesets. □

Then there exists the minimum subgraph satisfying the above condition. Let $H_m \subset H$ be the set of the *minimum subgraphs*. Then we can prove the following lemma.

Lemma 3.5.2.3(2). Let h_m be a *maximal tiesetfree and cutsetfree* subgraph of G. Then $h_m \in H_m$.

Proof. In general, let g_t be a tiesetfree subgraph in G and $g_t \cup e_i$ make a tie where $e_i \notin g_t$,then the graph derived from G short-circuiting the edges of a tiesetfree subgraph g_t includes self-loop (tie) e_i, Let g_c be a cutsetfree subgraph in G and $g_t \cup e_i$ make a cut where $e_i \notin g_t$, then the graph derived from open-circuiting the edges of a cutsetfree subgraph g_c includes self-cut e_i. Since h_m is a maximal tiesetfree and cutsetfree subgraph of G, the subgraph $h_m \cup e_i$ for any edge e_i $(E - h_m)$, contains a tieset or a cutset of G, where the tieset and cutset may match. Let ε_1 be the subset of h_m contained in such tiesets and ε_2 be the subset of h_m contained in such cutsets, where let the edges containing in both a tieset and a cutset belong ε_1 and $\varepsilon_1 \cup \varepsilon_2 = h_m$, $\varepsilon_1 \cap \varepsilon_2 = \emptyset$. Clearly, the graph derived from by short-circuiting all the edges of ε_1, and open-circuiting all the edges of ε_2 is connected and contains only self cutsets and self tiesets, that is, $h_m \in H$.

Consider a subgraph g of G which is tiesetfree and cutsetfree and is not maximal. Then there exists at least one edge e_j of $E - g$ such that $g \cup e_j$ contains neither tiesets nor cutsets of G. Therefore, g is not an element of

H, because the graph derived from G by short-circuiting some edges of g and open-circuiting all the other edges of g includes subgraphs except self cutsets and self tiesets. This means that if a subgraph $h \in H$ is tiesetfree and cutsetfree, it must be a maximal tiesetfree and cutsetfree subgraph of G.

In order to show that $h_m \in H_m$, we assume that $h_m \notin H_m$ and g_k is an element of H_m, such that $g_k = g_1 \cup g_2, g_1 \cap g_2 = \emptyset$ and the graph derived from G by short-circuiting all the edges of g_1 and open-circuiting all the edges of g_2 contains only self cutsets and self tiesets. From the above explanations, g_k must contain tiesets or cutsets. First, suppose tiesets are contained in g_k and let L_p be one of them. Then, L_p contains at least two edges of g_2 because if L_p consists of only edges of g_1 or, one edge of g_2 and edges of g_1, the subgraph $g_k - e_p$ $(e_p g_1)$ belongs to H since $G_{os}[g_1 - e_p; g_2]$ is connected and contains only self cutsets and self tiesets. That is g_k is not a minimum subgraph which means $g_k \notin H_m$. Thus L_p contains at least two edges of g_2. Let $L_{p2} = L_p \cap g_2$ and $G_p = G_{os}[g_1; g_2 - L_{p2}]$. In G_p, L_{p2} forms a tieset and let $e_x = (v_i, v_j)$ be an edge L_{p2}. Since $g_k \in H_m$, $G_{os}[g_1; g_2]$ is connected and contains only self cutsets and self tiesets and there exists an path P_x between v_i and v_j which is a tieset $P_x + e_x$ in G_p because $G_{os}[g_1; g_2]$ is the graph derived from G_p by open-circuiting L_{p2}. Since every edge of P_x belongs to self tiesets in $G_{os}[g_1; g_2]$, let $L'_{p2} = (L_{p2} - e_x) + e_y$ where $e_y P_x$ then the graph derived from G_p by open-circuiting L'_{p2}, is connected and contains only self cutsets and self tiesets. Let $g'_2 = (g_2 - e_x) + e_y$, then $g_1 \cup g'_2$ belongs to H and does not contain L_p. Applying this procedure to all the tiesets of g_k we obtain the subgraph that contains no tiesets and consists of the same number of edges as g_k. Next, suppose cutsets are contained in g_k and let F_c be one of them. Since F_c is a cutset of G, F_c means a partition of V and $F_c = E\langle V_1, V_2\rangle$. If F_c consists of only the edges of g_2, the subgraph $g_k - e_c$ where $e_c \in F_c$ belongs to H since $G_{os}[g_1; g_2 - e_c]$ is connected and contains only self cutsets and self tiesets. That is g_k is not a minimum subgraph which means $g_k \notin H_m$. Thus, F_c consists of the edges of g_1 or g_1 and g_2. $F_c = E\langle V_1, V_2\rangle$ is an edge set of a partition $V = V_1 \cup V_2, V_1 \cap V_2 = \emptyset$, $V_1 \neq \emptyset$, $V_2 \neq \emptyset$ such that one end vertex in every edge of F_c is in V_1 and the other end vertex is in V_2 and every path from a vertex of V_1 to a vertex of V_2 go through necessarily only the edges of F_c. Let $e_x = (v_i, v_j) \in g_1$ be an edge of F_c.

Since we have no path between $v_i \in V_1$ and $v_j \in V_2$ except e_x, there is no tie including e_x. This means that the graph derived from G short-circuiting e_x does not make new self tiesets and g_k is not a minimum subgraph. That is, cutsets are not contained in g_k. These contradict the assumption that $h_m \notin H_m$. □

From these two lemmas, we obtain the following theorem.

Theorem 3.5.2.4. *The minimum value m_f of m in the classification (3.5.2.4) of $HT(G)$ equals the number of edges contained in a maximal tiesetfree and cutsetfree subgraph in G.*

Proof. Let ht_m be a hybrid tree of hT_{mf} with respect to the partition (E_y, E_z) of E which gives the minimum value m_f of m. Then from Lemmas 3.5.2.3(1) and 3.5.2.3(2), ht_m belongs to H and $|ht_m| \geq |h_m|$, where h_m is a maximal tiesetfree and cutsetfree subgraph. Therefore, we have only to prove that h_m can be a hybrid tree in G. For any edge e_i of $\bar{h}_m(= E - h_m)$, the subgraph $h_m \cup e_i$ contains a tieset or a cutset, where the tieset and cutset may match. Let E_y be the set of all the edges contained in such tiesets which includes the case where the tieset and cutset may match and E_z be $E - E_y$. And let $h_m = \varepsilon_y \cup \varepsilon_z, \varepsilon_y \subset E_y, \varepsilon_z \subset E_z, \bar{\varepsilon_y} = E_y - \varepsilon_y$ and $\bar{\varepsilon_z} = E_z - \varepsilon_z$. Since h_m is a maximal tiesetfree and cutsetfree subgraph and ε_z is the subset of h_m, ε_z contains no cutsets. Since $h_m \cup e_j$ for any $e_j \in \bar{\varepsilon_z}$ contains a cutset of G, $\bar{\varepsilon_z}$ contains no tiesets because if there is a tieset including $e_{j1}, e_{j2} \in \bar{\varepsilon_z}$, $h_m \cup e_{j1}$ and $h_m \cup e_{j2}$ do not contain cutsets. In the graph G_y derived from G by open-circuiting all the edges of ε_z and short-circuiting all the edges of $\bar{\varepsilon_z}$, ε_y contains no tiesets and since $h_m \cup e_i$ for any $e_i \in \bar{\varepsilon_y}$ contains a tieset of G_y, $\bar{\varepsilon_y}$ contains no cutsets because if there a cutset including $e_{i1}, e_{i2} \in \bar{\varepsilon_y}$, $h_m \cup e_{i1}$ and $h_m \cup e_{i2}$ do not contain tiesets. This means that h_m is a hybrid tree of G with respect to the partition (E_y, E_z). □

In the proof of this theorem, a method of partition of E minimizing "m" was shown. The method, however, is not unique. Because if the subgraph $h_m \cup e_i$ for $e_i \in \bar{h}_m = E - h_m$, contains both a tieset and a cutset of G, e_i may be an element of either E_y or E_z.

Example. Figure 3.5.2.2 is a connected graph $G = (V, E)$, where $V = \{v_1, v_2, v_3, v_4, v_5\}$, $E = \{e_1, e_2, e_3, e_4, e_5, e_6, e_7\}$. Let $h_m = \{e_2, e_5\}$ be a maximal tiesetfree and cutsetfree subgraph. For any edge e_i of $\bar{h}_m (= E - h_m)$, the subgraph $h_m \cup e_i$ contains a tieset or a cutset. Let E_y be the set of all the edges contained in such tiesets and E_z be $E - E_y$, then we obtain $E_y = \{e_2, e_4, e_7\}$ and $E_z = \{e_1, e_3, e_5, e_6\}$. The set of hybrid trees of G with respect to the partition (E_y, E_z) is given as follows:

$$\begin{aligned} HT(G) = \{&\{e_1, e_2\}, \{e_1, e_4\}, \{e_1, e_7\}, \{e_2, e_3\}, \\ &\{e_2, e_5\}, \{e_2, e_6\}, \{e_3, e_4\}, \{e_3, e_7\}, \\ &\{e_4, e_5\}, \{e_4, e_6\}, \{e_5, e_7\}, \{e_6, e_7\}, \\ \{\emptyset\}\} &= \{hT_2, hT_0\}. \end{aligned} \tag{3.5.2.9}$$

Using (3.15) and (3.16), we obtain $T(G)$ and $C(G)$.

$$\begin{aligned} T(G) &= \{t | t = ht \oplus E_z, ht \in HT(G)\} \\ &= \{\{e_2, e_3, e_5, e_6\}, \{e_3, e_4, e_5, e_6\}, \{e_3, e_5, e_6, e_7\}, \\ &\quad \{e_1, e_2, e_5, e_6\}, \{e_1, e_2, e_3, e_6\}, \{e_1, e_2, e_3, e_5\}, \\ &\quad \{e_1, e_4, e_5, e_6\}, \{e_1, e_5, e_6, e_7\}, \{e_1, e_3, e_4, e_6\}, \\ &\quad \{e_1, e_3, e_4, e_5\}, \{e_1, e_3, e_6, e_7\}, \{e_1, e_3, e_5, e_7\}, \\ &\quad \{e_1, e_3, e_5, e_6\}\}, \end{aligned} \tag{3.5.2.10}$$

$$\begin{aligned} C(G) &= \{c | c = ht \oplus E_y, ht \in HT(G)\} \\ &= \{\{e_1, e_4, e_7\}, \{e_1, e_2, e_7\}, \{e_1, e_2, e_4\}, \{e_3, e_4, e_7\}, \{e_4, e_5, e_7\}, \\ &\quad \{e_4, e_6, e_7\}, \{e_2, e_3, e_7\}, \{e_2, e_3, e_4\}, \{e_2, e_5, e_7\}, \{e_2, e_6, e_7\}, \\ &\quad \{e_2, e_4, e_5\}, \{e_2, e_4, e_6\}, \{e_2, e_4, e_7\}\}. \end{aligned} \tag{3.5.2.11}$$

(End of Example)

In a graph $G = (V, E)$ of Fig. 3.5.2.2, the rank $\rho(G)$ and the nullity $\mu(G)$ are 4 and 3 respectively, which are equal to the numbers of edges of a tree and a cotree of G respectively, shown in (3.5.2.10) and (3.5.2.11). In a graph $G = (V, E)$ of Fig. 3.5.2.2, the minimum value m_f of m in

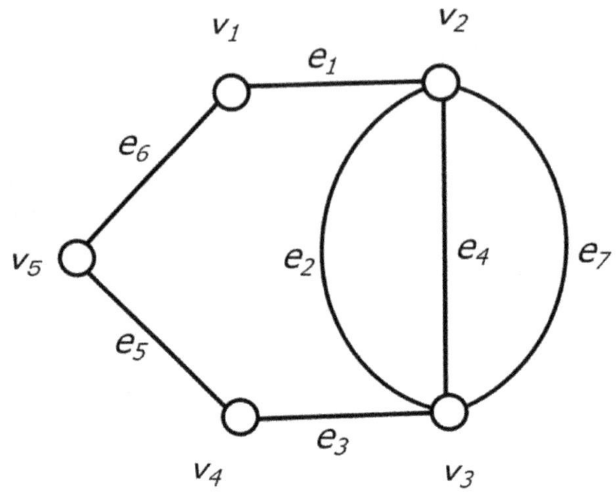

Fig. 3.5.2.2. A graph $G = (V, E)$ with a maximal tiesetfree and cutsetfree subgraph $h_m = \{e_2, e_5\}$.

the classification (3.5.2.4) of $HT(G)$ is 2 which is shown in (3.5.2.9). The minimum value m_f of m may be called *hybrid rank* which coincides with the *topological degrees of freedom* of G [11,18,19].

3.5.3. *Hybrid Tree Graphs*

The tree graph or the cotree graph of G is a graph which represents the relations among the elements of $T(G)$ or $C(G)$ respectively, and they are useful concepts to investigate the properties of the trees or the cotrees of G. In this section, the hybrid tree graph of G is defined and its some fundamental properties are investigated.

Let $HT(G)$ be the set of hybrid trees of G with respect to the partition (E_y, E_z) of E. From Theorem 3.5.2.1, $HT(G)$ is classified as (3.5.2.4). Then the hybrid tree graph of G is defined as follows.

Definition of hybrid tree Graphs: The hybrid tree graph $HT_g = (X, B)$ of G is a graph in which each vertex corresponds to an element of $HT(G)$ and an edge $b_k = (x_l, x_n), (x_l, x_n \in X, b_k \in B)$ joins vertices x_l and x_n whenever $ht_l = ht_n \oplus \{e_i, e_j\}, (ht_l, ht_n \in HT(G), e_i, e_j \in E)$. And the set

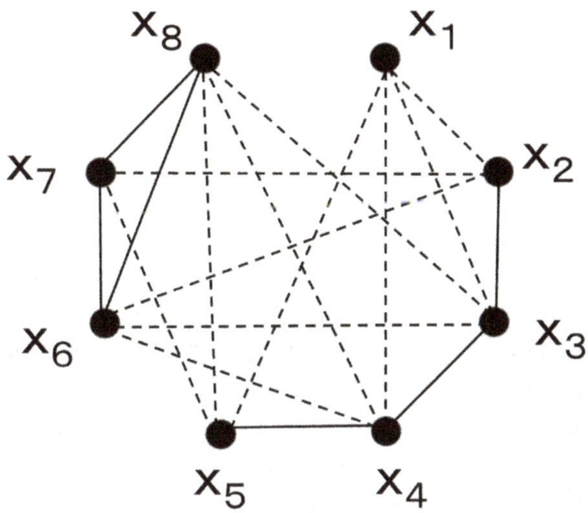

Fig. 3.5.3.1. The hybrid tree graph $HT_g = (X, B)$ of G in Fig. 3.5.2.1.

of edges B of HT_g is partitioned into two subsets such that $B = P \cup N$, $P \cap N = \emptyset$ and $b_k = (x_l, x_n) \epsilon P$, if $|ht_l| = |ht_n|$, otherwise $b_k \epsilon N$.

Example. The hybrid tree graph $HT_g = (X, B)$ of G in Fig. 3.5.2.1 is shown in Fig. 3.5.3.1 in which G has eight hybrid trees as (3.5.2.5). In Fig. 3.5.3.1, full lines are edges of P and broken lines are edges of N. And

$$P = \{(x_2, x_3), (x_3, x_4), (x_4, x_5), (x_6, x_7), (x_6, x_8), (x_7, x_8)\}, \tag{3.5.3.1}$$

$$N = \{(x_1, x_2), (x_1, x_3), (x_1, x_4), (x_1, x_5), (x_2, x_6), (x_2, x_7), (x_3, x_6), (x_3, x_8), (x_4, x_6), (x_4, x_8), (x_5, x_7), (x_5, x_8)\}. \tag{3.5.3.2}$$

(End of Example)

The edges of P and N are called positive edges and negative edges, respectively. And let the signs of their edges be positive (+) and negative (−), respectively. The number of edges of every tree in G is all the same. Also, the number of edges of every cotree and every cotree in G is all the same. In the definition of hybrid tree graphs, the graph obtained from $HT_g(X, B) = (X; P, N)$ by removing the discrimination between P and N and by changing the signs of all the edges into positive is called the tree

Table 3.5.3.1. Multiplication Table.

•	+	−
+	+	−
−	−	+

(cotree) graph of G. This means that a hybrid tree graph is a generalized concept of a tree graph or a cotree graph. Thus a hybrid tree graph HT_g has the same topological properties as a tree graph and furthermore it has some peculiar properties. In the rest of this section, we shall investigate them.

We define the sign of a pathset or a tieset as the product of the signs of the components edges, where the product is defined by the multiplication table of Table 3.5.3.1. For example, let $s(x_i, x_j)$ be the sign of an edge (x_i, x_j), and consider the sign of a pathset $\{(x_1, x_2), (x_2, x_3), (x_3, x_4)\}$ of Fig. 3.5.3.1. Then, the sign of the pathset is

$$s(x_1, x_2) \cdot s(x_2, x_3) \cdot s(x_3, x_4) = (-) \cdot (+) \cdot (+) = (-).$$

We also define the balance of a graph as follows:

Definition of the balance of a graph:
A graph G is balanced if and only if the signs of all of its tiesets are positive.

Then we obtain the following theorem.

Theorem 3.5.3.1. *The hybrid tree graph HT_g of G is balanced.*

Proof. Let $L = \{(x_{l_1}, x_{l_2}), (x_{l_2}, x_{l_3}), \ldots, (x_{l_k}, x_{l_1})\}$ be a tieset of HT_g, and ht_{l_i} be an element of hT_n. Then the hybrid tree ht_{li+1} obtained from ht_{li} by the operation $ht_{li+1} = ht_{l_i} \oplus \{e_{j_i}, e_{j_{i+1}}\}$, $(1 \leq i \leq k-1)$, where $e_{j_i}, e_{ji+1} \in E$ and $ht_{l_1} = ht_{l_k} \oplus \{e_{j_k}, e_{j_1}\}$ when $i = k$, belongs to hT_n, hT_{n-2} or hT_{n+2}. If $ht_{l_{i+1}} \in hT_{n-2}$, then the sign of the edge $(x_{l_i}, x_{l_{i+1}})$ is negative. Since L is a tieset of HT_g, there exists an edge $(x_{l_{i'}}, x_{l_{i'+1}})$ in L satisfying $ht_{l_{i'}} \in hT_{n-2}$ and $ht_{l_{i'+1}} \in hT_n$. Generally, there exists the same number of edges $(x_{l_{i'}}, x_{l_{i'+1}})$ in L satisfying $ht_{l_{i'}} \in hT_{n-2}$ and $ht_{l_{i'+1}} \in hT_n$ as the number of edges $(x_{l_i}, x_{l_{i+1}})$ satisfying $ht_{l_i} \in hT_n$ and $ht_{l_{i+1}} \in hT_{n-2}$. When $ht_{l_i} \in hT_{n+2}$, it is similar. These show that L has an even number of negative edges. □

From Theorem 3.5.3.1, since the sign of a tieset is positive, we can tell the sign of the one edge if the signs of the edges except one of a tieset are known. The subgraph consisting of a tree t and one edge of the cotree c contains a tieset which is called a fundamental tieset (see Section 3.3). This shows that if the signs of all the edges of a tree t of HT_g are known, then the signs of all the edges of the cotree c are obtained.

Corollary 3.5.3.1. *The signs of all the edges of a tree of HT_g give the signs of all the edges of HT_g.*

A balanced graph has a property as follows.

Lemma 3.5.3.1. *Let $G_b = (V; P, N)$ be a balanced graph. Then there exists a bipartition of V into V^1 and V^2 (one of these subsets can be empty) such that the sign of a path between any two vertices of V^1 (or V^2) is positive and the sign of a path between any two vertices of V^1 and V^2 is negative.*

Proof. From Theorem 2.7.1, a tieset and a cutset have an even number (including 0) of common edges. We traverse around a tieset $L = \{(v_{l1}, v_{l2}), (v_{l2}, v_{l3}), \ldots, (v_{l_k}, v_{l_1})\}$ in G_b by starting at a vertex v_{l_1} going through points $v_{l_2}, v_{l_3}, \ldots$ and coming back to v_{l_1} as the proof of Theorem 2.7.1. And let $v_{l_1} \in V^1$, then $v_{l_2} \in V^1$ if the sign of (v_{l_1}, v_{l_2}) is positive and $v_{l_2} \in V^2$ if the sign of (v_{l_1}, v_{l_2}) is negative. Similarly let $v_{l_i} \in V^p$, $(p = 1, 2)$, then $v_{l_{i+1}}$ belongs to the same V^p as v_{l_i} if the sign of $(v_{l_i}, v_{l_{i+1}})$ is positive and $v_{l_{i+1}}$ belongs to the another V^p from v_{l_i}. Since G_b is balanced, the number of negative edge of every tieset is even. Thus, $\langle V^1, V^2\rangle$ is a partition of V and the edges between V^1 and V^2 are sum of cutsets. Since a tieset and a cutset have an even number (including 0) of common edges and the sign of the edges between V^1 and V^2 is negative, the sign of a path between any two vertices of V^1 (or V^2) is positive and the sign of a path between any two vertices of V^1 and V^2 is negative. □

Since HT_g is balanced, we obtained the following theorem.

Theorem 3.5.3.2. *The set of vertices X of a hybrid tree graph $HT_g = (X; P, N)$ can be partitioned into two subsets X^1 and X^2 such that $X = X^1 \cup X^2$, $X^1 \cap X^2 = \emptyset$ and the sign of a path between any two vertices of*

X^1 (or X^2) is positive and the sign of a path between any two vertices of X^1 and X^2 is negative. And if $ht_i \in hT_n$ and the vertex x_i of HT_g corresponding to ht_i belongs to X^1, then all the vertices corresponding to the hybrid trees of hT_n, belong to X^1 and all the vertices corresponding to the hybrid trees of hT_{n+2} and hT_{n-2} belong to X^2.

Proof. Since HT_g is balanced, from Lemma 3.5.3.2, X is partitioned into two subsets as stated above. Let ht_i, $ht_j \in hT_n$, $ht_k \in hT_{n+2}$, $ht'_k \in hT_{n-2}$ and x_i, x_j, x_k and x'_k be vertices corresponding to ht_i, ht_j, ht_k and ht'_k, respectively. Then it can be proved in similar manner as the proof of Theorem 3.5.3.1 that the sign of a path between x_i and x_j is positive and the sign of a path between x_i and x_k and also x_i and x'_k is negative. □

In this theorem, clearly when $HT(G) = T(G)$ or $C(G)$, X^1 or X^2 is empty.

For $HT(G) = \{hT_m, hT_{m-2}, \ldots, hT_{l+2}, hT_l\}$, let X_n ($n = m, m-2, \ldots, l$) be the subset of X corresponding to hT_n. Then, from Theorem 3.5.3.1, we obtain

$$X^1 = \{X_m, X_{m-4}, X_{m-8}, \ldots\}, \tag{3.5.3.3}$$

$$X^2 = \{X_{m-2}, X_{m-6}, X_{m-10}, \ldots\}. \tag{3.5.3.4}$$

For example, for a hybrid tree graph of Fig. 3.5.3.1, we obtain

$$X^1 = \{X_5, X_1\} = \{x_1, x_6, x_7, x_8\},$$

$$X^2 = \{X_3\} = \{x_2, x_3, x_4, x_5\}.$$

The problem of finding a graph that realizes a given set of trees or its some subsets is known as a synthesis problem in graph theory. The problem has been solved by Hakimi [15,16] for labeled trees, that is, all the trees are expressed in terms of edges. Kishi and Kajitani [17] extended the problem to unlabeled trees, that is, they gave a method for obtaining a graph whose tree graph has a subgraph (local subgraph) isomorphic to a prescribed graph. These problems can be extended to hybrid trees. In [21], a method for finding a graph which realizes a given set of hybrid trees is presented. Theorems 3.5.3.1 and 3.5.3.2 are used for obtaining the partition (E_y, E_z) of edges E

of a graph that realizes a given hybrid tree graph. These problems are of interest in *topological synthesis* of electrical networks.

- **Summary:** The subgraphs, trees and cotrees are defined. The subgraph hybrid tree is defined as a generalization of a tree and a cotree.
- **Interesting challenges:** Consider what role the concepts of trees, cotrees and hybrid trees play in matroid theory.

Bibliography

[1] B. Bollobás, *Modern Graph Theory*, Springer-Verlag, 2002.

[2] W. K. Chen, *Applied Graph Theory*, North-Holland, Amsterdam, The Netherlands, 1976.

[3] N. Deo, *Graph Theory with Applications to Engineering and Computer Science*, Prentice-Hall, 1974.

[4] F. Harary, *Graph Theory*, Addison-Wesley Reading, MA., 1969.

[5] O. Ore, *Theory of Graphs*, American Mathematical Society Colloquium Publications XXXVIII, 1962.

[6] K. Ruohonen, *Graph Theory* (Translation by Janne Tamminen, Kung-Chung Lee and Robert Piché), 2013.

[7] S. Seshu and M. B. Reed, *Linear Graphs and Electrical Networks*, Addison-Wesley, 1961.

[8] M. N. S. Swamy and K. Thulasiraman, *Graphs: Theory and Algorithms*, Wiley, 1992.

[9] M. Iri, Network flow, *Transportation and Scheduling, Theory and Algorithms*, Academic Press, 1969.

[10] Y. Kajitani, *Graph Theory for Circuits*, Shokodo Pub., 1979.

[11] M. Iri, I. Shirakawa, Y. Kajitani, S. Shinoda, H. Ariyoshi, K. Onaga, T. Ozawa, T. Ohtsuki, T. Nishi, N. Tomizawa, M. Sengoku, R. Ishii, T. Kashiwabara, T. Nishizeki, S. Fujishige and T. Asano, *Graph Theory with Exercise*, Corona Pub. 1983.

[12] G. Chartrand, L. Lesniak and P. Zhang, *Graphs & Digraphs*, Sixth Edition, Chapman and Hall/CRC, 2015.

[13] L. Novak and A. Gibbons, *Hybrid Graph Theory and Network Analysis*, Cambridge University Press, 1999.

[14] J. G. Oxley, *Matroid Theory*, Oxford University Press, 2006.

[15] S. L. Hakimi, "On the realizability of a set of trees", *JRE Trans. Circuit Theory*, vol. CT-8, pp. 11–17, 1960.

[16] S. L. Hakimi and D. G. Green, "Generation and realization of trees and k-trees", *IEEE Trans. Circuit Theory*, vol. CT-11, pp. 247–255, 1964.

[17] G. Kishi and Y. Kajitani, "On realization of tree graphs", *IEEE Trans. Circuit Theory*, vol. CT-15, pp. 271–273, 1968.

[18] G. Kishi and Y. Kajitani, "Maximally distant trees and principal partition of a linear graph," *IEEE Trans. Circuit Theory*, vol. CT-16, pp. 323–330, 1969.

[19] T. Ohtsuki, Y. Ishizaki, and H. Watanabe, "Topological degrees of freedom and mixed analysis of electrical networks, *IEEE Trans. Circuit Theory,* vol. CT-17, pp. 491–499, 1970.

[20] G. Kishi and Y. Kajitani, "On Hamilton circuits in tree graphs," *IEEE Trans. Circuit Theory*, vol. CT-15, pp. 42–50, 1968.

[21] M. Sengoku, T. Kurobe, and Y. Ogawa, "On the realization of a set of hybrid trees," *IEICE,* vol. 55-A, no. 2, pp. 60–66, 1972.

[22] M. Sengoku, "On hybrid tree graphs," *IEICE*, vol. 57-A, no. 5, pp. 353–358, 1974.

[23] M. Sengoku, "Hybrid Trees and Hybrid Tree Graphs", *IEEE Trans. Circuits Syst.*, vol. CAS-22, no. 10, pp. 786–790, 1975.

[24] H. Tamura, K. Nakano and M. Sengoku, *Network Engineering*, IEICE, Corona Pub., 2020.

Chapter 4

Matrix and Application of Trees

4.1. Matrices of Graphs

4.1.1. *Adjacency matrix*

The *adjacency matrix* of a connected graph $G = (V, E)$ with $|V|$ vertices is a $|V| \times |V|$ matrix $X = [x_{ij}]$ defined such that

x_{ij} = number of edges between $v_i \epsilon V$ and $v_j \epsilon V$, and $x_{ij} = 0$ if there is no edge between them.

The *adjacency matrix* of a *directed graph* a graph $G = (V, E)$ with $|V|$ vertices is a $|V| \times |V|$ matrix $X_d = [x_{ij}]$ defined such that

x_{ij} = number of edges $(v_{i,} v_j)$ where $v_i \epsilon V$ and $v_j \epsilon V$ are the initial vertex and the terminal vertex, respectively.

Let us consider $X^2 = X \times X$, then *ij*th entry of X^2 is number of different edge sequences of two edges between $v_i \epsilon V$ and $v_j \epsilon V$. In a directed graph, then *ij*th entry of $X_d^2 = X_d \times X_d$ is number of different directed edge sequences of two edges from $v_i \epsilon V$ to $v_j \epsilon V$. Similarly, the *ij*th entry in $X^r (X_d^r)$ is the number of different edge sequences (different directed edge sequences) of r edges between vertices $v_i \epsilon V$ and $v_j \epsilon V$.

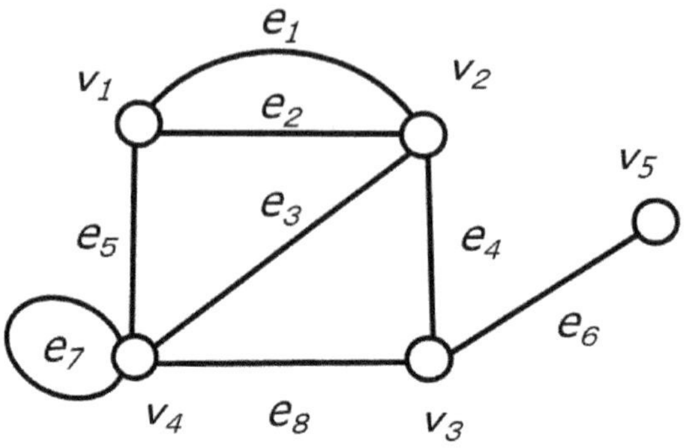

Fig. 4.1.1.1. A graph G.

Example. The *adjacency matrix* X of a graph of Fig. 4.1.1.1 is as follows:

$$X = \begin{array}{c} \\ v_1 \\ v_2 \\ v_3 \\ v_4 \\ v_5 \end{array} \begin{array}{c} \begin{array}{ccccc} v_1 & v_2 & v_3 & v_4 & v_5 \end{array} \\ \begin{bmatrix} 0 & 2 & 0 & 1 & 0 \\ 2 & 0 & 1 & 1 & 0 \\ 0 & 1 & 0 & 1 & 1 \\ 1 & 1 & 1 & 1 & 0 \\ 0 & 0 & 1 & 0 & 0 \end{bmatrix} \end{array}. \tag{4.1.1.1}$$

The *adjacency matrix* X_d of a directed graph of Fig. 4.1.1.2 is as follows:

$$X_d = \begin{array}{c} \\ v_1 \\ v_2 \\ v_3 \\ v_4 \\ v_5 \end{array} \begin{array}{c} \begin{array}{ccccc} v_1 & v_2 & v_3 & v_4 & v_5 \end{array} \\ \begin{bmatrix} 0 & 1 & 0 & 0 & 0 \\ 1 & 0 & 1 & 1 & 0 \\ 0 & 0 & 0 & 0 & 1 \\ 1 & 0 & 1 & 1 & 0 \\ 0 & 0 & 0 & 0 & 0 \end{bmatrix} \end{array}. \tag{4.1.1.2}$$

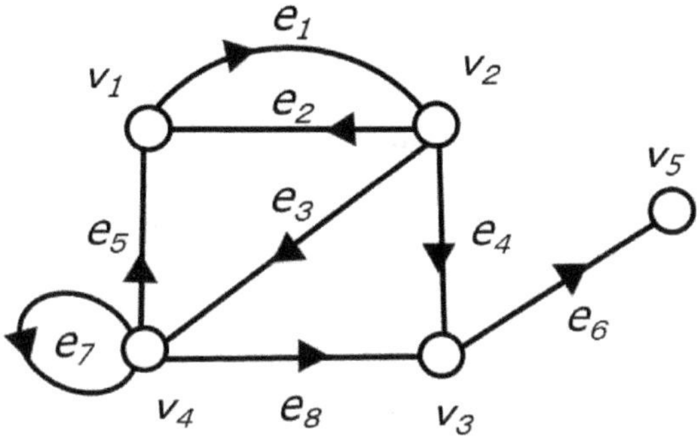

Fig. 4.1.1.2. A directed graph G.

Moreover,

$$X^2 = X \times X = \begin{array}{c} \\ v_1 \\ v_2 \\ v_3 \\ v_4 \\ v_5 \end{array} \begin{array}{c} \begin{array}{ccccc} v_1 & v_2 & v_3 & v_4 & v_5 \end{array} \\ \begin{bmatrix} 5 & 1 & 3 & 3 & 0 \\ 1 & 5 & 1 & 4 & 1 \\ 3 & 1 & 3 & 2 & 0 \\ 3 & 4 & 2 & 4 & 1 \\ 0 & 1 & 0 & 1 & 1 \end{bmatrix} \end{array}. \quad (4.1.1.3)$$

The *ij*th entry of X^2 is number of different edge sequences of two edges between $v_i \epsilon V$ and $v_j \epsilon V$.

For example, 11th entry of X^2 is 5: (e_1, e_1), (e_2, e_2), (e_5, e_5), $(e_1.e_2)$, (e_2, e_1). 31st entry of X^2 is 3: (e_4, e_1), (e_4, e_2), (e_8, e_5).

$$X_d^2 = X_d \times X_d = \begin{array}{c} \\ v_1 \\ v_2 \\ v_3 \\ v_4 \\ v_5 \end{array} \begin{array}{c} \begin{array}{ccccc} v_1 & v_2 & v_3 & v_4 & v_5 \end{array} \\ \begin{bmatrix} 1 & 0 & 1 & 1 & 0 \\ 1 & 1 & 1 & 1 & 1 \\ 0 & 0 & 0 & 0 & 0 \\ 1 & 1 & 1 & 2 & 1 \\ 0 & 0 & 0 & 0 & 0 \end{bmatrix} \end{array}. \quad (4.1.1.4)$$

The *ij*th entry of X_d^2 is number of different edge sequences of two edges from $v_i \epsilon V$ to $v_j \epsilon V$. For example, 11th entry of X_d^2 is 1: $(e_1.e_2)$. 41st entry of X_d^2 is 1: (e_7, e_5). **(End of Example)**

4.1.2. *Incidence matrix*

The *incidence matrix* of a connected graph $G = (V, E)$ is a $|V| \times |E|$ matrix $\boldsymbol{A} = [a_{ij}]$ defined such that

$$a_{ij} = \begin{cases} 1 & e_j \text{ is not a self tieset and } v_i \text{ is an end vertex of } e_j, \\ 0 & \text{otherwise} \end{cases} \tag{4.1.2.1}$$

The *incidence matrix* of a connected *directed graph* a graph $G = (V, E)$ is a $|V| \times |E|$ matrix $\boldsymbol{A_d} = [a_{ij}]$ defined such that

$$a_{ij} = \begin{cases} +1 & e_i \text{ is not a self tieset and } v_i \text{ is an initial vertex of } e_j, \\ -1 & e_j \text{ is not a self tieset and } v_i \text{ is a terminal vertex of } e_j, \\ 0 & \text{otherwise} \end{cases} \tag{4.1.2.2}$$

Example. The *incidence matrix A* of a graph of Fig. 4.1.1.1 is as follows:

$$\boldsymbol{A} = \begin{array}{c} \\ v_1 \\ v_2 \\ v_3 \\ v_4 \\ v_5 \end{array} \begin{array}{c} \begin{array}{cccccccc} e_1 & e_2 & e_3 & e_4 & e_5 & e_6 & e_7 & e_8 \end{array} \\ \begin{bmatrix} 1 & 1 & 0 & 0 & 1 & 0 & 0 & 0 \\ 1 & 1 & 1 & 1 & 0 & 0 & 0 & 0 \\ 0 & 0 & 0 & 1 & 0 & 1 & 0 & 1 \\ 0 & 0 & 1 & 0 & 1 & 0 & 0 & 1 \\ 0 & 0 & 0 & 0 & 0 & 1 & 0 & 0 \end{bmatrix} \end{array}. \tag{4.1.2.3}$$

The *incidence matrix* A_d of a directed graph of Fig. 4.1.1.2 is as follows:

$$\boldsymbol{A_d} = \begin{array}{c} \\ v_1 \\ v_2 \\ v_3 \\ v_4 \\ v_5 \end{array} \begin{array}{c} \begin{array}{cccccccc} e_1 & e_2 & e_3 & e_4 & e_5 & e_6 & e_7 & e_8 \end{array} \\ \begin{bmatrix} 1 & -1 & 0 & 0 & -1 & 0 & 0 & 0 \\ -1 & 1 & 1 & 1 & 0 & 0 & 0 & 0 \\ 0 & 0 & 0 & -1 & 0 & 1 & 0 & -1 \\ 0 & 0 & -1 & 0 & 1 & 0 & 0 & 1 \\ 0 & 0 & 0 & 0 & 0 & -1 & 0 & 0 \end{bmatrix} \end{array}. \tag{4.1.2.4}$$

(End of Example)

Since every column of an incidence matrix contains exactly two non-zero numbers (two ones), we can remove a row and still have enough information to define the graph. The *incidence matrix* of a graph is obtained by removing a row from the incidence matrix. It is not unique because there are $|V|$ possible rows to remove. The vertex corresponding to the row removed is called the *reference vertex*. Similarly, every column in the incidence matrix of a directed graph contains exactly two non-zero numbers, $+1$ and -1. We can remove a row from the incidence matrix and obtain the *incidence matrix* A_r called *reduced incidence matrix*. Notice that the rows of incidence matrix are linearly dependent because the sum of rows is a zero vector.

In an undirected graph, Galois field modulo 2 is used as the operation of the vector space. The modulo 2 is as follows:

- operation *addition* modulo 2 written as $+$ such that

$$\begin{aligned} &0+0 \equiv 0 \pmod 2 \\ &1+0 \equiv 0+1 \equiv 1 \pmod 2 \\ &1+1 \equiv 0 \pmod 2 \end{aligned} \tag{4.1.2.5}$$

- operation *multiplication* modulo 2 written as $\cdot$ such that

$$\begin{aligned} &0 \cdot 0 \equiv 0 \pmod 2 \\ &1 \cdot 0 \equiv 0 \cdot 1 \equiv 0 \pmod 2 \\ &1 \cdot 1 \equiv 1 \pmod 2. \end{aligned} \tag{4.1.2.6}$$

Theorem 4.1.2.1. *The rank of incidence matrix of a graph G is $\rho(G)$.*

Proof. Let us consider a connected *directed graph* a graph $G = (V, E)$. Since the incidence matrix is $|V| \times |E|$, every column in the incidence matrix contains exactly two non-zero numbers ($+1$ and -1) and the sum of rows is a zero vector, the rank of the incidence matrix $\leq |V| - 1$. Next let us proof by induction that the determinant of a incidence matrix of a tree is ± 1. If $|V| = 2$ and a tree consists of one edge of G, the a incidence matrix is 1×1 and the determinant is ± 1. Suppose that the determinant of of a tree in the graph of $|V| \leq k$ is ± 1. Let us consider a incidence matrix of a tree T in a graph with $k+1$ vertices. The tree T with $k+1$ vertices has a leaf from Lemma 3.1.1. Let the pendant vertex (whose degree is one) of the leaf e_j be v_i, then row of an incidence matrix of the tree contains exactly one non-zero numbers a_{ij}. We expand (cofactor expansion) the determinant of

$\det A$ by the ith row:

$$\det A_d = (\pm 1)(-1)^{i+j} \det A'_d, \qquad (4.1.2.7)$$

where A'_d is the minor corresponding to a_{ij}. Let $T' = T - v_i$ which is also a tree (see the proof of Theorem 3.1.1). From the induction hypothesis, $\det \boldsymbol{A}' = \pm 1$ because $\boldsymbol{A}'$ is obviously an incidence matrix of T'. Thus, the rank of the incidence matrix is $|V| - 1$. That is, $|V| - 1 = \rho(G)$.

Similarly, we can proof the result using mod 2 in an undirected graph. □

4.1.3. *Tieset matrix*

Let $F_i (i = 1, 2, \ldots, l)$ be tiesets of of a connected graph $G = (V, E)$. The *tieset matrix* of G is an $l \times |E|$ matrix $\boldsymbol{B} = [b_{ij}]$ defined such that

$$b_{ij} = \begin{cases} 1 & \text{if the edge } e_j \text{ is in the tieset } F_i, \\ 0 & \text{otherwise.} \end{cases} \qquad (4.1.3.1)$$

There are some ways of thinking about subgraphs in a directed graph. In Theorem 2.7.6, we consider that every edge in a directed tieset has same direction. As an another thinking about a tieset in directed graph, we consider the tiesets consisting of the tieset of the graph that ignores edge direction of a directed graph and every tieset is given an arbitrary direction (which is called that the tiesets in the directed graph G are *oriented*) for the sake of defining a tieset (tieset matrix) of a directed graph where every edge is contained in a tieset or not in the tieset and if the edge is contained in the tieset, the direction of the edge is in the same direction or in the opposite direction of the tieset.

Let F_i $(i = 1, 2, \ldots, l)$ be tiesets (which are *oriented*) of a connected directed graph $G = (V, E)$. The *tieset matrix* of a connected *directed graph* a graph $G = (V, E)$ is an $l \times |E|$ matrix $\boldsymbol{B_d} = [b_{ij}]$ defined such that

$$b_{ij} = \begin{cases} +1 & \text{if the edge } e_j \text{ is in the tieset } F_i \text{ in the same direction,} \\ -1 & \text{if the edge } e_j \text{ is in the tieset } F_i \text{ in the opposite direction,} \\ 0 & \text{otherwise.} \end{cases}$$

(4.1.3.2)

Example. The *tieset matrix* $\boldsymbol{B}$ of a graph of Fig. 4.1.1.1 is as follows:

$$\boldsymbol{B} = [b_{ij}] = \begin{array}{c} \\ F_1 \\ F_2 \\ F_3 \\ F_4 \\ F_5 \\ F_6 \\ F_7 \end{array} \begin{array}{c} \begin{array}{cccccccc} e_1 & e_2 & e_3 & e_4 & e_5 & e_6 & e_7 & e_8 \end{array} \\ \begin{bmatrix} 1 & 1 & 0 & 0 & 0 & 0 & 0 & 0 \\ 1 & 0 & 1 & 0 & 1 & 0 & 0 & 0 \\ 0 & 1 & 1 & 0 & 1 & 0 & 0 & 0 \\ 1 & 0 & 0 & 1 & 1 & 0 & 0 & 1 \\ 0 & 1 & 0 & 1 & 1 & 0 & 0 & 1 \\ 0 & 0 & 1 & 1 & 0 & 0 & 0 & 1 \\ 0 & 0 & 0 & 0 & 0 & 0 & 1 & 0 \end{bmatrix} \end{array}.$$

The *tieset matrix* $\boldsymbol{B_d}$ of a directed graph of Fig. 4.1.1.2 is as follows:

$$\boldsymbol{B_d} = [b_{ij}] = \begin{array}{c} \\ F_1 \\ F_2 \\ F_3 \\ F_4 \\ F_5 \\ F_6 \\ F_7 \end{array} \begin{array}{c} \begin{array}{cccccccc} e_1 & e_2 & e_3 & e_4 & e_5 & e_6 & e_7 & e_8 \end{array} \\ \begin{bmatrix} 1 & 1 & 0 & 0 & 0 & 0 & 0 & 0 \\ 1 & 0 & 1 & 0 & 1 & 0 & 0 & 0 \\ 0 & -1 & 1 & 0 & 1 & 0 & 0 & 0 \\ 1 & 0 & 0 & 1 & 1 & 0 & 0 & -1 \\ 0 & 1 & 0 & -1 & -1 & 0 & 0 & 1 \\ 0 & 0 & 1 & -1 & 0 & 0 & 0 & 1 \\ 0 & 0 & 0 & 0 & 0 & 0 & 1 & 0 \end{bmatrix} \end{array}.$$

(End of Example)

Let t and c be a tree and the cotree of a connected graph $G = (V, E)$. $t = \{e_{t1}, e_{t2}, \ldots, e_{t\rho(G)}\}$, $c = \{e_{c1}, e_{c2}, \ldots, e_{c\mu(G)}\}$ where $E = t + c$. Let us consider a fundamental tieset of the tree t and cotree c.

The *fundamental tieset matrix* $B_f$is a submatrix which is obtained by rearranging the rows or columns of B.

$$\boldsymbol{B_f} = [\overbrace{\,\mathbf{1}_\mu\,}^{c} \;\vdots\; \overbrace{\,\boldsymbol{B_{ft}}\,}^{t}]. \tag{4.1.3.3}$$

In (4.1.13), $\mathbf{1}_\mu$ is the *identity matrix* of order μ. $\boldsymbol{B_{ft}}$ is called the *principal part* of $\boldsymbol{B_f}$. Since $\boldsymbol{B_f}$ includes $\mathbf{1}_\mu$, the rank of $\boldsymbol{B_f}$ is $\mu(G)$ and the rank of $\boldsymbol{B} \geq$ the rank of $\boldsymbol{B_f} = \mu(G)$.

Theorem 4.1.3.1. *Let* $\boldsymbol{A}$ *and* $\boldsymbol{B}$ *be, respectively, the incidence matrix and the tieset matrix whose columns are arranged using the same order of edges. Then every row of* $\boldsymbol{A}$ *is orthogonal to every row* $\boldsymbol{B}$*; that is,*

$$\boldsymbol{A}\boldsymbol{B}^{\mathrm{T}} = \boldsymbol{0}, \quad \boldsymbol{B}\boldsymbol{A}^{\mathrm{T}} = \boldsymbol{0} \quad (\mathrm{mod}\ 2) \tag{4.1.3.4}$$

$$\boldsymbol{A}_d\boldsymbol{B}_d^{\mathrm{T}} = \boldsymbol{0}, \quad \boldsymbol{B}_d\boldsymbol{A}_d^{\mathrm{T}} = \boldsymbol{0} \quad (\text{for digraph}) \tag{4.1.3.5}$$

where superscript T denotes the transposed matrix.

Proof. Let us consider a directed graph first. Let a row of $\boldsymbol{A_d}$ and a row of $\boldsymbol{B_d}$ be $\boldsymbol{a}_i = [a_{i1}, a_{i2}, \ldots, a_{i|E|}]$ and $\boldsymbol{b}_j = [b_{j1}, b_{j2}, \ldots, b_{j|E|}]$, respectively, then $\boldsymbol{A_d}\boldsymbol{B_d^T} = [x_{ij}]$ where $x_{ij} = \boldsymbol{a}_i\boldsymbol{b}_j^T = \sum_{k=1}^{|E|} a_{ik}b_{jk}$. $\boldsymbol{a}_i$ represents the edges which are *incident out* of v_i and *incident into* v_i. $\boldsymbol{b}_j$ represents the edges which are included in the tieset F_j and their directions in the F_j. If the edge e_k is not in the tieset F_j, $a_{ik}b_{jk} = 0$. Since F_i is a tieset, there is another edge of v_i in F_i if an edge incident of v_i is in F_i. If an edge $e_k = (v_m, v_i)$ which is incident into v_i is in the tieset F_j and F_j include a $e_p = (v_i, v_n)$ which is incident out v_i, then e_k and e_p are the same direction but they are incident into v_i and incident out v_i respectively, which means $\boldsymbol{a}_i\boldsymbol{b}_j^T = a_{ik}b_{jk} + a_{ip}b_{jp} = 0$. If an edge $e_k = (v_m, v_i)$ which is incident into v_i is in the tieset F_j and F_j include a $e_p = (v_n, v_i)$ which is incident into v_i, then e_k and e_p are the opposite direction but they are incident into v_i, which means $\boldsymbol{a}_i\boldsymbol{b}_j^T = a_{ik}b_{jk} + a_{ip}b_{jp} = 0$. That is, $\boldsymbol{a}_i\boldsymbol{b}_j^T = 0$ and $\boldsymbol{A_d}\boldsymbol{B_d^{\mathrm{T}}} = \boldsymbol{0}$. Obviously, $\boldsymbol{B_d}\boldsymbol{A_d^{\mathrm{T}}} = \boldsymbol{0}$.

Similarly, we can proof the result using mod 2 in an undirected graph. □

According to *Sylvester's theorem*: In $p \times q$ Matrix X, $q \times r$ Matrix Y, if $XY = 0$, rank $X+$ rank $Y \leq q$, the rank of tieset matrix $\boldsymbol{B} \leq |E| -$ rank $\boldsymbol{A} = |E| - \rho(G) = \mu(G)$ from Theorem 4.1.2.1.

Therefore

$$\text{rank}\ \boldsymbol{B} = \mu(G). \tag{4.1.3.6}$$

Let $\boldsymbol{A_r}$ be the reduced incidence matrix which is obtained by rearranging the rows or columns of the incidence matrix.

$$\boldsymbol{A_r} = [\overbrace{\boldsymbol{A}_c}^{c} \ \vdots \ \overbrace{\boldsymbol{A}_t}^{t}]. \tag{4.1.3.7}$$

From (4.1.3.3)–(4.1.3.5) and (4.1.3.7),

$$\boldsymbol{A}_r\boldsymbol{B}_f^T = \mathbf{A}_c + \mathbf{A}_t\boldsymbol{B}_{ft}^T = \mathbf{0} \quad \text{(for digraph)}, \tag{4.1.3.8}$$

$$\boldsymbol{A}_r\boldsymbol{B}_f^T = \mathbf{A}_c + \mathbf{A}_t\boldsymbol{B}_{ft}^T \equiv \mathbf{0} \quad \text{(mod 2)}, \tag{4.1.3.9}$$

$$\boldsymbol{B}_{ft}^T = -\mathbf{A}_t^{-1}\boldsymbol{A}_c \quad \text{(for digraph)}, \tag{4.1.3.10}$$

$$\boldsymbol{B}_{ft}^T = \mathbf{A}_t^{-1}\mathbf{A}_c \quad \text{(mod 2)}. \tag{4.1.3.11}$$

Example. In Fig. 4.1.1.1,

$$\boldsymbol{A} = \begin{array}{c} \\ v_1 \\ v_2 \\ v_3 \\ v_4 \\ v_5 \end{array} \begin{array}{c} \begin{array}{cccccccc} e_1 & e_2 & e_3 & e_4 & e_5 & e_6 & e_7 & e_8 \end{array} \\ \begin{bmatrix} 1 & 1 & 0 & 0 & 1 & 0 & 0 & 0 \\ 1 & 1 & 1 & 1 & 0 & 0 & 0 & 0 \\ 0 & 0 & 0 & 1 & 0 & 1 & 0 & 1 \\ 0 & 0 & 1 & 0 & 1 & 0 & 0 & 1 \\ 0 & 0 & 0 & 0 & 0 & 1 & 0 & 0 \end{bmatrix} \end{array}, \tag{4.1.3.12}$$

$$\boldsymbol{B} = \begin{array}{c} \\ F_1 \\ F_2 \\ F_3 \\ F_4 \\ F_5 \\ F_6 \\ F_7 \end{array} \begin{array}{c} \begin{array}{cccccccc} e_1 & e_2 & e_3 & e_4 & e_5 & e_6 & e_7 & e_8 \end{array} \\ \begin{bmatrix} 1 & 1 & 0 & 0 & 0 & 0 & 0 & 0 \\ 1 & 0 & 1 & 0 & 1 & 0 & 0 & 0 \\ 0 & 1 & 1 & 0 & 1 & 0 & 0 & 0 \\ 1 & 0 & 0 & 1 & 1 & 0 & 0 & 1 \\ 0 & 1 & 0 & 1 & 1 & 0 & 0 & 1 \\ 0 & 0 & 1 & 1 & 0 & 0 & 0 & 1 \\ 0 & 0 & 0 & 0 & 0 & 0 & 1 & 0 \end{bmatrix} \end{array}, \tag{4.1.3.13}$$

$$\boldsymbol{A}\boldsymbol{B}^T = \begin{bmatrix} 0 & 0 & 0 & 0 & 0 \\ 0 & 0 & 0 & 0 & 0 \\ 0 & 0 & 0 & 0 & 0 \\ 0 & 0 & 0 & 0 & 0 \\ 0 & 0 & 0 & 0 & 0 \end{bmatrix}.$$

In Fig. 4.1.12,

$$\boldsymbol{A_d} = \begin{array}{c} \\ v_1 \\ v_2 \\ v_3 \\ v_4 \\ v_5 \end{array} \begin{array}{c} \begin{array}{cccccccc} e_1 & e_2 & e_3 & e_4 & e_5 & e_6 & e_7 & e_8 \end{array} \\ \begin{bmatrix} 1 & -1 & 0 & 0 & -1 & 0 & 0 & 0 \\ -1 & 1 & 1 & 1 & 0 & 0 & 0 & 0 \\ 0 & 0 & 0 & -1 & 0 & 1 & 0 & -1 \\ 0 & 0 & -1 & 0 & 1 & 0 & 0 & 1 \\ 0 & 0 & 0 & 0 & 0 & -1 & 0 & 0 \end{bmatrix} \end{array}, \tag{4.1.3.14}$$

$$\boldsymbol{B_d} = \begin{array}{c} \\ F_1 \\ F_2 \\ F_3 \\ F_4 \\ F_5 \\ F_6 \\ F_7 \end{array} \begin{array}{c} \begin{array}{cccccccc} e_1 & e_2 & e_3 & e_4 & e_5 & e_6 & e_7 & e_8 \end{array} \\ \begin{bmatrix} 1 & 1 & 0 & 0 & 0 & 0 & 0 & 0 \\ 1 & 0 & 1 & 0 & 1 & 0 & 0 & 0 \\ 0 & -1 & 1 & 0 & 1 & 0 & 0 & 0 \\ 1 & 0 & 0 & 1 & 1 & 0 & 0 & -1 \\ 0 & 1 & 0 & -1 & -1 & 0 & 0 & 1 \\ 0 & 0 & 1 & -1 & 0 & 0 & 0 & 1 \\ 0 & 0 & 0 & 0 & 0 & 0 & 1 & 0 \end{bmatrix} \end{array}, \tag{4.1.3.15}$$

$$\boldsymbol{A_d B_d^T} = \begin{bmatrix} 0 & 0 & 0 & 0 & 0 \\ 0 & 0 & 0 & 0 & 0 \\ 0 & 0 & 0 & 0 & 0 \\ 0 & 0 & 0 & 0 & 0 \\ 0 & 0 & 0 & 0 & 0 \end{bmatrix}.$$

Let $t = \{e_4, e_5, e_6, e_8\}$ and $c = \{e_1, e_2, e_3, e_7\}$ in a graph of Fig. 4.1.1.1. Then

$$\boldsymbol{B_f} = [\overbrace{\mathbf{1}_\mu}^{c} \;\vdots\; \overbrace{\boldsymbol{B}_{ft}}^{t}] = \begin{array}{c} \\ F_4 \\ F_5 \\ F_6 \\ F_7 \end{array} \begin{array}{c} \begin{array}{cccc:cccc} e_1 & e_2 & e_3 & e_7 & e_4 & e_5 & e_6 & e_8 \end{array} \\ \left[\begin{array}{cccc:cccc} 1 & 0 & 0 & 0 & 1 & 1 & 0 & 1 \\ 0 & 1 & 0 & 0 & 1 & 1 & 0 & 1 \\ 0 & 0 & 1 & 0 & 1 & 0 & 0 & 1 \\ 0 & 0 & 0 & 1 & 0 & 0 & 0 & 0 \end{array}\right] \end{array}, \tag{4.1.3.16}$$

$$\mathbf{A}_r = [\overbrace{\mathbf{A}_c}^{c} \;\vdots\; \overbrace{\mathbf{A}_t}^{t}] = \begin{matrix} & e_1 & e_2 & e_3 & e_7 & e_4 & e_5 & e_6 & e_8 \\ v_1 \\ v_2 \\ v_3 \\ v_4 \end{matrix} \left[\begin{array}{cccc|cccc} 1 & 1 & 0 & 0 & 0 & 1 & 0 & 0 \\ 1 & 1 & 1 & 0 & 1 & 0 & 0 & 0 \\ 0 & 0 & 0 & 0 & 1 & 0 & 1 & 1 \\ 0 & 0 & 1 & 0 & 0 & 1 & 0 & 1 \end{array}\right], \tag{4.1.3.17}$$

$$\boldsymbol{B}_{ft}^{T} = \mathbf{A}_t^{-1}\mathbf{A}_c \quad (\text{mod } 2). \tag{4.1.3.18}$$

Let $t = \{e_4, e_5, e_6, e_8\}$ and $c = \{e_1, e_2, e_3, e_7\}$ in a graph of Fig. 4.1.1.2. Then

$$\boldsymbol{B}_f = [\overbrace{\mathbf{1}_\mu}^{c} \;\vdots\; \overbrace{\boldsymbol{B}_{ft}}^{t}] = \begin{matrix} & e_1 & e_2 & e_3 & e_7 & e_4 & e_5 & e_7 & e_8 \\ F_4 \\ F_5 \\ F_6 \\ F_8 \end{matrix} \left[\begin{array}{cccc|cccc} 1 & 0 & 0 & 0 & 1 & 1 & 0 & -1 \\ 0 & 1 & 0 & 0 & -1 & -1 & 0 & 1 \\ 0 & 0 & 1 & 0 & -1 & 0 & 0 & 1 \\ 0 & 0 & 0 & 1 & 0 & 0 & 0 & 0 \end{array}\right], \tag{4.1.3.19}$$

$$\boldsymbol{A}_r = [\overbrace{\mathbf{A}_c}^{c} \;\vdots\; \overbrace{\boldsymbol{A}_t}^{t}] = \begin{matrix} & e_1 & e_2 & e_3 & e_7 & e_4 & e_5 & e_6 & e_8 \\ v_1 \\ v_2 \\ v_3 \\ v_4 \end{matrix} \left[\begin{array}{cccc|cccc} 1 & -1 & 0 & 0 & 0 & -1 & 0 & 0 \\ -1 & 1 & 1 & 0 & 1 & 0 & 0 & 0 \\ 0 & 0 & 0 & 0 & -1 & 0 & 1 & -1 \\ 0 & 0 & -1 & 0 & 0 & 1 & 0 & 1 \end{array}\right], \tag{4.1.3.20}$$

$$\boldsymbol{B}_{ft}^{\mathrm{T}} = -\boldsymbol{A}_{\mathbf{t}}^{-1}\boldsymbol{A}_c. \tag{4.1.3.21}$$

(End of Example)

4.1.4. *Cutset matrix*

Let H_i $(i = 1, 2, \ldots, m)$ be cutsets of of a connected graph $G = (V, E)$. The *cutset matrix* of G is an $m \times |E|$ matrix $\boldsymbol{Q} = [q_{ij}]$ defined such that

$$q_{ij} = \begin{cases} 1 & \text{if the edge } e_j \text{ is in the ticset } H_i, \\ 0 & \text{otherwise.} \end{cases} \tag{4.1.4.1}$$

There are some ways of thinking about subgraphs in a directed graph. In Theorem 2.7.6, we consider that every edge in a directed cutset has same direction. As an another thinking about a cutset in directed graph, we consider the cutsets consisting of the cutset of the graph that ignores edge direction of a directed graph and every cutset is given an arbitrary direction (which is called that the cutsets in the directed graph G are *oriented*) for the sake of defining a cutset (cutset matrix) of a directed graph where every edge is contained in a cutset or not in the cutset and if the edge is contained in the cutset, the direction of the edge is in the same direction or in the opposite direction of the cutset.

Let H_i $(i = 1, 2, \ldots, m)$ be cutsets (which are *oriented*) of a connected directed graph $G = (V, E)$. Then the *cutset matrix* of a connected *directed graph* a graph $G = (V, E)$ is an $m \times |E|$ matrix $\boldsymbol{Q_d} = [q_{ij}]$ defined such that

$$b_{ij} = \begin{cases} +1 & \text{if the edge } e_j \text{ is in the cutset } H_i \text{ in the same direction,} \\ -1 & \text{if the edge } e_j \text{ is in the cutset } H_i \text{ in the opposite direction,} \\ 0 & \text{otherwise.} \end{cases} \tag{4.1.4.2}$$

Example. The cutset matrix $\boldsymbol{Q}$ of a graph of Fig. 4.1.1.1 is as follows:

$$\boldsymbol{Q} = [q_{ij}] = \begin{array}{c} \\ H_1 \\ H_2 \\ H_3 \\ H_4 \\ H_5 \\ H_6 \\ H_7 \end{array} \begin{array}{c} \begin{array}{cccccccc} e_1 & e_2 & e_3 & e_4 & e_5 & e_6 & e_7 & e_8 \end{array} \\ \begin{bmatrix} 1 & 1 & 1 & 1 & 0 & 0 & 0 & 0 \\ 1 & 1 & 0 & 0 & 1 & 0 & 0 & 0 \\ 1 & 1 & 1 & 0 & 0 & 0 & 0 & 1 \\ 0 & 0 & 0 & 0 & 0 & 1 & 0 & 0 \\ 0 & 0 & 1 & 1 & 1 & 0 & 0 & 0 \\ 0 & 0 & 0 & 1 & 0 & 0 & 0 & 1 \\ 0 & 0 & 1 & 0 & 1 & 0 & 0 & 1 \end{bmatrix} \end{array}. \tag{4.1.4.3}$$

The cutset matrix $\boldsymbol{Q}_d$ of a directed graph of Fig. 4.1.1.2 is as follows:

$$\boldsymbol{Q}_d = [q_{ij}] = \begin{matrix} \\ H_1 \\ H_2 \\ H_3 \\ H_4 \\ H_5 \\ H_6 \\ H_7 \end{matrix} \begin{matrix} \begin{matrix} e_1 & e_2 & e_3 & e_4 & e_5 & e_6 & e_7 & e_8 \end{matrix} \\ \begin{bmatrix} -1 & 1 & 1 & 1 & 0 & 0 & 0 & 0 \\ 1 & -1 & 0 & 0 & 1 & 0 & 0 & 0 \\ 1 & -1 & -1 & 0 & 0 & 0 & 0 & 1 \\ 0 & 0 & 0 & 0 & 0 & 1 & 0 & 0 \\ 0 & 0 & 1 & 1 & -1 & 0 & 0 & 0 \\ 0 & 0 & 0 & 1 & 0 & 0 & 0 & 1 \\ 0 & 0 & -1 & 0 & 1 & 0 & 0 & 1 \end{bmatrix} \end{matrix}. \tag{4.1.4.4}$$

(End of Example)

Let t and c be a tree and the cotree of a connected graph $G = (V, E)$. $t = \{e_{t1}, e_{t2}, \ldots, e_{t\rho(G)}\}$, $c = \{e_{c1}, e_{c2}, \ldots, e_{c\mu(G)}\}$ where $E = t + c$. Let us consider a fundamental cutset of the tree t and cotree c.

The *fundamental cutset matrix* $\boldsymbol{Q}_f$ is a submatrix which is obtained by rearranging the rows or columns of $\boldsymbol{Q}$.

$$\boldsymbol{Q}_f = [\overbrace{\boldsymbol{Q}_{fc}}^{c} \ \vdots \ \overbrace{\boldsymbol{1}_\rho}^{t}]. \tag{4.1.4.5}$$

In (4.1.4.5), $\boldsymbol{1}_\rho$ is the *identity matrix* of order ρ. $\boldsymbol{Q}_{fc}$ is called the *principal part* of $\boldsymbol{Q}_f$. Since $\boldsymbol{Q}_f$ includes $\boldsymbol{1}_\rho$, the rank of $\boldsymbol{Q}_f$ is $\rho(G)$ and the rank of $\boldsymbol{Q} \geq$ the rank of $\boldsymbol{Q}_f = \rho(G)$.

Theorem 4.1.4.1. *Let $\boldsymbol{Q}$ and $\boldsymbol{B}$ be, respectively, the cutset matrix and the tieset matrix whose columns are arranged using the same order of edges. Then every row of $\boldsymbol{Q}$ is orthogonal to every row $\boldsymbol{B}$, that is,*

$$\boldsymbol{Q}\boldsymbol{B}^{\mathrm{T}} = \boldsymbol{0}, \boldsymbol{B}\boldsymbol{Q}^{\mathrm{T}} = \boldsymbol{0} \quad (\text{mod } 2), \tag{4.1.4.6}$$

$$\boldsymbol{Q}_d\boldsymbol{B}_d^{\mathrm{T}} = \boldsymbol{0}, \boldsymbol{B}_d\boldsymbol{Q}_d^{\mathrm{T}} = \boldsymbol{0} \quad (\text{for digraph}), \tag{4.1.4.7}$$

where superscript T denotes the transposed matrix.

Proof. Let us consider a directed graph first. Let a row of $\boldsymbol{Q}_d$ and a row of $\boldsymbol{B}_d$ be $\boldsymbol{q}_i = [q_{i1}, q_{i2}, \ldots, q_{i|E|}]$ and $\boldsymbol{b}_j = [b_{j1}, b_{j2}, \ldots, b_{j|E|}]$, respectively, then $\boldsymbol{Q}_d\boldsymbol{B}_d^{\mathrm{T}} = [x_{ij}]$ where $x_{ij} = \boldsymbol{q}_i\boldsymbol{b}_j^T = \sum_{k=1}^{|E|} q_{ik}b_{jk}$. $\boldsymbol{q}_i$ represents the edges which are included in the cutset H_i and their directions in the H_i. $\boldsymbol{b}_j$ represents the edges which are included in the tieset F_j and their directions in the F_j. Form Theorem 2.7.1, a tieset and a cutset have an even number (including 0) of common edges. If a cutset H_i and a tieset F_j have no common edges, $\boldsymbol{q}_i\boldsymbol{b}_j^T = 0$. If a cutset H_i and a tieset F_j have common edges, the number of common edges is even and, the number of same direction and the number of opposite direction in a cutset H_i and a tieset F_j are equal. That is, $\boldsymbol{q}_i\boldsymbol{b}_j^T = 0$, thus $\boldsymbol{Q}_d\boldsymbol{B}_d^{\mathrm{T}} = \boldsymbol{0}$. Obviously $\boldsymbol{B}_d\boldsymbol{Q}_d^{\mathrm{T}} = \boldsymbol{0}$.

Similarly, we can proof the result using mod 2 in an undirected graph. □

According to *Sylvester's theorem*, rank $\boldsymbol{Q} \leq |E| - \text{rank}\boldsymbol{B} = |E| - \mu(G) = \rho(G)$ from (4.1.3.6).

Therefore,

$$\text{rank } \boldsymbol{Q} = \rho(G) \tag{4.1.4.8}$$

From (4.1.3.3), (4.1.3.15)–(4.1.3.17),

$$\boldsymbol{Q}_f\boldsymbol{B}_f^T = \boldsymbol{Q}_{\mathrm{fc}} + \boldsymbol{B}_{ft}^{\mathrm{T}} = 0, \tag{4.1.4.9}$$

$$\boldsymbol{Q}_{fc} = -\boldsymbol{B}_{ft}^{\mathrm{T}} \quad \text{(for digraph)}, \tag{4.1.4.10}$$

$$\boldsymbol{Q}_{\mathrm{fc}} = \boldsymbol{B}_{ft}^{\mathrm{T}} \quad \text{(mod 2)}. \tag{4.1.4.11}$$

Example. Let $t = \{e_4, e_5, e_6, e_8\}$ and $c = \{e_1, e_2, e_3, e_7\}$ in a graph of Fig. 4.1.1.1.

$$\boldsymbol{Q}_f = [\overbrace{\boldsymbol{Q}_{fc}}^{c} \;\vdots\; \overbrace{\boldsymbol{1}_\rho}^{t}] = \begin{matrix} \\ H_1 \\ H_2 \\ H_4 \\ H_3 \end{matrix} \begin{matrix} \begin{matrix} e_1 & e_2 & e_3 & e_7 & e_4 & e_5 & e_6 & e_8 \end{matrix} \\ \left[\begin{array}{cccc|cccc} 1 & 1 & 1 & 0 & 1 & 0 & 0 & 0 \\ 1 & 1 & 0 & 0 & 0 & 1 & 0 & 0 \\ 0 & 0 & 0 & 0 & 0 & 0 & 1 & 0 \\ 1 & 1 & 1 & 0 & 0 & 0 & 0 & 1 \end{array}\right] \end{matrix}. \tag{4.1.4.12}$$

From (4.1.3.16),

$$\boldsymbol{Q}_{fc} = \boldsymbol{B}_{ft}^{\mathrm{T}} \quad \text{(mod 2)}. \tag{4.1.4.13}$$

Let $t = \{e_4, e_5, e_6, e_8\}$ and $c = \{e_1, e_2, e_3, e_7\}$ in a graph of Fig. 4.1.1.2 .

$$Q_f = [\overbrace{Q_{fc}}^{c} \;\vdots\; \overbrace{\mathbf{1}_\rho}^{t}] = \begin{array}{c} \\ H_1 \\ H_2 \\ H_4 \\ H_3 \end{array}\begin{array}{c} \begin{array}{cccc|cccc} e_1 & e_2 & e_3 & e_7 & e_4 & e_5 & e_6 & e_8 \end{array} \\ \left[\begin{array}{cccc|cccc} -1 & 1 & 1 & 0 & 1 & 0 & 0 & 0 \\ -1 & 1 & 0 & 0 & 0 & 1 & 0 & 0 \\ 0 & 0 & 0 & 0 & 0 & 0 & 1 & 0 \\ 1 & -1 & -1 & 0 & 0 & 0 & 0 & 1 \end{array}\right] \end{array}. \tag{4.1.4.14}$$

From (4.1.3.19),

$$Q_{fc} = -B_{ft}^{\mathrm{T}} \quad \text{(for digraph)}. \tag{4.1.4.15}$$

(End of Example)

4.2. Application to Linear Networks

As an example of a network, let us consider an electrical network. The weighted edges and/or points of the graph are called a *network*. In an electrical network, every edge e_k in directed connected graph $G(V, E)$ is associated with a number i_k called *current* or *edge current* which may be thought of the *through-quantity* or *flow* and every vertex v_i in G is associated with a number p_i called the *potential* or *voltage* which may be thought of the *across-quantity, potential difference* or *tension* of the edge $e_k = (v_{i1}, v_{i2})$ where $u_k = p_{i1} - p_{i2}$ and u_k called *edge voltage* of e_k. Thus an electrical network is a connected directed graph G in which each edge e_k is assigned two variables u_k and i_k. Let the values of currents flowing through edges be represented by a column vector (called the *edge current vector* or *through vector*) $\boldsymbol{I}$. Similarly, *potential vector* $\boldsymbol{P}$ and *edge voltage vector*, *potential difference vector* or *across vector* $\boldsymbol{U}$ are defined as follows:

$$\boldsymbol{I} = \begin{bmatrix} i_1 \\ \vdots \\ i_{|E|} \end{bmatrix}, \tag{4.2.1}$$

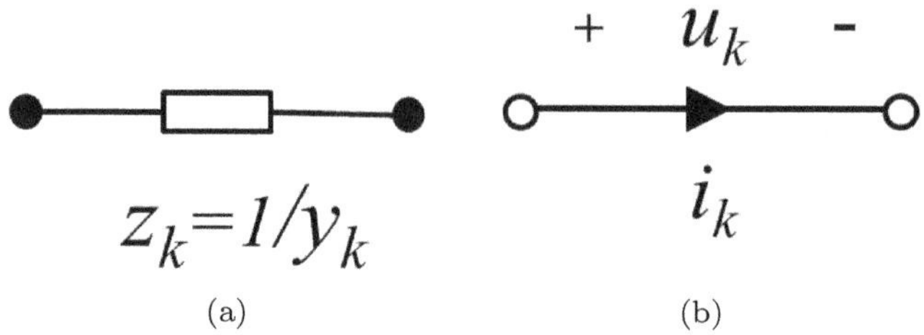

Fig. 4.2.1. Electrical element and its representation as an edge of a directed graph. (a) An electrical element. (b) An edge e_k.

$$\boldsymbol{P} = \begin{bmatrix} p_1 \\ \vdots \\ p_{|V|} \end{bmatrix}, \tag{4.2.2}$$

$$\boldsymbol{U} = \begin{bmatrix} u_1 \\ \vdots \\ u_{|E|} \end{bmatrix}. \tag{4.2.3}$$

There are some electrical components in an electrical network: An *impedance* element, an *admittance* element, an *across source* and a *through source*.

(1) In Fig. 4.2.1, an *impedance* element (an *admittance* element) and its representation as an edge e_k of a directed graph are shown where z_k and y_k are called *impedance* and *admittance* respectively, and the voltage $+$ is always at tail of current arrow.

The impedance z_k and admittance y_k of e_k are constant and the following equations link the edge current and voltage (linear):

$$u_k = z_k i_k, \tag{4.2.4}$$

$$i_k = y_k v_k, \tag{4.2.5}$$

where $z_k = 1/y_k$ and $z_k \neq 0$, $y_k \neq 0$.

(2) In Fig. 4.2.2(a), an across source (voltage source) is shown, where the voltage s_u is fixed and the current of the element is arbitrary value.

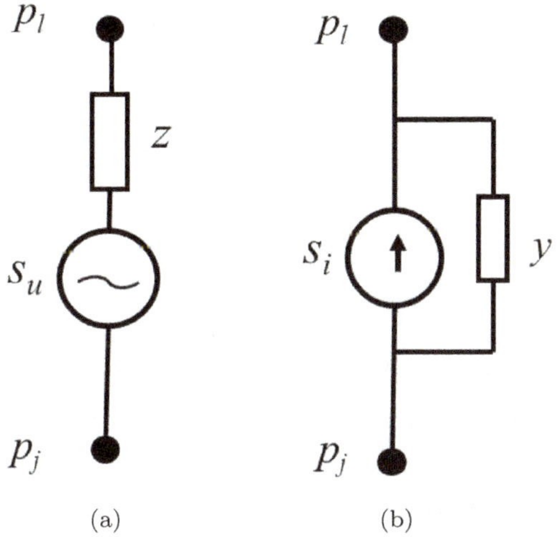

Fig. 4.2.2. An across source and a through source electrical elements. (a) An across source. (b) A through source.

(3) In Fig. 4.2.2(b), a through source (current source) is shown, where the current s_i is fixed and the voltage of the element is arbitrary value.

Let $\boldsymbol{S_u}$ and $\boldsymbol{S_i}$ be *cross source and a through source vectors*, respectively.

$$\boldsymbol{S_u} = \begin{bmatrix} s_{u1} \\ \vdots \\ s_{u|E|} \end{bmatrix}, \tag{4.2.6}$$

$$\boldsymbol{S_i} = \begin{bmatrix} s_{i1} \\ \vdots \\ s_{i|E|} \end{bmatrix}. \tag{4.2.7}$$

It was mentioned above that each element in an electrical network is governed by a specific relationship imposed upon its two edge variables. When the elements are interconnected to form a network, there are additional relationships called *Kirchhoff 's Law* imposed on these edge variables.

- *Kirchhoff's Current Law (KCL)*: The sum (taking into account the orientations) of all the currents leaving any node (or vertex) is zero. This means that the sum (taking into account the orientations) of all the currents of a cutset is zero.

- *Kirchhoff's Voltage Law (KVL)*: The sum (taking into account the orientations) of the voltages around a tieset is zero.

The *electrical network analysis* in which the electrical components (above (1)–(3)) and the connected directed graph are given is to find all of edge currents, voltages, using *KCL* and *KVL*. Let us show a method of the electrical network analysis called nodal analysis. In the analysis method, only the admittance elements and through sources are used as electrical components in an electrical network. If impedance elements and across sources are included, the impedance elements are transformed to admittance elements by Fig. 4.2.3:

$$z_k = 1/y_k. \tag{4.2.8}$$

The across sources are transformed to through sources admittance elements by Fig. 4.2.3. In Fig. 4.2.3, an across source and a through source electrical elements are shown. Let us find the condition to have the same currents and voltages of these source electrical elements. Let the current of the elements be i_k. In the across source, since z_k and s_{uk} are in series,

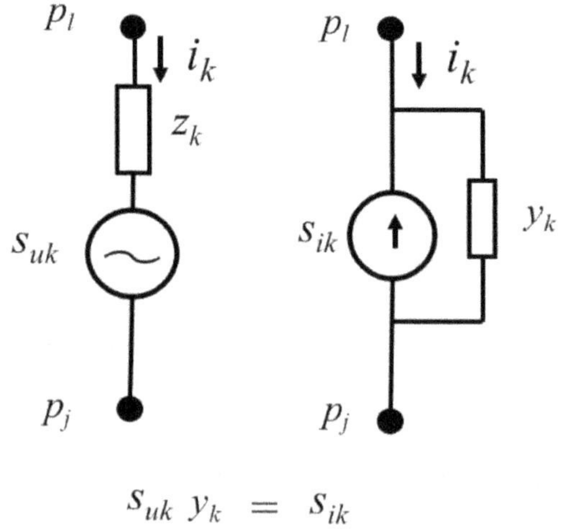

Fig. 4.2.3. Transformation between an across source and a through source.

every tieset including the edge e_k have z_k and s_{uk} in series. From *KVL*, $u_k = (p_l - p_k)$ is the sum of the voltage of z_k and s_{uk}. That is,

$$u_k = (p_l - p_k) = i_k z_k + s_{uk}. \tag{4.2.9}$$

In the through source, since y_k and s_{ik} are in parallel, the current of the edge e_k equals the sum of the current of y_k and s_{ik} from *KCL*.

That is,

$$u_k = (p_l - p_k) = (i_k + s_{ik}) \cdot 1/y_k. \tag{4.2.10}$$

From (4.2.9) and (4.2.10), we obtain

$$s_{uk} y_k = s_{jk}. \tag{4.2.11}$$

That is, if $s_{uk} y_k = s_{jk}$, an across source and a through source in Fig. 4.2.3 are equivalent.

Thus, if impedance elements and across sources are included, the impedance elements are transformed to admittance elements and the across sources are transformed to through sources using (4.2.8) and (4.2.11).

Let us consider a directed graph where every admittance of edges and through sources (current sources) are given.

From *KCL*,

$$\boldsymbol{A_r I} = \boldsymbol{0}, \tag{4.2.12}$$

$$\boldsymbol{Q_f I} = \boldsymbol{0}. \tag{4.2.13}$$

From *KVL*,

$$\boldsymbol{B_f U} = \boldsymbol{0}. \tag{4.2.14}$$

From (4.2.5) and (4.2.7),

$$\boldsymbol{I} = \boldsymbol{YU} + \boldsymbol{S_i}, \tag{4.2.15}$$

where

$$Y = \begin{bmatrix} y_1 & 0 & \cdots & 0 \\ 0 & y_2 & \cdots & 0 \\ \vdots & \vdots & \ddots & \vdots \\ 0 & 0 & \cdots & y_{|E|} \end{bmatrix}. \tag{4.2.16}$$

Since the edge voltage u_k of $e_k = (v_{i1}, v_{i2})$ is $u_k = p_{i1} - p_{i2}$,

$$\boldsymbol{U} = \boldsymbol{A}_r^T \boldsymbol{P}. \tag{4.2.17}$$

Multiplying $\boldsymbol{B}_f$ to $\boldsymbol{U}$ in (4.2.17) and using Theorem 4.1.2, we obtain

$$\boldsymbol{B}_f \boldsymbol{U} = \boldsymbol{B}_f \boldsymbol{A}_r^T \boldsymbol{P} = \boldsymbol{0}. \tag{4.2.18}$$

This satisfies *KVL* of (4.2.14).

Substituting (4.2.17) in (4.2.15) and substituting (4.2.15) in *KCL* of (4.2.12), we obtain

$$\boldsymbol{A}_r \boldsymbol{Y} \boldsymbol{A}_r^T \boldsymbol{P} = -\boldsymbol{A}_r \boldsymbol{S}_i \tag{4.2.19}$$

In (4.2.19), let $\boldsymbol{Y}_n = \boldsymbol{A}_r \boldsymbol{Y} \boldsymbol{A}_r^T$ and $\boldsymbol{J}_n = -\boldsymbol{A}_r \boldsymbol{S}_i$ which are called the node admittance matrix and the node current column vector, respectively,

$$\boldsymbol{Y}_n \boldsymbol{P} = \boldsymbol{J}_n. \tag{4.2.20}$$

In a method of the electrical network analysis called nodal analysis, we solve equation (4.2.20) for $\boldsymbol{P}$ and substituting $\boldsymbol{P}$ in (4.2.17) and substituting $\boldsymbol{U}$ of (4.2.17) in (4.2.15), we obtain all of edge currents and voltages.

Similarly, there are some analytical methods using $\boldsymbol{Q}_f \boldsymbol{I} = \boldsymbol{0}$ of (4.2.13) as *KCL* or edge currents and edge voltages as unknown quantities in place of vertex potentials in (4.2.19).

Example. Figure 4.2.4 is an electrical network in which impedance elements, admittance elements, an across source and a through source are included. Figure 4.2.5 is an electrical network with only admittance elements and the across sources which are transformed from impedance elements and through sources using (4.2.8) and (4.2.11). Figure 4.2.6 is a connected graph $G = (V, E)$ of an electrical network in Fig. 4.2.5, where

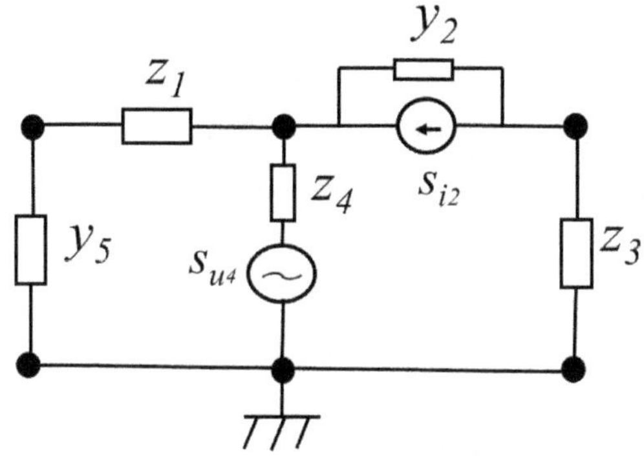

Fig. 4.2.4. An electrical network.

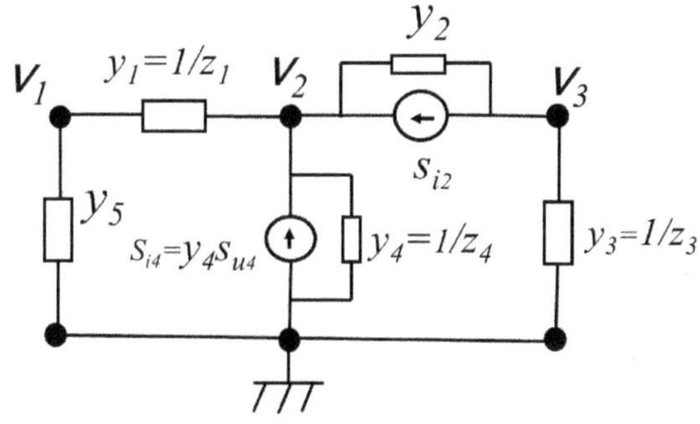

Fig. 4.2.5. An electrical network with admittance elements and the across sources transformed from impedance elements and through sources.

$V = \{v_1, v_2, v_3, v_4\}$, $E = \{e_1, e_2, e_3, e_4, e_5\}$, $e_1 = (v_2, v_1)$, $e_2 = (v_3, v_2)$, $e_3 = (v_3, v_4)$, $e_4 = (v_4, v_2)$ and $e_5 = (v_1, v_4)$.

Then

$$\boldsymbol{I} = \begin{bmatrix} i_1 \\ i_2 \\ i_3 \\ i_4 \\ i_5 \end{bmatrix}. \tag{4.2.21}$$

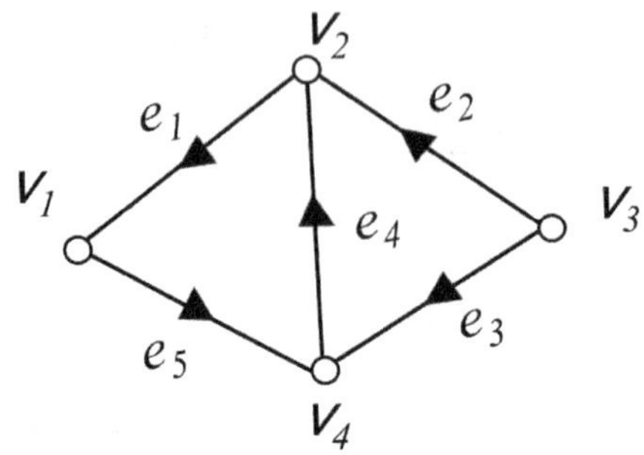

Fig. 4.2.6. The directed graph of an electrical network of Fig. 4.2.5.

Let $p_4 = 0$, then

$$\boldsymbol{P} = \begin{bmatrix} p_1 \\ p_2 \\ p_3 \end{bmatrix}, \tag{4.2.22}$$

$$\boldsymbol{U} = \begin{bmatrix} u_1 \\ u_2 \\ u_3 \\ u_4 \\ u_5 \end{bmatrix}, \tag{4.2.23}$$

$$\boldsymbol{S}_i = \begin{bmatrix} 0 \\ s_{i2} \\ 0 \\ s_{i4} \\ 0 \end{bmatrix}, \tag{4.2.24}$$

$$\boldsymbol{Y} = \begin{bmatrix} y_1 & 0 & 0 & 0 & 0 \\ 0 & y_2 & 0 & 0 & 0 \\ 0 & 0 & y_3 & 0 & 0 \\ 0 & 0 & 0 & y_4 & 0 \\ 0 & 0 & 0 & 0 & y_5 \end{bmatrix}, \tag{4.2.25}$$

$$\boldsymbol{A}_r = [\overbrace{\boldsymbol{A}_c}^{c} \ \vdots \ \overbrace{\boldsymbol{A}_t}^{t}] = \begin{array}{c} \\ v_1 \\ v_2 \\ v_3 \end{array} \begin{array}{c} \begin{array}{ccccc} e_1 & e_2 & e_3 & e_4 & e_5 \end{array} \\ \begin{bmatrix} -1 & 0 & 0 & 0 & 1 \\ 1 & -1 & 0 & -1 & 0 \\ 0 & 1 & 1 & 0 & 0 \end{bmatrix} \end{array}, \tag{4.2.26}$$

$$\boldsymbol{B}_{Bf} = [\overbrace{\boldsymbol{1}_\mu}^{c} \ \vdots \ \overbrace{\boldsymbol{B}_{ft}}^{t}] = \begin{array}{c} \\ F_1 \\ F_2 \end{array} \begin{array}{c} \begin{array}{ccccc} e_1 & e_2 & e_3 & e_4 & e_5 \end{array} \\ \begin{bmatrix} 1 & 0 & 0 & 1 & 1 \\ 0 & 1 & -1 & -1 & 0 \end{bmatrix} \end{array}. \tag{4.2.27}$$

From (4.2.26) and (4.2.27),

$$\boldsymbol{B}_f \boldsymbol{A}_r^T \boldsymbol{P} = \begin{bmatrix} 1 & 0 & 0 & 1 & 1 \\ 0 & 1 & -1 & -1 & 0 \end{bmatrix} \begin{bmatrix} -1 & 1 & 0 \\ 0 & -1 & 1 \\ 0 & 0 & 1 \\ 0 & -1 & 0 \\ 1 & 0 & 0 \end{bmatrix} \boldsymbol{P} = \begin{bmatrix} 0 & 0 \\ 0 & 0 \end{bmatrix} \quad \boldsymbol{P} = \boldsymbol{0}. \tag{4.2.28}$$

From (4.2.25) and (4.2.26),

$$\boldsymbol{Y}_n = \boldsymbol{A}_r \boldsymbol{Y} \boldsymbol{A}_r^T = \begin{bmatrix} -1 & 0 & 0 & 0 & 1 \\ 1 & -1 & 0 & -1 & 0 \\ 0 & 1 & 1 & 0 & 0 \end{bmatrix} \begin{bmatrix} y_1 & 0 & 0 & 0 & 0 \\ 0 & y_2 & 0 & 0 & 0 \\ 0 & 0 & y_3 & 0 & 0 \\ 0 & 0 & 0 & y_4 & 0 \\ 0 & 0 & 0 & 0 & y_5 \end{bmatrix}$$

$$\times \begin{bmatrix} -1 & 1 & 0 \\ 0 & -1 & 1 \\ 0 & 0 & 1 \\ 0 & -1 & 0 \\ 1 & 0 & 0 \end{bmatrix} = \begin{bmatrix} y_1 + y_5 & -y_1 & 0 \\ -y_1 & y_1 + y_2 + y_4 & -y_2 \\ 0 & -y_2 & y_2 + y_3 \end{bmatrix}. \tag{4.2.29}$$

From (4.2.26) and (4.2.24),

$$\boldsymbol{J}_n = -\boldsymbol{A}_r\boldsymbol{S}_i = \begin{bmatrix} -1 & 0 & 0 & 0 & 1 \\ 1 & -1 & 0 & -1 & 0 \\ 0 & 1 & 1 & 0 & 0 \end{bmatrix} \begin{bmatrix} 0 \\ s_{i2} \\ 0 \\ s_{i4} \\ 0 \end{bmatrix} = \begin{bmatrix} 0 \\ s_{i2} + s_{i4} \\ -s_{i2} \end{bmatrix}. \quad (4.2.30)$$

From (4.2.20), (4.2.29) and (4.2.30),

$$\boldsymbol{Y}_n\boldsymbol{P} = \boldsymbol{J}_n \begin{bmatrix} y_1 + y_5 & -y_1 & 0 \\ -y_1 & y_1+y_2+y_4 & -y_2 \\ 0 & -y_2 & y_2+y_3 \end{bmatrix} \boldsymbol{P} = \begin{bmatrix} 0 \\ s_{i2} + s_{i4} \\ -s_{i2} \end{bmatrix}. \quad (4.2.31)$$

From (4.2.31), we solve the equation for $\boldsymbol{P}$ and substituting $\boldsymbol{P}$ in (4.2.17) and substituting $\boldsymbol{U}$ of (4.2.17) in (4.2.15), we can obtain all of edge currents, voltages, and all of vertex potentials. **(End of Example)**

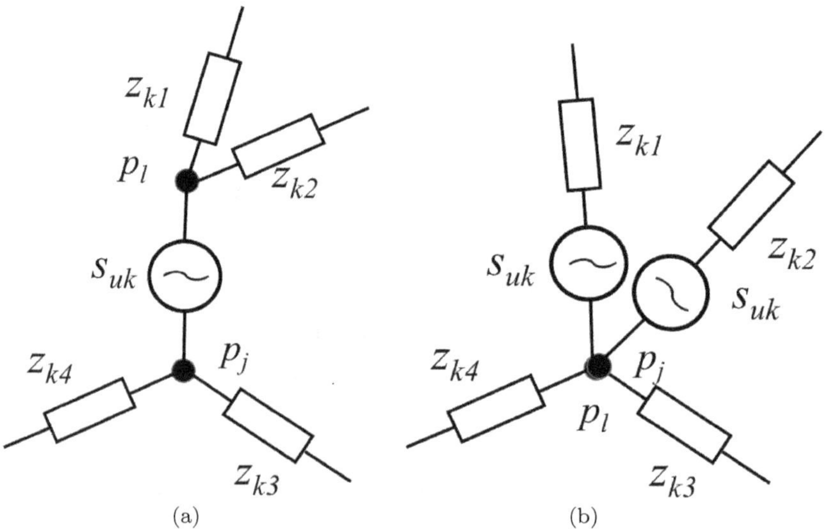

Fig. 4.2.7. An equivalent transformation to an across source with no impedance to an across source with impedance. (a) An across source with no impedance ($z_{k\cdot} = 0$). (b) An equivalent transformation to an across source with impedance.

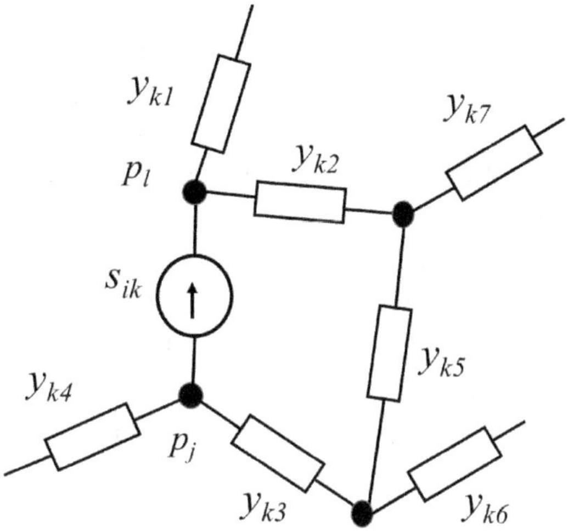

(a)

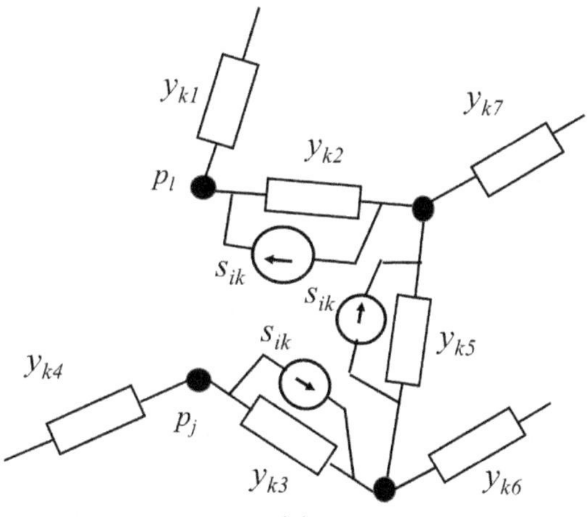

(b)

Fig. 4.2.8. An equivalent transformation to a through source with no admittance to a through source with admittance. (a) A through source with no admittance ($y_{k\cdot} = 0$). (b) An equivalent transformation to a through source with admittance.

In the example, we show an electrical network with admittance elements and the across sources transformed from impedance elements and through sources. Figure 4.2.3 shows how to transform sources but if $z_k = 0$ and $y_k = \infty$, the transformation is impossible. In a real electrical network, the case of $z_k = 0(y_k = \infty)$ is rare. However for the theory of analysis of an electrical network, we will show the analytical method in the case of $z_k = 0(y_k = \infty)$.

Figure 4.2.7(a) is an example of an across source with no impedance ($z_{k\cdot} = 0$). Figure 4.2.7(b) is an equivalent transformation an across source with no impedance to an across source with impedance, where the source s_{uk} is moved to every incident edge of p_l, and p_l and p_j are identified. Obviously, the transformation keeps *KCL* and *KVL*.

Figure 4.2.8(a) is an example of a through source with no admittance ($y_{k\cdot} = 0$). Figure 4.2.8(b) is an equivalent transformation a through source with no admittance to a through source with admittance, where the source s_{ik} is moved to every edge of a tieset including s_{ik}, and s_{ik} is removed. Obviously, the transformation keeps *KCL* and *KVL*.

4.3. Tellegen's Theorem in Electrical Networks

The electrical network analysis in which the electrical components (impedance elements, admittance elements, across sources (voltage sources) and through sources (current sources)) and the connected directed graph are given is to find all of edge currents, voltages, and all of vertex potentials, using *KCL* and *KVL*. The electrical components, the connected directed graph and *KCL* and *KVL* are also considered for studying electrical network analysis and electrical network synthesis. The process of the study has examined various properties of electrical network. Among the properties, there is a unique property called *Tellegen's theorem*.

Theorem 4.3.1. *If two electrical networks have the same connected directed graph $G = (V, E)$ with corresponding through-vectors $\boldsymbol{I}$ and $\boldsymbol{I}'$ as well as corresponding across-vectors $\boldsymbol{U}$ and $\boldsymbol{U}'$, then*

$$\boldsymbol{U}^{\mathrm{T}}\boldsymbol{I}' = 0 \quad \text{and} \quad \boldsymbol{U}'^{\mathrm{T}}\boldsymbol{I} = 0. \tag{4.3.1}$$

That is

$$\sum_{k=1}^{|E|} u_k i'_k = 0 \quad \text{and} \quad \sum_{k=1}^{|E|} u'_k i_k = 0. \tag{4.3.2}$$

If we apply this to the case of an electrical network of a connected directed graph $G = (V, E)$ with through-vectors $\boldsymbol{I}$ and across-vectors $\boldsymbol{U}$, then

$$\boldsymbol{U}^T\boldsymbol{I} = 0 \quad \text{that is} \sum_{k=1}^{|E|} u_k i_k = 0. \tag{4.3.3}$$

This is known as the Law of Conservation of Energy.

Proof. Let $\mathbf{A_r}$ be the reduced incidence matrix of G and from KCL,

$$\boldsymbol{A_r I} = \mathbf{0}. \tag{4.3.4}$$

Since the edge voltage u_k of $e_k = (v_{i1}, v_{i2})$ is $u_k = p_{i1} - p_{i2}$ where p_{i1} and p_{i2} are potential elements of *potential vector* $\boldsymbol{P}$,

$$\boldsymbol{U} = \boldsymbol{A_r^T P}. \tag{4.3.5}$$

Therefore, from (4.3.5) and (4.3.4),

$$\boldsymbol{U^T I} = \boldsymbol{P^T A_r I} = 0. \tag{4.3.6}$$

This result only depends on the structure of the connected directed graph G. That is, this result holds regardless of the electrical components (impedance elements, admittance elements, across sources (voltage sources) and through sources (current sources)). This means that if two electrical networks have the same connected directed graph $G = (V, E)$ with corresponding through-vectors $\boldsymbol{I}$ and $\boldsymbol{I}'$ as well as corresponding across-vectors $\boldsymbol{U}$ and $\boldsymbol{U}'$, then $\boldsymbol{U^T I'} = \mathbf{0}$ and $\boldsymbol{U'^T I} = \mathbf{0}$. □

Example. Figure 4.3.1 is an example of electrical networks which was shown in Fig. 1.2(a) and Fig. 4.3.2 is a generalized model of Fig. 4.3.1. Figure 4.3.2 shows two electrical networks (a) and (b) which have the same connected directed graph $G = (V, E)$ where impedance elements z_k, z'_k, admittance elements y_k, y'_k and an across source s_{uk}. s'_{uk} are included and they may have different values and types in two electrical networks (a) and (b). This means that the impedance elements of (a) may be the admittance elements in (b) and an across source of (a) may be the through sources in (b), and sources may be included in the some elements in (b). Figure 4.3.3 is the directed graph $G = (V, E)$ of Figs. 4.3.2(a) and 4.3.2(b).

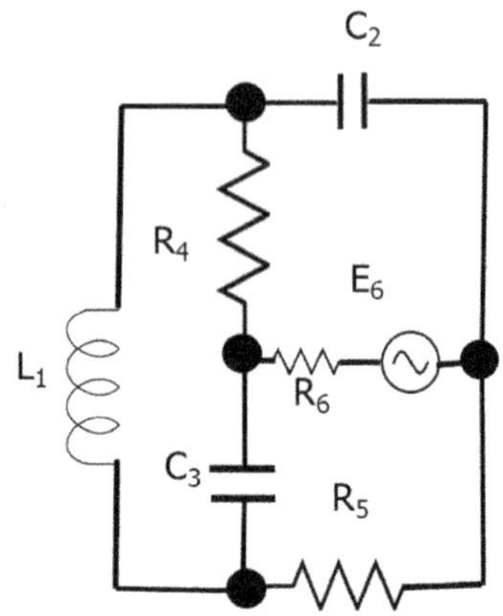

Fig. 4.3.1. An electrical network (Fig. 1.2(a)).

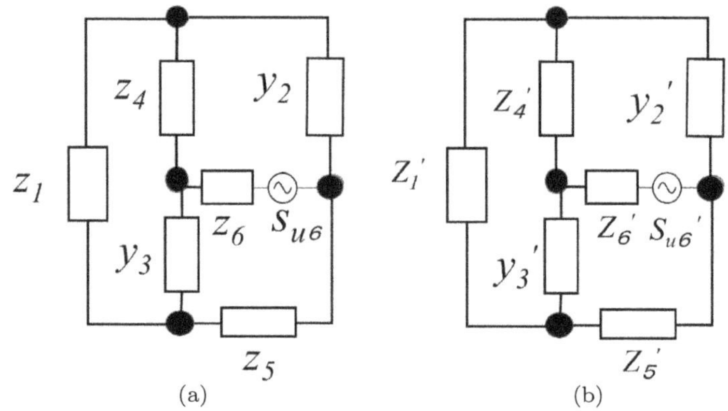

Fig. 4.3.2. Two electrical networks (a) and (b) which have the same connected directed graph $G = (V, E)$. (a) An electrical network with element y_k, z_k etc. (b) An electrical network with element y'_k, z'_k etc.

Let v_4 be the reference vertex of G.

$$\boldsymbol{A}_r = \begin{matrix} & \begin{matrix} e_1 & e_2 & e_3 & e_4 & e_5 & e_6 \end{matrix} \\ \begin{matrix} v_1 \\ v_2 \\ v_3 \end{matrix} & \begin{bmatrix} 1 & 1 & 0 & -1 & 0 & 0 \\ 0 & 0 & -1 & 1 & 0 & 1 \\ -1 & 0 & 1 & 0 & 1 & 0 \end{bmatrix} \end{matrix}. \quad (4.3.7)$$

Since v_4 is reference, $p_4 = 0$. Let the potential vector, voltage vector and current vector of (a) in Fig. 4.3.2 be $\boldsymbol{P}$, $\boldsymbol{U}$, and $\boldsymbol{I}$. Let the potential vector,

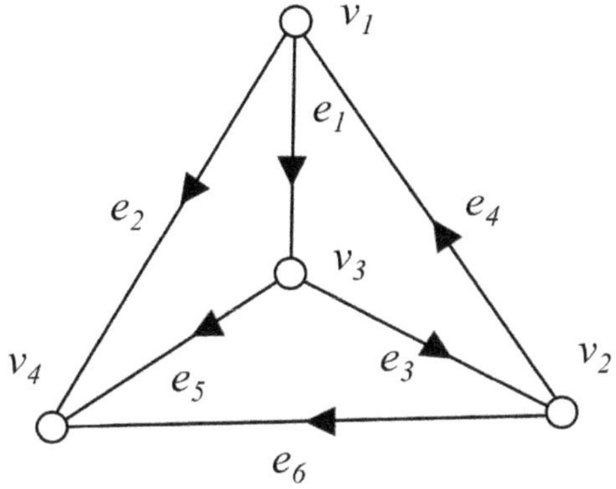

Fig. 4.3.3. The directed graph $G = (V, E)$ of two electrical networks (a) and (b) in Fig. 4.3.2.

voltage vector and current vector of (b) in Fig. 4.3.2 be $\boldsymbol{P}'$, $\boldsymbol{U}'$, and $\boldsymbol{I}'$.

$$\boldsymbol{P} = \begin{bmatrix} p_1 \\ p_2 \\ p_3 \end{bmatrix}, \tag{4.3.7}$$

$$\boldsymbol{U} = \begin{bmatrix} u_1 \\ u_2 \\ u_3 \\ u_4 \\ u_5 \\ u_6 \end{bmatrix}, \tag{4.3.8}$$

$$\boldsymbol{I} = \begin{bmatrix} i_1 \\ i_2 \\ i_3 \\ i_4 \\ i_5 \\ i_6 \end{bmatrix}, \tag{4.3.9}$$

$$\boldsymbol{P}' = \begin{bmatrix} p_1' \\ p_2' \\ p_3' \end{bmatrix}, \tag{4.3.10}$$

$$\boldsymbol{U}' = \begin{bmatrix} u_1' \\ u_2' \\ u_3' \\ u_4' \\ u_5' \\ u_6' \end{bmatrix}, \tag{4.3.11}$$

$$\boldsymbol{I}' = \begin{bmatrix} i_1' \\ i_2' \\ i_3' \\ i_4' \\ i_5' \\ i_6' \end{bmatrix}. \tag{4.3.12}$$

Since (4.3.7) is incident matrix of a directed graph and for example $e_1 = (v_1, v_2)$, $u_1 = p_1 - p_2$, all of edge voltages are given as follows:

$$\boldsymbol{U} = \boldsymbol{A}_r^T \boldsymbol{P}. \tag{4.3.13}$$

Similarly,

$$\boldsymbol{U}' = \boldsymbol{A}_r^T \boldsymbol{P}'. \tag{4.3.14}$$

From *KCL*,

$$\boldsymbol{A}_r \boldsymbol{I} = \boldsymbol{0} \tag{4.3.15}$$

and

$$\boldsymbol{A}_r \boldsymbol{I}' = \boldsymbol{0}. \tag{4.3.16}$$

Then,

$$\boldsymbol{U}^T \boldsymbol{I}' = (\boldsymbol{A}_r^T \boldsymbol{P})^T \boldsymbol{I}'.$$

Using (4.3.16),

$$U^T I' = (A_r^T P)^T I' = P^T A_r I' = P^T \mathbf{0} = 0. \tag{4.3.17}$$

Similarly, $U'^T I = 0$. **(End of Example)**

4.4. Tree and Computer Algorithm

4.4.1. *Binary tree and the amount of calculation*

The trees are defined in Section 3.1 and have a hierarchical structure which has many applications in computer *algorithms*. Figure 4.4.1.1 shows a tree with a hierarchical structure in which one vertex at the top is called the *root* and the other vertices branching down from it. In Fig. 4.4.1.1, the length of a path between v_1 (root) and v_4 is 2 then the number 2 is called the *depth* of v_4. The maximum depth of a tree is called the *height* of the tree. Vertices of a particular "depth" are sometimes collectively referred to as "*levels*" in a tree (see Fig. 4.4.1.1). v_3 belongs to level 1 and v_3 is adjacent to v_7 which belongs to level 2, then v_3 is called the *parent* of v_7 which is called the *child* of v_3. v_5 is also a child of v_3 which has no child, then v_5 is called *leaf* or *pendant vertex* (see Section 3.1).

Let us consider an example of linear search and binary search algorithms. In Fig. 4.4.1.2, suppose that 17 same boxes contain various fruits (a banana,

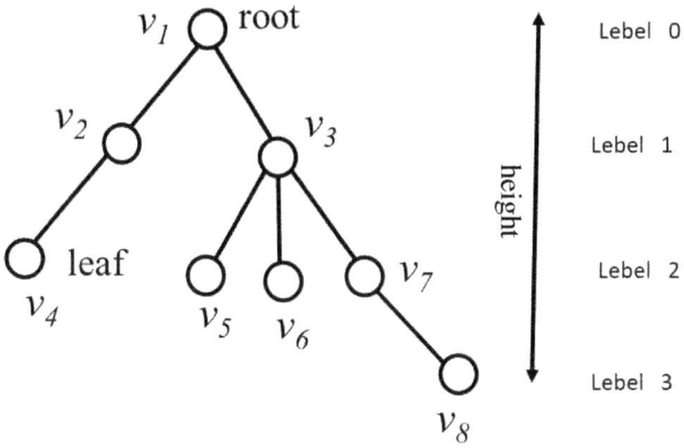

Fig. 4.4.1.1. A tree with a hierarchical structure.

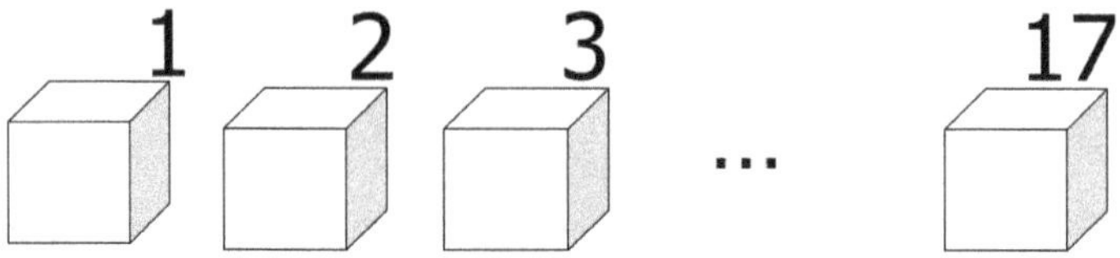

Fig. 4.4.1.2. 17 boxes contain various fruits (a banana, an apple, an orange, etc.).

an apple, an orange, etc.) and each box contains one different fruit. Find the box containing an apple. It is completely indistinguishable from the outside, and you cannot tell what is inside unless you open the box. The method of searching for an apple by opening them one by one from the first box is called *linear search.* If there is an apple in the first box, you can open it once, and if it is in the last box, you can open it 17 times. The worst number of searches is 17 times. The linear search is a simple method, and when the number n of boxes is as small as 17, the worst is 17 times, but when n is large, the worst is n, and the average is $n/2$. The amount of calculation is "$O(n)$" (called in the O notation of the *order*, and the amount of calculation is determined by the order of n).

In Fig. 4.4.1.3, suppose that 17 same boxes contain apples of different weights. Find a box containing 300 grams of apples. It is completely indistinguishable from the outside, and you cannot tell unless you open the box and weigh it. The weights are arranged in order from the lightest one. Let us show a more efficient way to weigh the apples by opening them one by one from the first box like a linear search. Open the eighth box in the middle and weigh 250 grams. As a result, there are 300 grams of an apple in boxes 9–17. Next, open the box 12 in the middle of boxes 9–17. If we weigh 270 grams, then 300 grams of apples are in boxes 13–17. Next, open the box 15 in the middle of boxes 13–17. If we weigh 320 grams, then 300 grams of apples are in boxes 13 or 14. At this time, the number of times the box is opened is 4 or 5 times, and the worst number is 5 times.

If the number of boxes is n, the maximum number of searches is $(O(\log_2 n)) + 1$. The amount of calculation when n becomes large is $O(\log_2 n)$. This method is called *binary search.* Table 4.4.1.1 shows that the binary search is extremely effective when the number of data n becomes large.

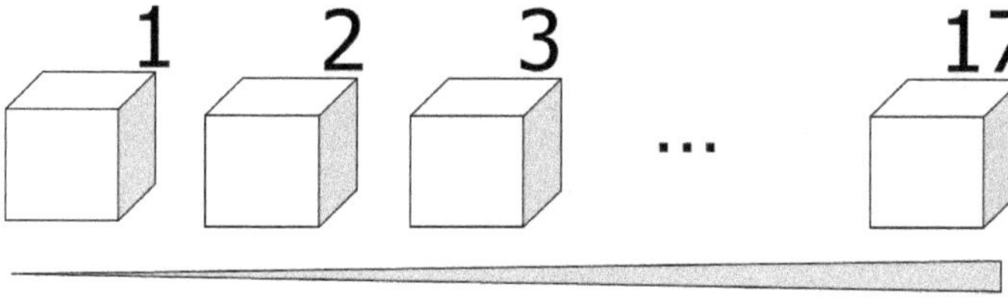

Fig. 4.4.1.3. 17 boxes contain apples in order from the lightest.

Table 4.4.1.1. Number of data n and $O(\log_2 n)$.

Number of Data n	$O(\log_2 n)$
100	7
1,000	10
10,000	14
100,000	17
1,000,000	20
10,000,000	24
100,000,000	27
1,000,000,000	30
10,000,000,000	34

If a parent in a tree has one or two children, the tree is called a *binary tree.* In binary trees, if every parent has two children, the tree is called a *full binary tree* and in full binary trees, if the depth of every leaf is the same, the tree is called a *perfect binary tree.*

Example. Figure 4.4.1.4 shows an example of a binary tree. Figure 4.4.1.5 shows an example of a full binary tree. Figure 4.4.1.6 shows an example of a perfect binary tree. **(End of Example)**

Theorem 4.4.1.1. *Let the number of leafs in a full binary tree* $G = (V, E)$ *be* l. *Then* $l = (|V| + 1)/2$ *and the number of parents is one less than the number of leafs.*

Proof. There are three kind of degrees of each vertices in a full binary tree $G = (V, E)$, that is, one vertex with degree 2 called root, l vertices with degree 1 called leaf and other vertices with degree 3.

The sum of degrees of all vertices $= 2 + l + 3(|V| - 2 - l) = 2|E|$. Since G is a tree, $|E| = |V| - 1$. Therefore $l = (|V| + 1)/2$.

Since the vertices except for leafs in a full binary tree $G = (V, E)$ are parents, the number of parents is $|V| - l$. Since $l - (|V| - l) = 2l - |V| = 1$, the number of parents is one less than the number of leafs. □

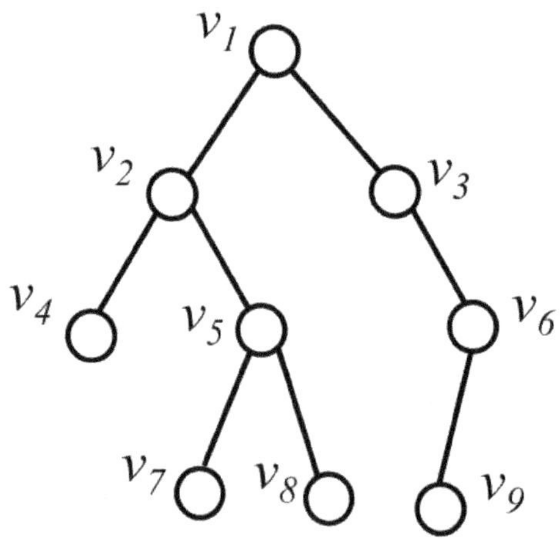

Fig. 4.4.1.4. A binary tree.

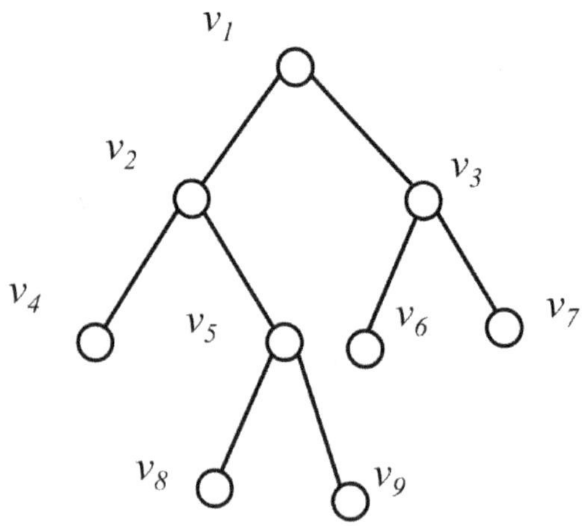

Fig. 4.4.1.5. A full binary tree.

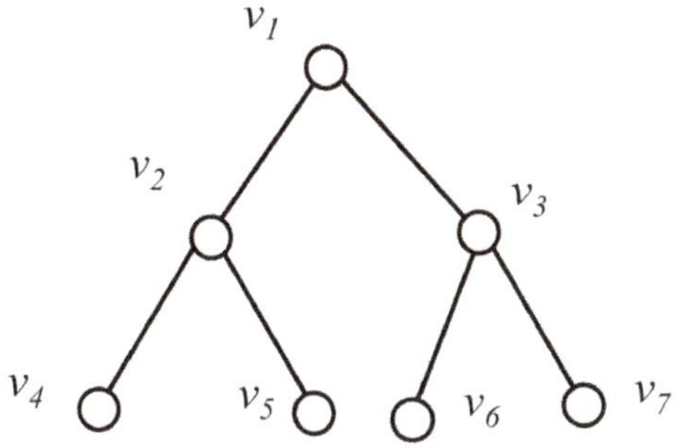

Fig. 4.4.1.6. A perfect binary tree.

Theorem 4.4.1.2. *Let $G = (V, E)$ be a binary tree with $|V| = n$. Then the maximum height $= (n - 1)$ and the minimum height $= \lceil \log_2(n+1) - 1 \rceil$, where $\lceil x \rceil$ is the smallest integer greater than or equal to x for a real number x.*

Proof. Since a binary tree has one or two children, the binary tree with maximum height has one child. That is, the maximum height $= (n - 1)$.

Since the maximum number of vertices of a binary tree with height h is $2^{h+1} - 1$, $2^h - 1 < n \leq 2^{h+1} - 1$, that is, $\log_2(n+1) - 1 \leq h < \log_2(n) - 1$. Therefore, the minimum height = ceiling $(\log_2(n+1) - 1) = \lceil \log_2(n+1) - 1 \rceil$. □

Example. Figure 4.4.1.7(a) shows a binary tree with $|V| = n$ and the maximum height$= (n - 1)$. Figure 4.4.1.7(b) shows a binary tree with $|V| = n$ and the minimum height = ceiling $(\log_2(n+1) - 1) = \lceil \log_2(n+1) - 1 \rceil$.
(End of Example)

A *binary search tree* is a binary tree in which the value of the child node on the left is smaller or equal to the value of the child node on the right. The binary search tree is one of the data structures suitable for searching data. To search for a value, start from the root node, move to the right child if the value of each node is greater than the value being searched, move to the left child if it is smaller, and end. The search ends when you trace to the end node (leaf node). Only the number of depths (heights) from the root to the leaf need to be examined, and the values can be searched more efficiently than simply examining the values in order from the first end.

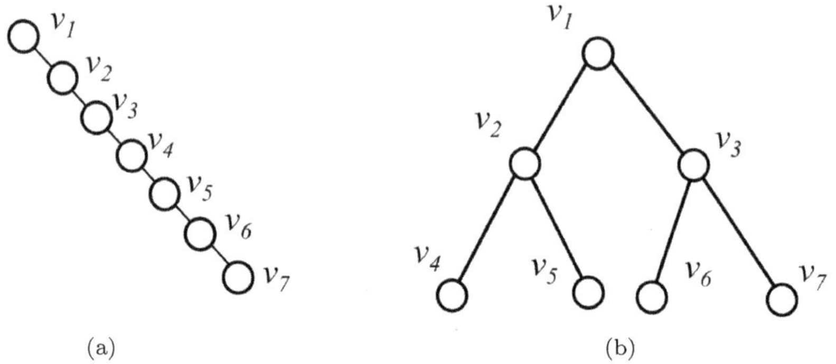

Fig. 4.4.1.7. (a) A binary tree with $|V| = n$ and the maximum height $= (n-1)$. (b) A binary tree with $|V| = n$ and the minimum height = ceiling $(\log_2(n+1)-1) = \lceil \log_2(n+1)-1 \rceil$. (a) a binary tree with the maximum height. (b) A binary tree with the minimum height.

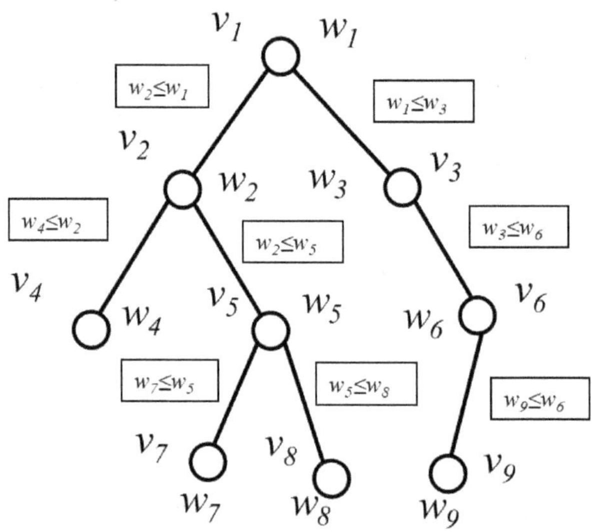

Fig. 4.4.1.8. An example of A *binary search tree*.

Figure 4.4.1.8 is an example of a binary search tree in which the value of v_i is $w_i (i = 1, 2, \ldots)$. The problem of Fig. 4.4.1.3 is an example of a binary search tree and the height shows the amount of calculation (complexity): $O(\log_2 n)$.

Let us consider some algorithms in graphs. The complexity of a problem in graphs is related to the resources required to compute a solution as a function of the size of the problem where the simple size of a graph $G = (V, E)$ is the number of vertices $|V| = n$ and the number of vertices $|E| = m$. So, for example, the complexity of a problem in graphs is sometimes expressed as $O(mn)$ or $O(m)$ etc. $O(mn)$ means the time obtained by multiplying mn by a constant at most. A problem that can be solved in polynomial time with respect to the input length is called a *P-problem* because it is a problem that can be solved efficiently. The problem for which the answer is required by the computational complexity $O(mn)$ is the P-problem. For many problems, there is no known way to find an answer quickly, but if one is provided with information showing what the answer is, it is possible to verify the answer quickly. The class of problems for which an answer can be verified in polynomial time is *NP*, which stands for "*nondeterministic polynomial time*". There are a group of problems that are believed to be unsolvable efficiently, some of which are called *NP-complete* and some of which are called *NP-hard* [17]. If the problem to be solved is found to be NP-complete or NP-hard, it is necessary to pursue an approximate solution to the problem. This approximation study is a meaningful and useful study because the problem is originally known to be NP-complete or NP-hard. The reason is that it seems just a difficult problem, and pursuing an approximate solution loses its effectiveness if the problem later turns out to be a non-difficult problem. This difficult problem is effectively used for *cryptography* (the security problems area) by taking advantage of its characteristic of being difficult. As mentioned above, there is no known way to quickly find the answer, but a function that has the property of being able to quickly see the answer if provided with information that indicates what the answer. Such function is called a *one-way function* (e.g. a *hash function*). This function is used in cryptography. The *blockchain* technology in cryptography (e.g. the area of *Bitcoin*) and the technology using a *directed acyclic graph* (DAG) that does not include directed tiesets are known [18].

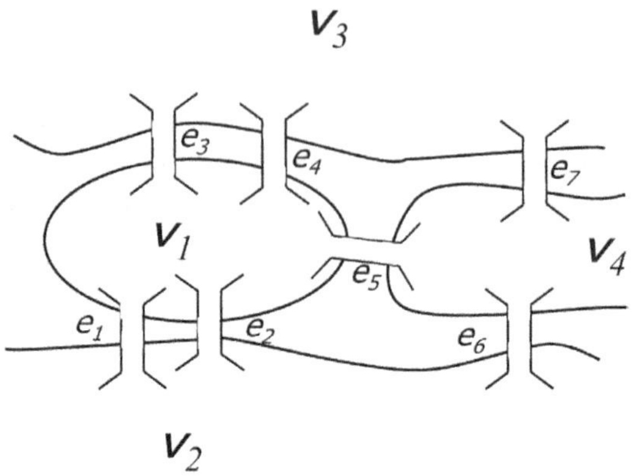

Fig. 4.4.2.1. Seven bridges on the Pregel River in Königsberg.

4.4.2. *Königsberg bridge problem*

The *Königsberg bridge problem* is considered as well known example in graph theory. It was a long-standing problem until solved by Leonhard Euler (1707–1783, a prominent mathematician and astrophysicist) in 1736, by means of a graph. In 18th century, there are seven bridges on the Pregel River in Königsberg as shown in Fig. 4.4.2.1 where $v_1, v_2, \ldots, v_4$ are land areas and $e_1, e_2, \ldots, e_7$ are bridges.

The problem was to start at any of the four land areas of v_1, v_2, and v_4, walk over each of the seven bridges exactly once, and return to the starting point. *Euler* represented this situation by means of a graph, as shown in Fig. 4.4.2.2 where the vertices represent the land areas and the edges represent the bridges.

In Section 2.4, a walk in the graph $G = (V, E)$ is defined as a finite sequence of the form: $v_{a0}, e_{b1}, v_{a1}, e_{b2}, \ldots, e_{bk}, v_{ak}$ which consists of alternating vertices and edges of G. A *trail* is a walk with no repeated edges. An *Eulerian trail* in a graph $G = (V, E)$ is a walk in G passing through every edge exactly once. If this walk is closed (starts and ends at the same vertex) it is called an *Eulerian tour*.

Lemma 4.4.2.1. *Every maximal trail in a graph $G = (V, E)$, where the degree $d(v_i)$ for any vertex $v_i \epsilon V$ is even, is a closed trail.*

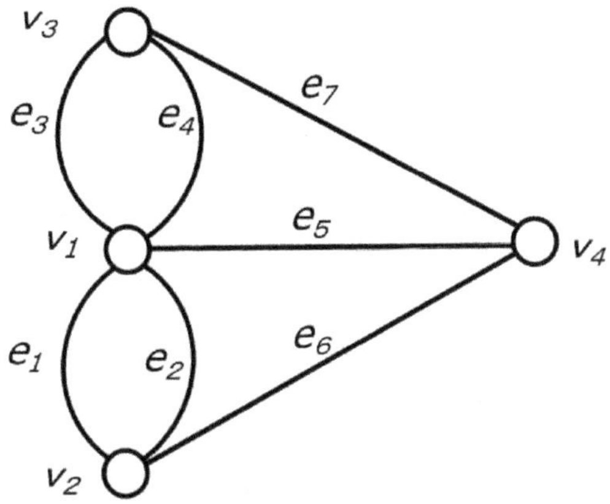

Fig. 4.4.2.2. The graph for seven bridges in Königsberg.

Proof. Let X be a maximal trail $v_{a0}, e_{b1}, v_{a1}, e_{b2}, \ldots, e_{bk}v_{ak}$ of G. If X is not closed that is $v_{a0} \neq v_{ak}$, then X has odd numbers of edges incident to the first and final vertices v_{a0} and v_{ak}, since X is a trail and there is a pair edges e_{bj}, e_{bj+1} for a vertex $v_{aj\,(j\neq 0,k)}$, thus v_{aj} has even numbers of edges of X. However, since v_{a0} and v_{ak} have even degrees, there is an edge incident to v_{a0} (and v_{ak}) which is not in X. This edge can be used to extend X to a longer trail, contradicting the maximality of X. □

Theorem 4.4.2.1. *A connected graph $G = (V, E)$ has an Eulerian tour if and only if each vertex has even degree.*

Proof. Suppose G has an Eulerian tour X. If a vertex v_i is visited k times in the tour X, then each visit uses 2 edges incident to v_i (one incoming edge and one outgoing edge). Thus, $d(v_i) = 2k$, which is even.

Let G be a connected graph with even degrees for every vertex. Let $X = v_{a0}, e_{b1}, v_{a1}, e_{b2}, \ldots, e_{bk}, v_{ak}$ be a longest trail in G. Then, by Lemma 4.4.3, X is closed, i.e., $v_{a0} = v_{ak}$. If X does not include all the edges of G, let G' be a graph removing all edges of X from G, then G' is a graph, possibly disconnected, with even degrees for every vertex. Since G is connected, there is an edge e_j outside of X such that $e_j = (v_{ai}, v_p)$ where v_{ai} is a vertex of X. Since G' is a graph, possibly disconnected, with even degrees for every vertex, there is a closed trail X' with $e_j = (v_{ai}, v_p)$ in a connected component of G' form Lemma 4.4.3. Since X' is a trail with first vertex

v_{ai} and final vertex v_{ai}, $X \bigcup X'$ is a trail in G which is longer than X, contradicting the fact that X is a longest trail in G. Thus, we conclude that X includes all the edges of G and so it is an Eulerian tour. □

Corollary 4.4.2.1. *A connected graph $G = (V, E)$ has an Eulerian trail if and only if it has either 0 or 2 vertices of odd degree.*

Proof. Suppose X is an Eulerian trail from vertex u to vertex v in G. If $u = v$ then X is an Eulerian tour and so by Theorem 4.4.2.1 it follows that all the vertices in G have even degree. If $u \neq v$, X is not an Eulerian tour but an Eulerian trail. Since an Eulerian trail is passing through every edge exactly once in G and not closed, the degree of every vertex except for u and v is even and the degrees of only two vertices u and v are odd.

If the number of vertices of odd degree in G is 0, that is, all the vertices in G have even degree, G has an Eulerian tour which is an Eulerian trail, from Theorem 4.4.2.1. Next, let the number of vertices of odd degree in G be 2 and let the two vertices be u and v, then the graph $G + \{e\}$, where $e = (u, v)$ is a new edge, has an Eulerian tour from Theorem 4.4.2.1, since all the degrees in $G + \{e\}$ are even. Thus, the graph removing $\{e\}$ from $G + \{e\}$ is the original graph G which has only two vertices u and v with odd degree and has an Eulerian trail from u to v. □

Example. Figure 4.4.2.2 shows the graph for Seven bridges in Königsberg. In the graph, $d(v_1) = 5$, $d(v_2) = 3$, $d(v_3) = 3$ and $d(v_4) = 3$, thus the number of even degree vertices is 0 and the number of odd degree vertices is 4. That is, the graph of Fig. 4.4.2.2 which shows the problem of Seven bridges in Königsberg have no Eulerian tour and also no Eulerian trail from Theorem 4.4.2.1 and Corollary 4.4.2.1.

In the graph of Fig. 4.4.2.3, $d(v_1) = d(v_4) = 2$, $d(v_2) = d(v_3) = d(v_5) = d(v_6) = 4$, thus the degree of all vertices is even. That is, the graph has an Eulerian tour from Theorem 4.4.2.1, which is $v_1 \rightarrow e_2, e_3, e_5, e_6, e_4, e_7, e_8, e_9, e_{10}, e_1 \rightarrow v_1$.

In the graph of Fig. 4.4.3.1, $d(v_1) = d(v_5) = 3$, $d(v_2) = d(v_3) = d(v_4) = 2$, thus the number of even degree vertices is 3 and the number

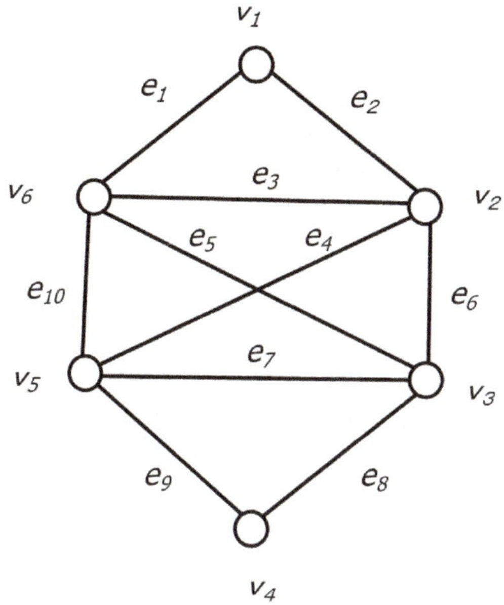

Fig. 4.4.2.3. A graph with an Eulerian tour and a Hamiltonian cycle.

of odd degree vertices is 2. That is, the graph has an Eulerian trail from Corollary 4.4.2.1, which is $v_1 \to e_3, e_{4,} e_5, e_{2,} e_1, e_6, \to v_5$.

(End of Example)

The problem to find an Eulerian tour or an Eulerian trail can be solved in polynomial time with respect to edges. That is, this is a P-problem and the complexity is $O(m)$.

4.4.3. *Hamiltonian cycle problem*

In Königsberg bridge problem mentioned above in Section 4.4.2, we discussed whether there exists a closed trail that includes every edge of a given connected graph G. A similar problem is to determine whether there exists a closed trail passing exactly once through each vertex of G. Such a cycle is a *Hamiltonian cycle* and G is a *Hamiltonian graph.*

In Fig. 4.4.2.3, $v_1, e_2, v_2, e_6, v_3, e_8, v_4, e_9, v_5, e_{10}, v_6, e_1, v_1$ is a Hamiltonian cycle of G. The graph of Fig. 4.4.3.1 has no Hamiltonian cycles.

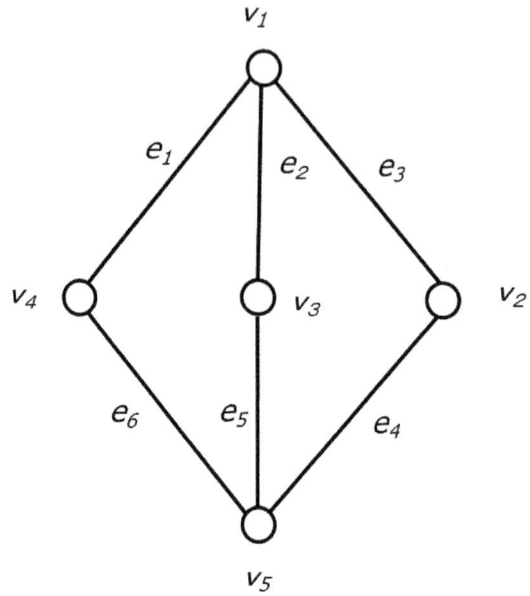

Fig. 4.4.3.1. A graph with an Eulerian trail and without a Hamiltonian cycle.

This problem is known as a problem that cannot be solved efficiently and *NP-complete.*

4.4.4. *Shortest path problem*

Let us consider a network with a digraph where each edge has weight which is the length. The distance of a directed path $P(v_i, v_j)$ from v_i to v_j is defined to be the sum of the lengths of the edges in $P(v_i, v_j)$. The problem is to find the shortest possible path and its length from a vertex v_i to a vertex v_j.

For example, Fig. 4.4.4.1 shows a network with a digraph where each edge has weight. The each weight of edges in the digraph is the length of the edge. Let us find a shortest path from v_1 to v_4. There are two directed paths from v_1 to v_4 where $P_1(v_1.v_4) = \{e_1, e_6, e_9\}$ and $P_2(v_1.v_4) = \{e_1, e_3, e_{4,}e_7, e_9\}$. The distance of $P_1(v_1.v_4) = \{e_1, e_6, e_9\}$ is 8 (= 2+5+1). The distance of $P_2(v_1.v_4) = \{e_1, e_3, e_{4,}e_7, e_9\}$ is $7(= 2 + 1 + 1 + 2 + 1)$. That is, the shortest path from v_1 to v_4 is $P_2(v_1.v_4) = \{e_1, e_3, e_{4,}e_7, e_9\}$ and the length is 7.

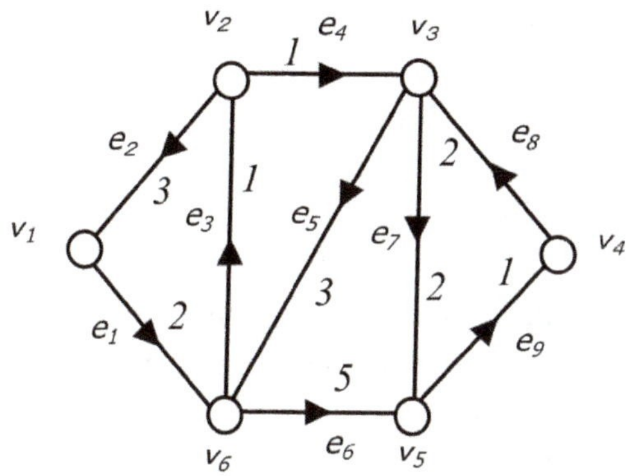

Fig. 4.4.4.1. A network with a digraph where each edge has length weight.

Among several algorithms that have been proposed for the shortest path between a specified vertex pair, perhaps the most efficient one is an algorithm due to [14] (Dijkstra) and the complexity is $O(n^2)$. Note that if G has an edge $e_{i'} = (v_k, v_j)$ with the same length w_i for $e_i = (v_j, v_k)$ with length w_i, G is considered as the undirected graph.

4.4.5. *Maximum flow problem*

Let us consider a network with a digraph where each edge has weight which is the capacity: The maximum amount of flow in the edge. It is assumed that at each intermediate vertex the total flow entering is equal to the flow leaving. In other words, there is no accumulation or generation of flow at any vertex along the way. Furthermore, the flow through a vertex is limited only by the capacities of the edges incident on it. In other words, the vertex itself can handle as much flow as allowed through the edges. The problem is to find the maximum flow possible through the network between a specified pair of vertices.

Let H_k be *a directed cutset with respect to a pair of vertices from* v_i *to* v_j in a connected directed graph G in which the direction of all edges of H_k is the same direction from v_i to v_j. The *capacity of cutset* H_k in a weighted connected graph G (in which the weight of each edge represents its flow capacity) is defined as the sum of the weights of all the edges in H_k.

Let H_k be a directed cutset with respect to a pair of vertices from v_i to v_j in a connected digraph G. In the graph $G - H_k$ which is derived from G by open-circuiting all edges of H_k, there is no directed path from v_i to v_j. Therefore, every directed path from v_i to v_j in G must contain at least one edge of H_k. Thus every flow from v_i to v_j must pass through one or more edges of S. Hence the total flow rate between these two vertices cannot exceed the capacity of H_k. Since this holds for all directed cutsets with respect to a pair of vertices from v_i to v_j, the flow cannot exceed the minimum of their capacities.

And also, it is proofed that we can obtain the maximum flow which equals the minimum of the capacities in all directed cutsets with respect to a pair of vertices from v_i to v_j using *augmenting paths* [15]. The complexity is $O(m$ maximum flow), where each edge capacity is not real number. If each edge capacity is real number, the problem may not be obtained by polynomial time. The improvement method to polynomial time for the problem is known [16].

The following theorem called *max-flow min-cut theorem* on the maximum flow is obtained from the above. Note that the theorem is applicable to a graph with non-directed edges in which each non-directed edge is replaced with two directed edges in both directions.

Theorem 4.4.5.1. *The maximum flow possible between two vertices from v_i to v_j in a network with a digraph G where each edge has capacity weight is equal to the minimum of the capacities in all directed cutsets with respect to vertices from v_i to v_j.*

Example. Figure 4.4.5.1 shows a network with a digraph where each edge has capacity weight. Let us consider the maximum flow from v_1 to v_4. The directed cutsets with respect to a pair of vertices from v_1 to v_2 are $\{e_1\}, \{e_1, e_8\}, \{e_3, e_5, e_6\}, \{e_3, e_5, e_9\}, \{e_5, e_6\}, \{e_5, e_9\}, \{e_6, e_8\}$ and $\{e_8, e_9\}$, where the capacities of the cutsets are 5, 6, 5, 9, 4, 8, 3 and 7, respectively. Then the minimum of the capacities in all directed cutsets with respect to vertices from v_1 to v_4 is 3 whose cutset is $\{e_{6,} e_8\}$ shown as a broken line in Fig. 4.4.5.1. Thus, the maximum flow from v_1 to v_4 is 3.

(End of Example)

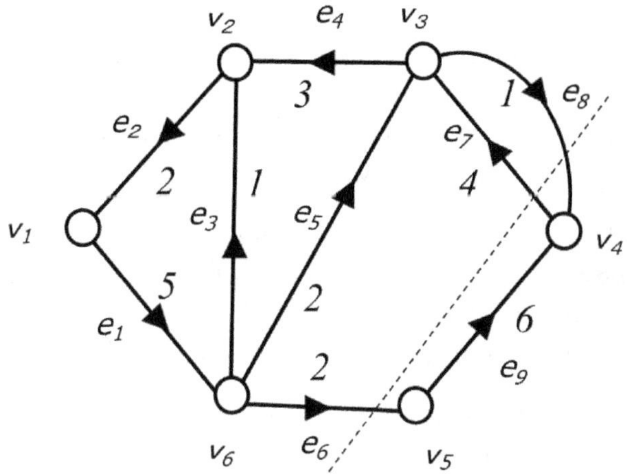

Fig. 4.4.5.2. A network with a digraph where each edge has capacity weight.

4.4.6. *Coloring problem*

Let us consider a graph $G = (V, E)$ and a painting its vertices such that no two adjacent vertices have the same color. When we paint the vertices, we can group them into different sets, that is, one set consisting of all blue vertices, another of green, and so on. This is a partitioning problem. The coloring and partitioning can, of course, be performed on edges or vertices of a graph. In the case of a planar graph, we may be interested in coloring the regions which consist of areas surrounded by vertices and edges drawn on a plane such that no two of its edges intersect. We say that a graph is *k-colorable* if it can be colored using (at most) k colors. The smallest number k for which the graph G is k-colorable, is called the *chromatic number* of G.

The problem is to find the minimum number of colors and the coloring. This problem is using in an application to mobile communication systems in this book.

Corollary 4.4.6.1. *The chromatic number of every tree with two or more vertices is* 2.

Proof. Select any vertex υ in the given tree t. Consider t as a rooted tree at vertex υ. Paint υ with color 1. Paint all vertices adjacent to υ with color 2. Next, paint the vertices adjacent to these (those that just have been colored with 2) using color 1. Continue this process till every vertex in t has been painted. Along any path in t the vertices are of alternating colors. Since there is one and only one path between any two vertices in a tree from Theorem 3.9, no two adjacent vertices have the same color. Thus, the chromatic number of every tree with two or more vertices is 2. □

Theorem 4.4.6.1. (1) *The chromatic number of K_n is n.*

(2) *The chromatic number of a graph with at least one edge is 2 if and only if it has no tiesets of odd length.*

(3) *The chromatic number of $K_{n,n}$ is* 2.

Proof. (1) Since all vertices in a complete graph are adjacent, the required number of colors is equal to the number of vertices of the graph. Therefore, the chromatic number of K_n is n.

(2) Let G be a connected graph with tiesets of only even lengths. Let us consider a tree t in G. Using the coloring procedure and the result of Corollary 4.4.6.1, the required number of colors of t is 2. Let us consider a fundamental tieset by $t + \{e_{ci}\}$, where e_{ci} is an edge of cotree of t. Since G had no tiesets of odd length, a fundamental tieset has odd length and the end vertices of e_{ci} are differently colored in t. Add the edges of cotree one by one, then G is colored with two colors. Conversely, if G has a tieset of odd length, we would need at least three colors just for that tieset.

(3) Since every bipartite graph has tiesets of only even lengths, the graph is colored with two colors from (2). Hence, the chromatic number of $K_{n,n}$ is 2. □

This coloring problem is known as a problem that cannot be solved efficiently and *NP-complete*.

Example. Figure 4.4.6.1 shows a tree colored by two colors blue (color C_1) and green (color C_2). Figure 4.4.6.2 shows a graph with a even length tieset whose length is 4 and a graph with a odd length tieset whose length is 5. A graph with a even length tieset is colored by two colors C_1 and C_2 shown in Fig. 4.4.6.2. A graph with a odd length tieset is colored by three colors C_1, C_2 and C_3 shown in Fig. 4.4.6.2.

Figure 4.4.6.3 shows Kuratowski's two graphs K_5 and $K_{3,3}$ and their coloring. Since K_5 is a complete graph and $K_{3,3}$ is a bipartite graph, K_5 is

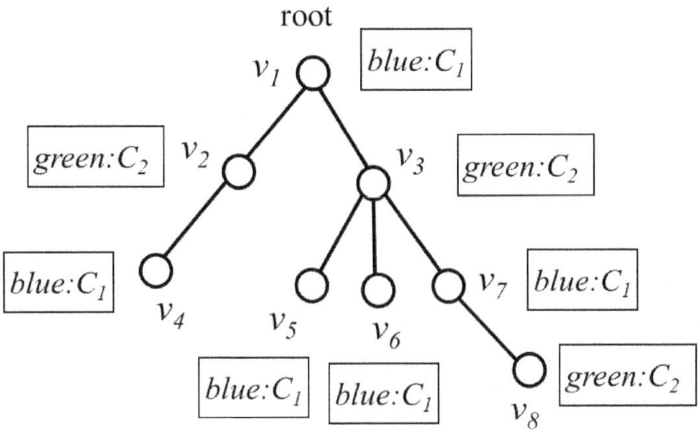

Fig. 4.4.6.1. A tree colored by two colors blue (color 1) and green (color 2).

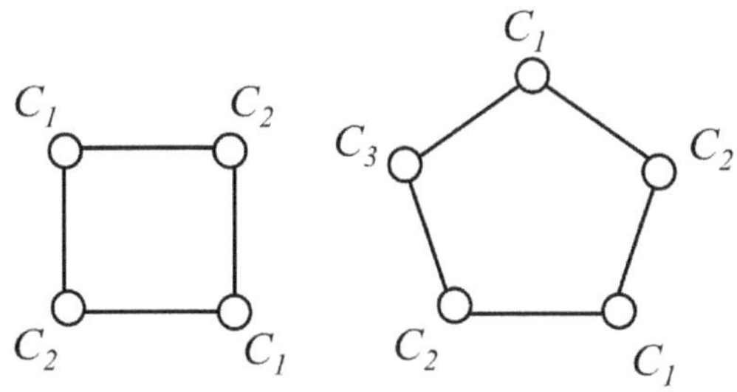

Fig. 4.4.6.2. A graph with a even length tieset and a graph with a odd length tieset.

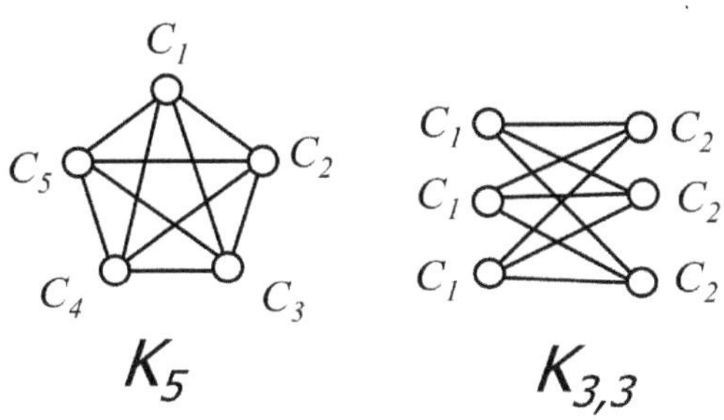

Fig. 4.4.6.3. Kuratowski's two graphs K_5 and $K_{3,3}$ and their coloring.

colored by five colors C_1, C_2, C_3, C_4 and C_5, and $K_{3,3}$ is colored by two colors C_1 and C_2 shown in Fig. 4.4.6.3. **(End of Example)**

- **Summary:** Some matrices of graphs are introduced and their application to linear networks and Tellegen's theorem in electrical networks are described. And also tree and computer algorithms: Königsberg bridge problem, Hamiltonian cycle problem, shortest path problem, Maximum flow problem and coloring problem, are described.
- **Interesting challenges:** (1) The electrical networks in this chapter include three elements: inductors, capacitors and resistors. Consider the electrical networks that also includes the fourth element: memristors (by Leon O. Chua).

  (2) Trees are useful for considering computer algorithms. Consider the use of cotrees and hybrid trees for considering computer algorithms.

Bibliography

[1] G. Kirchhoff, "Über die Auflösung der Gleichungen, auf welche man bei der Untersuchungen der Linearen Verteilung Galvanisher Ströme geführt wird," *Poggendorf Ann. Physik*, vol. 72, 497–508, 1847. English translation, *IRE Trans. Circuit Theory*, vol. CT-5, pp. 4–7, March 1958.

[2] L. O. Chua, C. A. Desoer, and E. S. Kuh, *Linear and Nonlinear Circuits*, McGraw-Hill, 1987.

[3] S. Seshu and M. B. Reed, *Linear Graphs and Electrical Networks*, Addison-Wesley, 1961.

[4] W. K. Chen, *Applied Graph Theory*, North-Holland, Amsterdam, The Netherlands, 1976.

[5] N. Deo, *Graph Theory with Applications to Engineering and Computer Science*, Prentice-Hall, 1974.

[6] O. Ore, *Theory of Graphs*, American Mathematical Society Colloquium Publications XXXVIII, 1962.

[7] K. Ruohonen, *Graph Theory* (Translation by Janne Tamminen, Kung-Chung Lee and Robert Piché) 2013.

[8] M. N. S. Swamy and K.Thulasiraman, *Graphs: Theory and Algorithms*. Wiley, 1992.

[9] M. Iri, I. Shirakawa, Y. Kajitani, S. Shinoda, H. Ariyoshi, K. Onaga, T. Ozawa, T. Ohtsuki, T. Nishi, N. Tomizawa, M. Sengoku, R. Ishii, T. Kashiwabara, T. Nishizeki, S. Fujishige and T. Asano, *Graph Theory with Exercise*, Corona Pub. 1983.

[10] Y. Kajitani, *Graph Theory for Circuits*, Shokodo Pub., 1979.

[11] P. Penfield, Jr., R. Spence and S. Duniker, *Tellegen's Theorem and Electric Networks*, MIT Press, Cambridge, MA, 1970.

[12] G. Kishi and T. Kida, Edge-port conservation in networks," *IEEE Trans. Circuit Theory*, vol. CT-15, pp. 274–276, Sept. 1968.

[13] P. Penfield, Jr., R. Spence and S. Duniker, "A generalized form of Tellegen's theorem," *IEEE Trans. Circuit Theory*, vol. CT-17, pp. 302–305, Aug. 1970.

[14] E. W. Dijkstra, A note on two problems in connexion with graphs, *Numer. Math.*, vol. 1, pp. 269–271, 1959.

[15] L. R. Ford, Jr. and D. R. Fulkerson, *Flows in Networks*, Princeton University Press, 1962.

[16] J. Edmonds and R. M. Karp, "Theoretical improvements in algorithmic efficiency for network flow problems," *J. ACM*, 19, 2, pp. 248–264, 1972.

[17] A. V. Aho, J. E. Hopcroft and J. D. Ullman, *The Design and Analysis of Computer Algorithms*, Addison-Wesley, 1974.

[18] Y. Sompolinsky and A. Zohar, "Accelerating bitcoin's transaction processing. Fast money grows on trees, not chains", *Cryptology ePrint Archive, eprint.iacr.org*, 2013.

[19] H. Tamura, K. Nakano and M. Sengoku, *Network Engineering*, IEICE, Corona Pub., 2020.

Chapter 5

Graphical Views of Electrical and Electronic Networks

From recent years to the present, information and communication technologies including the Internet (of course, including mobile communication technologies which have become particularly important) have formed the infrastructure or the basis of society in the world. In the base part of this information and communication technology, there are electrical networks including electronic device, new components, and electronic circuits and digital circuits. The various application technologies, for example mobile smart phone and games, etc., are mounted on the electrical networks. This chapter describes the graphical views of this electrical and electronic networks, and a *signal flow graph* for digital circuits.

5.1. Graphical Consideration of Electrical Networks Analysis

5.1.1. *Some analysis methods*

In Section 4.2, an electrical network and a method of the analysis are shown. The *electrical network analysis* in which the electrical components (Figs. 4.2.1 and 4.2.2) and the connected directed graph $G = (V, E)$ are given, is to find all of edge currents, edge voltages using *KCL* (*Kirchhoff's Current Law*) and *KVL*(*Kirchhoff's Voltage Law*). Let $\boldsymbol{I}$, $\boldsymbol{P}$ and $\boldsymbol{U}$ be *edge current vector*, *edge voltage vector* and *potential difference vector*, respectively as in Section 4.2.

$$\boldsymbol{I} = \begin{bmatrix} i_1 \\ \vdots \\ i_{|E|} \end{bmatrix}, \tag{5.1.1.1}$$

$$\boldsymbol{P} = \begin{bmatrix} p_1 \\ \vdots \\ p_{|V|-1} \end{bmatrix}, \tag{5.1.1.2}$$

moreover, where $p_{|V|} = 0$ called reference.

$$\boldsymbol{U} = \begin{bmatrix} u_1 \\ \vdots \\ u_{|E|} \end{bmatrix}. \tag{5.1.1.3}$$

The impedance z_k and admittance y_k of e_k are constant and the following equations link the edge current and voltage (linear):

$$u_k = z_k i_k, \tag{5.1.1.4}$$

$$i_k = y_k v_k, \tag{5.1.1.5}$$

where $z_k = 1/y_k$ and $z_k \neq 0$, $y_k \neq 0$.

As in Section 4.2, let $\boldsymbol{S_u}$ and $\boldsymbol{S_i}$ be a *cross source* (*voltage source*) *and a through source* (*current source*) *vectors*, respectively.

$$\boldsymbol{S_u} = \begin{bmatrix} s_{u1} \\ \vdots \\ s_{u|E|} \end{bmatrix}, \tag{5.1.1.6}$$

$$\boldsymbol{S_i} = \begin{bmatrix} s_{i1} \\ \vdots \\ s_{i|E|} \end{bmatrix}. \tag{5.1.1.7}$$

From *KCL*,

$$\boldsymbol{A}_r\boldsymbol{I} = \boldsymbol{0}, \tag{5.1.1.8}$$

$$\boldsymbol{Q}_f\boldsymbol{I} = \boldsymbol{0}, \tag{5.1.1.9}$$

where let t and c be a tree and the cotree of the graph $G = (V, E)$ and $\boldsymbol{Q}_f$ is fundamental cutset matrix with respective to t.

From *KVL*,

$$\boldsymbol{B}_f\boldsymbol{U} = \boldsymbol{0}, \tag{5.1.1.10}$$

where $\boldsymbol{B}_f$ is fundamental tieset matrix with respective to c.

From (5.1.1.5) and (5.1.1.7),

$$\boldsymbol{I} = \boldsymbol{Y}\boldsymbol{U} + \boldsymbol{S}_i \tag{5.1.1.11}$$

where

$$\boldsymbol{Y} = \begin{bmatrix} y_1 & 0 & \cdots & 0 \\ 0 & y_2 & \cdots & 0 \\ \vdots & \vdots & \ddots & \vdots \\ 0 & 0 & \cdots & y_{|E|} \end{bmatrix}. \tag{5.1.1.12}$$

From (5.1.1.4) and (5.1.1.6),

$$\boldsymbol{U} = \boldsymbol{Z}\boldsymbol{I} + \boldsymbol{S}_u, \tag{5.1.1.13}$$

where

$$\boldsymbol{Z} = \begin{bmatrix} z_1 & 0 & \cdots & 0 \\ 0 & z_2 & \cdots & 0 \\ \vdots & \vdots & \ddots & \vdots \\ 0 & 0 & \cdots & z_{|E|} \end{bmatrix}. \tag{5.1.1.14}$$

In the electrical network analysis, the connected directed graph and edge impedance, edge admittance, cross source (voltage source) and through source (current source) are given and the values to be obtained are all of edge currents and edge voltages.

There are some methods for network analysis and we show the following four methods.

(1) Nodal analysis: This method is introduced in Section 4.2, but here it is reprinted for ease of comparison with the other methods described below.

Since the edge voltage u_k of $e_k = (v_{i1}, v_{i2})$ is $u_k = p_{i1} - p_{i2}$,

$$\boldsymbol{U} = \boldsymbol{A}_r^T \boldsymbol{P}. \tag{5.1.1.15}$$

Multiplying $\boldsymbol{B}_f$ to $\boldsymbol{U}$ in (5.1.1.15) and using Theorem 4.1.3.1, we obtain

$$\boldsymbol{B}_f \boldsymbol{U} = \boldsymbol{B}_f \boldsymbol{A}_r^T \boldsymbol{P} = \boldsymbol{0}. \tag{5.1.1.16}$$

This satisfies *KVL* of (5.1.1.10).

Substituting (5.1.1.15) in (5.1.1.11) and substituting (5.1.1.11) in *KCL* of (5.1.1.8), we obtain

$$\boldsymbol{A}_r \boldsymbol{Y} \boldsymbol{A}_r^T \boldsymbol{P} = -\boldsymbol{A}_r \boldsymbol{S}_i. \tag{5.1.1.17}$$

In (5.1.1.17), let $\boldsymbol{Y}_n = \boldsymbol{A}_r \boldsymbol{Y} \boldsymbol{A}_r^T$ and $\boldsymbol{J}_n = -\boldsymbol{A}_r \boldsymbol{S}_i$ which are called the node admittance matrix and the node current source vector, respectively.

$$\boldsymbol{Y}_n \boldsymbol{P} = \boldsymbol{J}_n. \tag{5.1.1.18}$$

In a method of the electrical network analysis called nodal analysis, we solve Equation (5.1.1.18) for $\boldsymbol{P}$ and substituting $\boldsymbol{P}$ in (5.1.1.15) and substituting $\boldsymbol{U}$ of (5.1.1.15) in (5.1.1.11), we obtain all of edge currents and voltages.

(2) Cutset analysis: In nodal analysis, the potential difference $\boldsymbol{P}$ is the unknown of the equation to be solved first. In cutset analysis, let $\boldsymbol{U}_t$ be the edge voltage vector of t, then $\boldsymbol{U}_t$ is the unknown of the equation to be solved first.

$$\boldsymbol{U}_t = \begin{bmatrix} u_1 \\ \vdots \\ u_{\rho(G)} \end{bmatrix}, \tag{5.1.1.19}$$

$$\boldsymbol{U} = \boldsymbol{Q}_f^T \boldsymbol{U}_t. \tag{5.1.1.20}$$

Multiplying $\boldsymbol{B}_f$ to $\boldsymbol{U}$ in (5.1.1.20) and using Theorem 4.1.3.1, we obtain

$$\boldsymbol{B}_f\boldsymbol{U} = \boldsymbol{B}_f\boldsymbol{Q}_f^T\boldsymbol{U}_t = \boldsymbol{0}. \tag{5.1.1.21}$$

This satisfies *KVL* of (5.1.1.10).

Substituting (5.1.1.20) in (5.1.1.11) and substituting (5.1.1.11) in *KCL* of (5.1.1.9), we obtain

$$\boldsymbol{Q}_f\boldsymbol{Y}\boldsymbol{Q}_f^T\boldsymbol{U}_t = \boldsymbol{Q}_f\boldsymbol{S}_i. \tag{5.1.1.22}$$

In (5.1.1.22), let $\boldsymbol{Y}_c = \boldsymbol{Q}_f\boldsymbol{Y}\boldsymbol{Q}_f^T$ and $\boldsymbol{J}_c = \boldsymbol{Q}_f\boldsymbol{S}_i$ which are called the cutset admittance matrix and the cutset current source vector, respectively.

$$\boldsymbol{Y}_c\boldsymbol{U}_t = \boldsymbol{J}_c. \tag{5.1.1.23}$$

In a method of the electrical network analysis called cutset analysis, we solve Equation (5.1.1.23) for $\boldsymbol{U}_t$ and substituting $\boldsymbol{U}_t$ in (5.1.1.20) and substituting $\boldsymbol{U}$ of (5.1.1.20) in (5.1.1.11), we obtain all of edge currents and voltages.

(3) Tieset analysis: In tieset analysis, let $\boldsymbol{I}_c$ be the edge current vector of c, then $\boldsymbol{I}_c$ is the unknown of the equation to be solved first.

$$\boldsymbol{I}_c = \begin{bmatrix} i_1 \\ \vdots \\ i_{\mu(G)} \end{bmatrix}, \tag{5.1.1.24}$$

$$\boldsymbol{I} = \boldsymbol{B}_f^T\boldsymbol{I}_c. \tag{5.1.1.25}$$

Multiplying $\boldsymbol{Q}_f$ to $\boldsymbol{I}$ in (5.1.1.25) and using Theorem 4.1.3.1, we obtain

$$\boldsymbol{Q}_f\boldsymbol{I} = \boldsymbol{Q}_f\boldsymbol{B}_f^T\boldsymbol{I}_c = \boldsymbol{0}. \tag{5.1.1.26}$$

This satisfies *KCL* of (5.1.1.9).

Substituting (5.1.1.25) in (5.1.1.13) and substituting (5.1.1.13) in *KVL* of (5.1.1.10), we obtain

$$\boldsymbol{B}_f\boldsymbol{Z}\boldsymbol{B}_f^T\boldsymbol{I}_c = -\boldsymbol{B}_f\boldsymbol{S}_u. \tag{5.1.1.27}$$

In (5.1.1.27), let $\boldsymbol{Z}_t = \boldsymbol{B}_f\boldsymbol{Z}\ \boldsymbol{B}_f^T$ and $\boldsymbol{J}_t = -\boldsymbol{B}_f\boldsymbol{S}_u$ which are called the tieset impedance matrix and the tieset voltage source vector, respectively.

$$\boldsymbol{Z}_t\boldsymbol{I}_c = \boldsymbol{J}_t. \tag{5.1.1.28}$$

In a method of the electrical network analysis called tieset analysis, we solve Equation (5.1.1.28) for $\boldsymbol{I}_c$ and substituting $\boldsymbol{I}_c$ in (5.1.1.25) and substituting $\boldsymbol{I}$ of (5.1.1.25) in (5.1.1.13), we obtain all of edge currents and voltages.

(4) Hybrid analysis: In hybrid analysis, as the unknown of the equation to be solved first, some of the edge voltage vector of t and some of the edge current vector of c are chosen, where the edges consisting of elements of the vectors constitute a hybrid tree. In the graph $G = (V, E)$ where let the partition of E be (E_y, E_z), the edge of E_y is admittance and the edge of E_z is impedance, and each current source is included in E_y and each voltage source is included in E_z.

It is known that the minimum number of the unknown of the equation to be solved first is *hybrid rank* as mentioned in Section 3.5.2.

5.1.2. *Graphical considerations by trees*

Some analysis methods of electrical networks are introduced in Section 5.1.1. It is obvious that these methods are substantially capable of equivalent conversion using (4.2.4), (4.2.5) and (4.2.11). So, let us examine graphical considerations called *topological formulas* using nodal analysis.

In (5.1.1.18),

$$\boldsymbol{Y}_n\boldsymbol{P} = \boldsymbol{J}_n, \tag{5.1.2.1}$$

where $\boldsymbol{Y}_n = \boldsymbol{A}_r\boldsymbol{Y}\ \boldsymbol{A}_r^T$ and $\boldsymbol{J}_n = -\boldsymbol{A}_r\boldsymbol{S}_i$.

First, let us solve for $\boldsymbol{P}$ in (5.1.2.1) where $\boldsymbol{Y_n}$ and $\boldsymbol{J_n}$ are given by the electrical components (Figs. 4.2.1 and 4.2.2) and the connected directed graph.

$$\boldsymbol{P} = \boldsymbol{Y}_n^{-1}\boldsymbol{J}_n, \tag{5.1.2.2}$$

where $\boldsymbol{Y}_n^{-1}$ is the inverse matrix of $\boldsymbol{Y_n}$.

Then all edge voltages $\boldsymbol{U}$ are obtained from (5.1.1.15) using $\boldsymbol{P}$ and all edge currents $\boldsymbol{I}$ are obtained from (5.1.1.11) using $\boldsymbol{U}$.

Let

$$\boldsymbol{Y}_n^{-1} = [a_{ij}], \tag{5.1.2.3}$$

$$a_{ij} = \Delta_{ji}/\Delta_n, \tag{5.1.2.4}$$

$$\Delta_n = \det \boldsymbol{Y_n}, \tag{5.1.2.5}$$

$$\Delta_{ij} = (-1)^{i+j}(\det N_{ij}), \tag{5.1.2.6}$$

where N_{ij} is a matrix obtained by deleting the ith row and jth column from $\boldsymbol{Y}_n$ and Δ_{ij} is called (i, j) *cofactor* of $\boldsymbol{Y_n}$.

Before proving Theorem 5.1.2.1, let us explain a theorem of matrix theory known as the *Binet–Cauchy theorem.*

Binet–Cauchy Theorem. *If* $\boldsymbol{A}$ *is a* $k \times m$ *matrix and* $\boldsymbol{B}$ *is an* $m \times k$ *matrix* and let $\boldsymbol{C} = \boldsymbol{AB}$, then

(1) for $k = m$, $\det \boldsymbol{C} = (\det \boldsymbol{A})(\det \boldsymbol{B})$ (5.1.2.7)

(2) for $k < m$, $\det \boldsymbol{C} = \sum_P (\det \boldsymbol{A}_P)(\det \boldsymbol{B}_P)$, (5.1.2.8)

where $\boldsymbol{A}_P$ *is the* $k \times k$ *submatrix of* $\boldsymbol{A}$ *with column set* P, *and* $\boldsymbol{B}_P$ is the $k \times k$ *submatrix of* $\boldsymbol{B}$ *with the corresponding rows* P, *and the sum is over all* P *which is a set selecting* k *out of* m, *that is, there are* $\binom{m}{k}$ *such product terms.*

(3) for $k > m$, $\det \boldsymbol{C} = 0$. (5.1.2.9)

As an illustration of the theorem, let us show an example.

Let

$$A = \begin{bmatrix} 1 & 3 & -1 \\ 0 & 2 & 1 \end{bmatrix} \quad \text{and} \quad B = \begin{bmatrix} 2 & 0 \\ -1 & 3 \\ 0 & 5 \end{bmatrix}.$$

Since A is a 2×3 matrix and B is a 3×2 matrix, $\det C = \det(AB)$ is obtained using (5.1.2.8).

$$\det C = \det(AB) = \sum_P (\det A_P)(\det B_P) = \begin{vmatrix} 1 & 3 \\ 0 & 2 \end{vmatrix} \cdot \begin{vmatrix} 2 & 0 \\ -1 & 3 \end{vmatrix}$$

$$+ \begin{vmatrix} 1 & -1 \\ 0 & 1 \end{vmatrix} \cdot \begin{vmatrix} 2 & 0 \\ 0 & 5 \end{vmatrix} + \begin{vmatrix} 3 & -1 \\ 2 & 1 \end{vmatrix} \cdot \begin{vmatrix} -1 & 3 \\ 0 & 5 \end{vmatrix}$$

$$= 2 \cdot 6 + 1 \cdot 10 + 5 \cdot (-5) = -3.$$

The result can be verified by computing $AB = \begin{bmatrix} -1 & 4 \\ -2 & 11 \end{bmatrix}$ and finding its determinant directly.

Theorem 5.1.2.1. *The determinant* Δ_n *of a matrix* Y_n *is*

$$\Delta_n = \sum_{\text{all trees}} (\text{tree-admittance product of tree of } G) \quad (5.1.2.10)$$

where a tree-admittance product *is the product of the admittance of the edges of a tree in G.*

Proof. From (5.1.2.5),

$$\Delta_n = \det Y_n = \det(A_r Y A_r^T) = \det((A_r Y)(A_r^T)). \quad (5.1.2.11)$$

In (5.1.2.11), $A_r Y$ and A_r^T are $\rho(G) \times |E|$ and $|E| \times \rho(G)$ matrices, respectively. Let us apply *Binet–Cauchy theorem* to Equation (5.1.2.11). Let A_{rs} be a $\rho(G) \times \rho(G)$ submatrix of A_r with column set S and A_{rs}^T be the $\rho(G) \times \rho(G)$ submatrix of A_r^T with the corresponding rows S, then det $A_{rs} = \det A_{rs}^T = \pm 1$ if and only if submatrix of A_r with column set S corresponds a tree from the proof of Theorem 4.1.2.1. Since Y of (5.1.1.12) is the diagonal matrix consisting of the admittance of edges in G, det $((\boldsymbol{A_r Y})(\boldsymbol{A_r^T})) = \sum_{\text{all trees}}$ (tree-admittance product of tree of G). □

Theorem 5.1.2.2. *The determinant* Δ_{ij} *is*

$$\Delta_{ij} = \sum_{\text{all 2-trees}(T_{ij,r})} (\text{2-tree-admittance product of 2-tree of } T_{ij,r} \ of \ G), \tag{5.1.2.12}$$

where a 2-tree of $T_{ij,r}$ *consists of a subgraph obtained by removing an edge from a tree in G which has only two components and the reference vertex* v_r *must be in one of the components, both* v_i *and* v_j *vertices must be in the other and a 2-tree-admittance product is the product of the admittance of the edges of a 2-tree of* $T_{ij,r}$ *of in G.*

Proof. From (5.1.2.6),

$$\Delta_{ij} = (-1)^{i+j}(\det N_{ij}) = (-1)^{i+j}\det(\boldsymbol{A_{r\text{-}i} Y A_{r\text{-}j}^T}), \tag{5.1.2.13}$$

where $\boldsymbol{A_{r\text{-}i}}$ denotes the submatrix of $\boldsymbol{A_r}$ remaining after its ith row has been deleted.

In (5.1.2.13), $\boldsymbol{A_{r\text{-}i}\,Y}$ and $\boldsymbol{A_{r\text{-}j}^T}$ are $(\rho(G)-1)\times|E|$ and $|E|\times(\rho(G)-1)$ matrices, respectively. Let us apply *Binet–Cauchy theorem* to Equation (5.1.2.13). Let $\boldsymbol{A}_{r-i,s}$ be a $(\rho(G)-1)\times(\rho(G)-1)$ submatrix of $\boldsymbol{A_{r\text{-}i}}$ with column set S and $\boldsymbol{A}^T_{r-j,s}$ be the $(\rho(G)-1)\times(\rho(G)-1)$ submatrix of $\boldsymbol{A_{r\text{-}j}^T}$ with the corresponding rows S. Since from the proof of Theorem 4.1.2.1, det $\boldsymbol{A}_{r-i,s} = \pm 1$ if and only if submatrix of $\boldsymbol{A_{r\text{-}i}}$ with column set S corresponds a tree of $G_{i,r}$ obtained by coinciding v_i and v_r (reference vertex) and det $\boldsymbol{A}^T_{r-j,s} = \pm 1$ if and only if submatrix of $\boldsymbol{A_{r\text{-}j}^T}$ with rows set S corresponds a tree of $G_{j,r}$ obtained by coinciding v_j and v_r (reference vertex), a tree in both $G_{i,r}$ and $G_{j,r}$ means that det $\boldsymbol{A}_{r-i,s}$ and det $\boldsymbol{A}^T_{r-j,s}$ are $+1$ or -1, and only the tree contributes to det $(\boldsymbol{A_{r\text{-}i} Y A_{r\text{-}j}^T})$. A tree in both $G_{i,r}$ and $G_{j,r}$ or a *common tree* of $G_{i,r}$ and $G_{j,r}$ shows a *2-tree* of $T_{ij,r}$. Since $\boldsymbol{Y}$ of (5.1.1.12) is the diagonal matrix consisting of the admittance of edges in G, $\Delta_{ij} = \sum_{\text{all 2-trees}(T_{ij,r})}$ (2-tree-admittance product of 2-tree of $T_{ij,r}$ of G). □

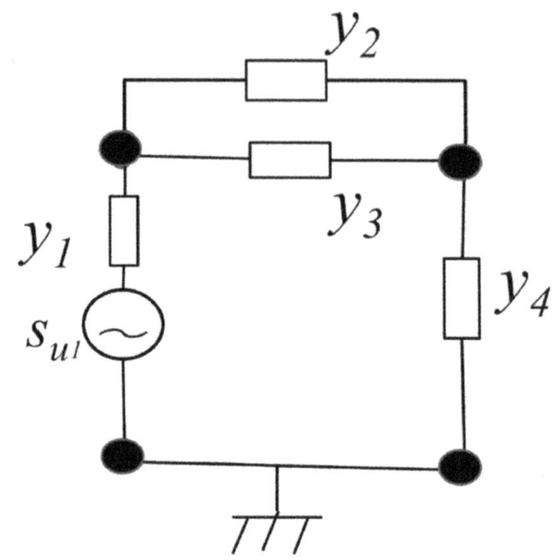

Fig. 5.1.2.1. An electrical network with a voltage source.

Theorems 5.1.2.1 and 5.1.2.2 show that the inverse matrix of $\boldsymbol{Y}_n$ is obtained by trees and 2-trees which are graphical considerations in the connected graph G. The formulas (5.1.2.10) and (5.1.2.12) in analysis methods are called *topological formulas.*

Example. Figure 5.1.2.1 shows an electrical network with a voltage source and Figure 5.1.2.2 shows the electrical network transformed to the current source from a voltage source of Fig. 5.1.2.1.

The graph of Fig. 5.1.2.3 is a directed graph G for the electrical network of Fig. 5.1.2.2.

Let $\boldsymbol{A}_r$ be an incidence matrix, then

$$\boldsymbol{A}_r = \begin{array}{c} \\ v_1 \\ v_2 \end{array} \begin{array}{c} \begin{array}{cccc} e_1 & e_2 & e_3 & e_4 \end{array} \\ \begin{bmatrix} -1 & 1 & -1 & 0 \\ 0 & -1 & 1 & 1 \end{bmatrix} \end{array}, \tag{5.1.2.14}$$

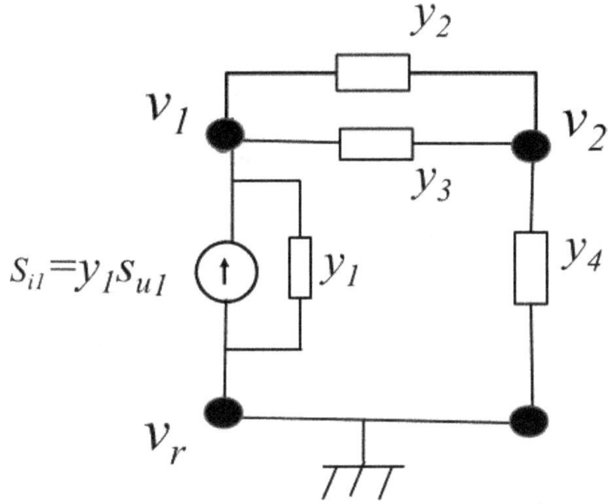

Fig. 5.1.2.2. The electrical network transformed to the current source from a voltage source of Fig. 5.1.2.1.

$$Y_n = A_r Y A_r^T = \begin{bmatrix} -1 & 1 & -1 & 0 \\ 0 & -1 & 1 & 1 \end{bmatrix} \begin{bmatrix} y_1 & 0 & 0 & 0 \\ 0 & y_2 & 0 & 0 \\ 0 & 0 & y_3 & 0 \\ 0 & 0 & 0 & y_4 \end{bmatrix} \begin{bmatrix} -1 & 0 \\ 1 & -1 \\ -1 & 1 \\ 0 & 1 \end{bmatrix}$$

$$= \begin{bmatrix} y1 + y2 + y3 & -y2 - y3 \\ -y2 - y3 & y2 + y3 + y4 \end{bmatrix}. \qquad (5.1.2.15)$$

From (5.1.1.17), $\boldsymbol{A_r Y A_r^T P} = -\boldsymbol{A_r S_i}$, that is,

$$\begin{bmatrix} y1 + y2 + y3 & -y2 - y3 \\ -y2 - y3 & y2 + y3 + y4 \end{bmatrix} \begin{bmatrix} p_1 \\ p_2 \end{bmatrix} = \begin{bmatrix} s_{i1} \\ 0 \end{bmatrix}. \qquad (5.1.2.16)$$

From (5.1.2.1), $\boldsymbol{P} = \boldsymbol{Y_n^{-1} J_n}$ that is,

$$\boldsymbol{P} = \boldsymbol{Y_n^{-1} J_n}.$$

Let

$$Y_n^{-1} = a_{ij} \tag{5.1.2.17}$$

then using the inverse matrix calculation method for matrix theory,

$$a_{11} = (y_2 + y_3 + y_4)/(y_1y_2 + y_1y_3 + y_1y_4 + y_2y_4 + y_3y_4),$$
$$a_{12} = (y_2 + y_3)/(y_1y_2 + y_1y_3 + y_1y_4 + y_2y_4 + y_3y_4),$$
$$a_{21} = (y_2 + y_3)/(y_1y_2 + y_1y_3 + y_1y_4 + y_2y_4 + y_3y_{4,})$$

and

$$a_{22} = (y_1 + y_2 + y_3)/(y_1y_2 + y_1y_3 + y_1y_4 + y_2y_4 + y_3y_4).$$

Next, let us obtain a_{ij} using Theorems 5.1.2.1 and 5.1.2.2.

Let $T(G)$ be the set of tree of G of Fig. 5.1.2.3.

$$T(G) = \{\{e_1,\ e_2\}, \{e_1, e_3\}, \{e_1, e_4\}, \{e_2, e_4\}, \{e_3, e_4\}\}. \tag{5.1.2.18}$$

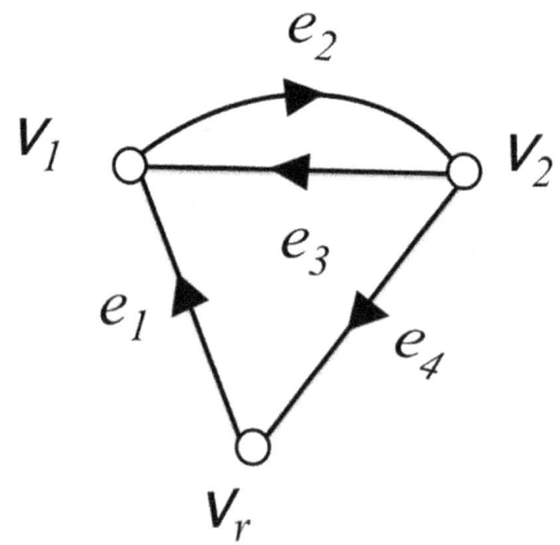

Fig. 5.1.2.3. The graph G of an electrical network of Fig. 5.1.2.2.

Let $T_{ij,r}$ be the set of 2-tree of G.

$$T_{11,r} = \{\{e_2\}, \{e_3\}, \{e_4\}\}, \tag{5.1.2.19}$$

$$T_{12,r} = \{\{e_2\}, \{e_3\}\}. \tag{5.1.2.20}$$

Using (5.1.2.10) and (5.1.2.18),

$$\Delta_n = \sum_{\text{all trees}} (\text{tree-admittance product of tree of G})$$
$$= (y_1y_2 + y_1y_3 + y_1y_4 + y_2y_4 + y_3y_4). \tag{5.1.2.21}$$

Using (5.1.2.12), (5.1.2.19) and (5.1.2.20),

$$\Delta_{11} = \sum_{\text{all 2-trees}} (T_{11}, r)\ (\text{2-tree-admittance product of 2-tree of } T_{11,r} \text{ of } G)$$
$$= (y_2 + y_3 + y_4). \tag{5.1.2.22}$$

$$\Delta_{12} = \sum_{\text{all 2-trees } (T_{12},r)} (\text{2-tree-admittance product of 2-tree of } T_{12,r} \text{ of } G)$$
$$= (y_2 + y_3). \tag{5.1.2.23}$$

Similarly, Δ_{21} and Δ_{22} are obtained.

Therefore, from $a_{ij} = -\Delta_{ji}/\Delta_n$ in (5.1.2.4),

$$a_{11} = (y_2 + y_3 + y_4)/(y_1y_2 + y_1y_3 + y_1y_4 + y_2y_4 + y_3y_4), \tag{5.1.2.24}$$

$$a_{12} = (y_2 + y_3)/(y_1y_2 + y_1y_3 + y_1y_4 + y_2y_4 + y_3y_4), \tag{5.1.2.25}$$

$$a_{21} = (y_2 + y_3)/(y_1y_2 + y_1y_3 + y_1y_4 + y_2y_4 + y_3y_4) \tag{5.1.2.26}$$

and

$$a_{22} = (y_1 + y_2 + y_3)/(y_1y_2 + y_1y_3 + y_1y_4 + y_2y_4 + y_3y_4). \tag{5.1.2.27}$$

The results (5.1.2.24)–(5.1.2.27) by graphical calculations can be verified by matrix computing (5.1.2.17). **(End of Example)**

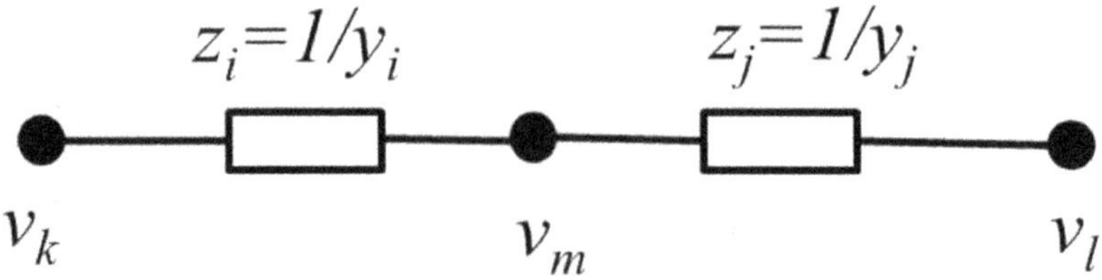

Fig. 5.1.3.1. The *series* of impedance or admittance elements.

5.1.3. *Graphical considerations by the series and parallel*

From (4.2.4) and (4.2.5), the voltage/current and the current/voltage mean impedance and admittance, respectively. The unit of a_{ij} in (5.1.2.17) means impedance. Then a_{ii} and a_{ij} are called the *input impedance* between v_i and v_r, and the *transfer impedance* from v_i to v_j, respectively.

Speaking of input impedance, it is often calculated using the *formula of the series and parallel* of impedance or admittance elements.

Figure 5.1.3.1 shows a *series* of impedance or admittance elements in which two elements $z_i = 1/y_i$ and $z_j = 1/y_j$ are connected in series. In Fig. 5.1.3.1, the currents of the two elements are same from *KCL* at v_m which is a cutset of two elements and let the current be i_{kl}. Let the voltage between v_k and v_l be u_{kl}, then u_{kl} is $u_i + u_j$ from *KVL*, where $u_i = z_i i_{kl} = i_{kl}/y_i$ and $u_j = z_j i_{kl} = i_{kl}/y_j$ from (4.2.4) and (4.2.5).

Therefore $u_{kl} = u_i + u_j = z_i i_{kl} + z_j i_{kl} = (z_i + z_j) i_{kl} = (1/y_i + 1/y_{j.})$ i_{kl} and $u_{kl}/i_{kl} = z_i + z_j = 1/y_i + 1/y_{j.}$ that is, the impedance z_{kl} between v_k and v_l is as follows:

$$z_{kl} = z_i + z_j = 1/y_i + 1/y_{j.} \tag{5.1.3.1}$$

Figure 5.1.3.2 shows a *parallel* of impedance or admittance elements in which two elements $z_i = 1/y_i$ and $z_j = 1/y_j$ are connected in *parallel*. In Fig. 5.1.3.2, the voltages of the two elements are same from *KVL* at a tieset of two elements and let the voltage be v_{kl}. Let the currents of z_i and z_j be i_i and i_j respectively, and the current between v_k and v_l be i_{kl}, then i_{kl} is $i_i + i_j$ from *KCL*, where $i_i = u_{kl}/z_i = y_i u_{kl}$ and $i_j = u_{kl}/z_j = y_j u_{kl}$ from (4.2.4) and (4.2.5).

Therefore $i_{kl} = i_i + i_j = u_{kl}/z_i + u_{kl}/z_j = (1/z_i + 1/z_j)u_{kl} = (y_i + y_j)u_{kl}$ and $u_{kl}/i_{kl} = 1/(1/z_i + 1/z_j) = 1/(y_i + y_j)$. that is, the

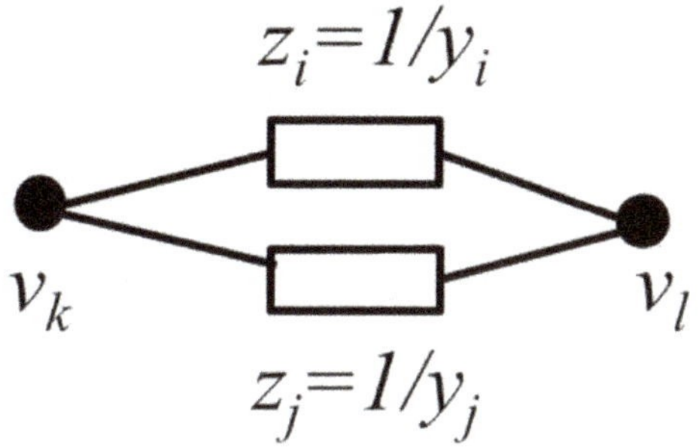

Fig. 5.1.3.2. The *parallel* of impedance or admittance elements.

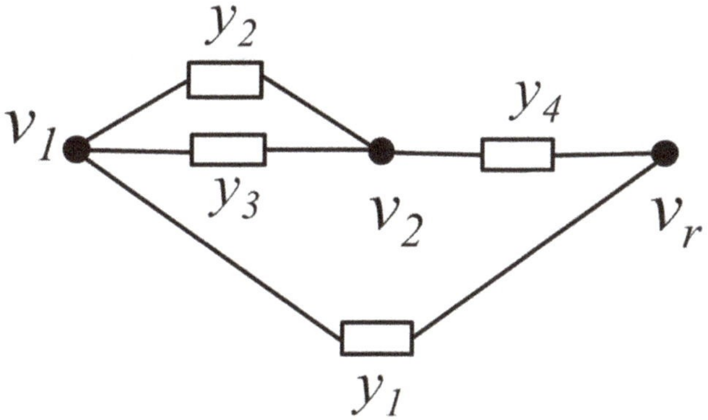

Fig. 5.1.3.3. The input impedance between v_1 and v_r in network of Fig. 5.1.2.2.

impedance z_{kl} between v_k and v_l is as follows:

$$z_{kl} = 1/(1/z_i + 1/z_j) = 1/(y_i + y_j). \qquad (5.1.3.2)$$

Example. Let us obtain input impedance between v_1 and v_r using the formula of the series and parallel of impedance or admittance elements, and verify the input impedance (5.1.2.24) which is obtained by topological formulas.

In Fig. 5.1.3.3, the network of Fig. 5.1.2.2 is redrawn so that we can see that the impedance between v_1 and v_r is composed in series and parallel elements. In Fig. 5.1.3.4, y_3' is obtained by (5.1.3.2) where y_2 and y_3 in

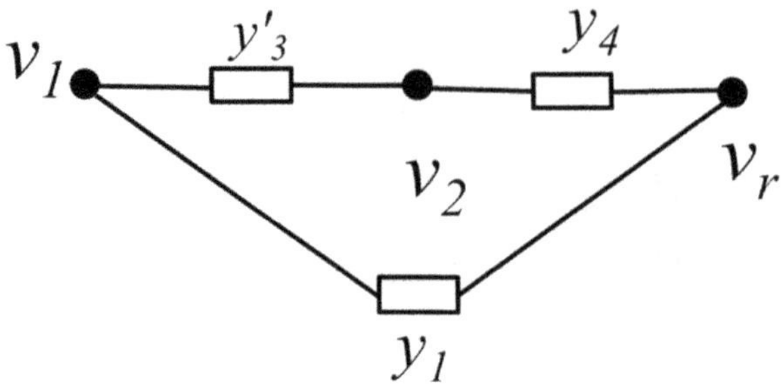

Fig. 5.1.3.4. The network by converting the parallel elements y_2 and y_3 into y'_3.

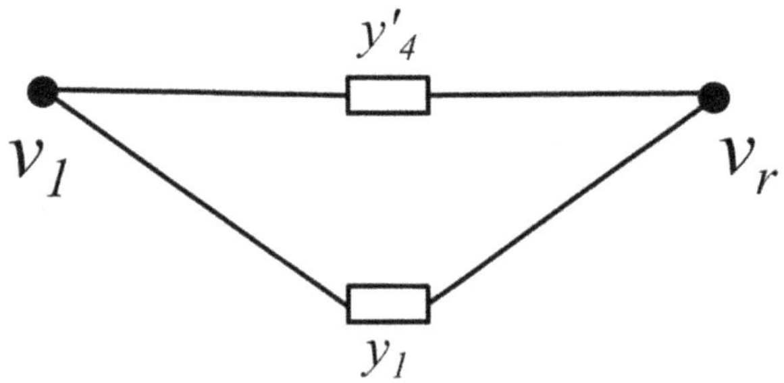

Fig. 5.1.3.5. The network by converting the series elements y'_3 and y_4 into y'_4.

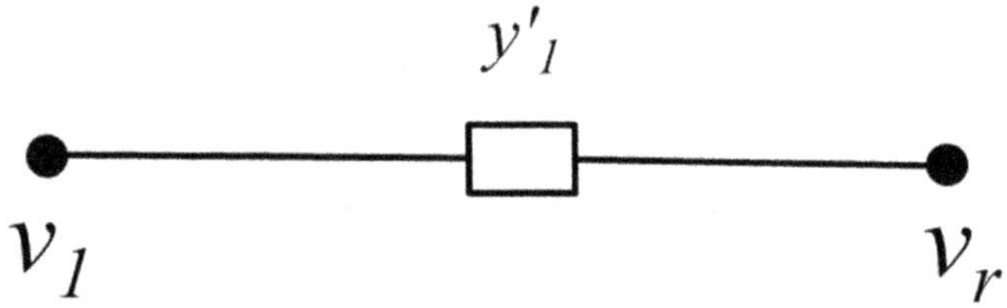

Fig. 5.1.3.6. The network by converting the parallel elements y'_4 and y_1 into y'_1.

Fig. 5.1.3.3 are parallel.

$$y'_3 = y_2 + y_3. \tag{5.1.3.3}$$

In Fig. 5.1.3.5, y'_4 is obtained by (5.1.3.1) where y'_3 and y_4 in Fig. 5.1.3.4 are series.

$$y'_4 = 1/(1/y'_3 + 1/y_4. \tag{5.1.3.4}$$

In Fig. 5.1.3.6, y'_1 is obtained by (5.1.3.2) where y'_4 and y_1 in Fig. 5.1.3.5 are parallel.

$$y'_1 = y'_4 + y_1. \tag{5.1.3.5}$$

From (5.1.3.3), (5.1.3.4) and (5.1.3.5),

$$y'_1 = (y_1y_2 + y_1y_3 + y_1y_4 + y_2y_4 + y_3y_4)/(y_2 + y_3 + y_4). \tag{5.1.3.6}$$

Therefore, the input impedance z_{1r} between v_1 and v_r is as follows:

$$z_{1r} = 1/y'_1 = (y_2 + y_3 + y_4)/(y_1y_2 + y_1y_3 + y_1y_4 + y_2y_4 + y_3y_4), \tag{5.1.3.7}$$

which coincides with (5.1.2.24) obtained by topological formulas.

(End of Example)

5.1.4. *Graphical considerations by hybrid trees*

From the definition of hybrid trees, Theorems (3.5.1.1) and (3.5.2.1), *the set of hybrid trees* $HT(G)$ of a connected graph $G = (V, E)$ where $E = E_y \bigcup E_z$ and $E_y \bigcap E_z = \emptyset$ is as follows:

$$HT(G) = \{ht | ht = t \oplus E_z, tT(G)\}, \tag{5.1.4.1}$$

where $T(G)$ is the set of trees of G.

Let us define *the set of hybrid 2-trees* $HT_{ij,r}$ of a connected graph $G = (V, E)$ where $E = E_y \bigcup E_z$ and $E_y \bigcap E_z = \emptyset$ is as follows:

$$HT_{ij,r} = \{ht_{ij,r} | ht_{ij,r} = t_{ij,r} \oplus E_z, t_{ij,r} \in T_{ij,r}\}. \tag{5.1.4.2}$$

Let us define a *hybrid tree immittance product* and a *hybrid 2-tree immittance product* in a connected graph $G = (V, E)$ where $E = E_y \bigcup E_z$ and $E_y \bigcap E_z = \emptyset$.

A *hybrid tree immittance product* and a *hybrid 2-tree immittance product* are the products of the admittance of the edges of E_y and the impedance of the edges of E_z of a hybrid tree and a hybrid 2-tree in G, respectively, where if a hybrid tree or a hybrid 2-tree is $\{\phi\}$, let the immittance product be 1.

Then, from (5.1.2.4), (5.1.2.10), (5.1.2.12), (5.1.4.1) and (5.1.4.2), the input impedance a_{ii} between v_i and v_r, and the transfer impedance a_{ij} from v_i to v_j are as follows:

$$\begin{aligned} a_{ij} &= \Delta_{h,ji}/\Delta_{h,n} \\ &= \sum_{\text{all hybrid 2-trees}(HT_{ij,r})} (\text{hybrid 2-tree immittance product of hybrid 2-tree of } HT_{ij,r} \text{ of } G) \\ &\quad \Big/ \sum_{\text{all hybrid trees}} (\text{hybrid tree immittance product of hybrid tree of } G), \end{aligned} \tag{5.1.4.3}$$

where if $i = j$ or $i \neq j$, a_{ij} is input impedance or transfer impedance, respectively.

Equation (5.1.4.3) is also called a *topological formula* as the formulas (5.1.2.10) and (5.1.2.12).

Example. Let us consider a network of Fig. 5.1.4.1 which is a network with impedance and admittance elements of Fig. 4.3.2(a). Figure 5.1.4.2 shows the graph G of Fig. 5.1.4.1, where $G = (V, E)$, $E = E_y \bigcup E_z$, $E_y \bigcap E_z = \emptyset$, $V = \{v_1, v_2, v_3, v_4\}$, $E = \{e_1, e_2, e_3, e_4, e_5, e_6\}$, $E_y = \{e_2, e_3\}$ and $E_z = \{e_1, e_4, e_5, e_6\}$.

The set of hybrid trees $HT(G)$ and the set of hybrid 2-trees $HT_{22,r}$ of G is given as follows:

$$\begin{aligned} HT(G) = \{&\{e_1, e_2, e_3, e_4, e_5\}, \{e_1, e_2, e_3, e_4, e_6\}, \{e_1, e_2, e_3, e_5, e_6\}, \\ &\{e_2, e_3, e_4, e_5, e_6\}, \{e_1, e_2, e_4\}, \{e_1, e_2, e_6\}, \{e_1, e_3, e_5\}, \\ &\{e_1, e_3, e_6\}, \{e_2, e_4, e_5\}, \{e_2, e_5, e_6\}, \{e_3, e_4, e_5\}, \\ &\{e_3, e_4, e_6\}, \{e_1\}, \{e_4\}, \{e_5\}, \{e_6\}\}, \end{aligned} \tag{5.1.4.4}$$

$$\begin{aligned} HT_{22,r} = \{&\{e_1, e_2, e_3, e_4, e_5, e_6\}, \{e_1, e_2, e_4, e_6\}, \{e_1, e_3, e_5, e_6\}, \\ &\{e_2, e_4, e_5, e_6\}, \{e_3, e_4, e_5, e_6\}, \{e_1, e_6\}, \{e_4, e_6\}, \{e_5, e_6\}\}. \end{aligned} \tag{5.1.4.5}$$

Then the sum $\Delta_{h,n}$ of hybrid tree immittance product of hybrid tree of G and the sum $\Delta_{h,22}$ of hybrid 2-tree immittance product of hybrid 2-tree

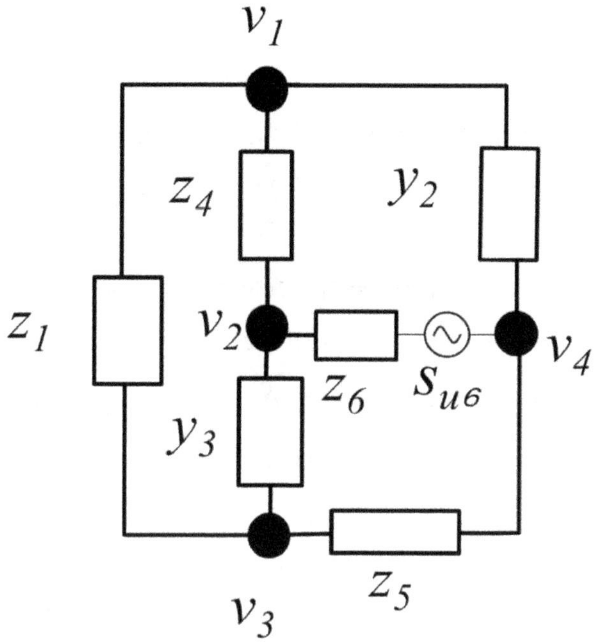

Fig. 5.1.4.1. A network with impedance and admittance elements of Fig. 4.3.2(a).

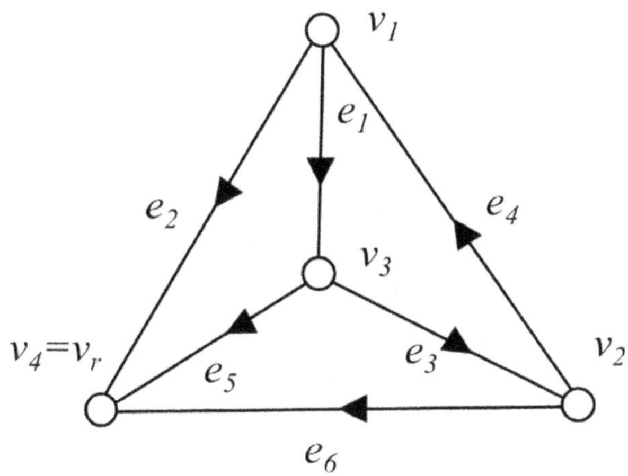

Fig. 5.1.4.2. The graph G of Fig. 5.1.4.1.

of G are as follows:

$$\begin{aligned}\Delta_{h,n} &= z_1y_2y_3z_4z_5 + z_1y_2y_3z_4z_6 + z_1y_2y_3z_5z_6 + y_2y_3z_4z_5z_6\\&\quad + z_1y_2z_4 + z_1y_2z_6 + z_1y_3z_5 + z_1y_3z_6 + y_2z_4z_5\\&\quad + y_2z_5z_6 + y_3z_4z_5 + y_3z_4z_6 + z_1 + z_4 + z_5 + z_6,\end{aligned} \tag{5.1.4.6}$$

$$\begin{aligned}\Delta_{h,22} &= z_1y_2y_3z_4z_5z_6 + z_1y_2z_4z_6 + z_1y_3z_5z_6 + y_2z_4z_5z_6\\&\quad + y_3z_4z_5z_6 + z_1z_6 + z_4z_6 + z_5z_6.\end{aligned} \tag{5.1.4.7}$$

Therefore, input impedance $a_{22} = \Delta_{h,22}/\Delta_{h,n}$ is as follows:

$$\begin{aligned}a_{22} = \Delta_{h,22}/\Delta_{h,n} &= (z_1y_2y_3z_4z_5z_6 + z_1y_2z_4z_6 + z_1y_3z_5z_6\\&\quad + y_2z_4z_5z_6 + y_3z_4z_5z_6 + z_1z_6 + z_4z_6 + z_5z_6)\\&\quad /(z_1y_2y_3z_4z_5 + z_1y_2y_3z_4z_6 + z_1y_2y_3z_5z_6\\&\quad + y_2y_3z_4z_5z_6 + z_1y_2z_4 + z_1y_2z_6 + z_1y_3z_5\\&\quad + z_1y_3z_6y_2z_4z_5 + y_2z_5z_6 + y_3z_4z_5 + y_3z_4z_6\\&\quad + z_1 + z_4 + z_5 + z_6).\end{aligned} \tag{5.1.4.8}$$

In the network of Fig. 5.1.4.1, note that input impedance between v_2 and $v_4(v_r)$ cannot obtain by the formula of the series and parallel of impedance or admittance elements, because the network cannot be decomposed in series and parallel.

Next, let us consider a network of Fig. 5.1.4.3 with impedance and admittance elements. Figure 5.1.4.4 is the graph G of Fig. 5.1.4.3, where $G = (V, E)$, $E = E_y \bigcup E_z$, $E_y \bigcap E_z = \emptyset$, $V = \{v_1, v_2, v_3, v_4, v_5\}$, $E = \{e_1, e_2, e_3, e_4, e_5, e_6\}$, e_7, $E_y = \{e_2, e_4, e_7\}$ and $E_z = \{e_1, e_3, e_5, e_6\}$. Since the graph G of Fig. 5.1.4.3 is as same as the graph of Fig. 3.5.2.2, the set of hybrid trees $HT(G)$, the set of trees $T(G)$ and the set of cotrees $C(G)$ are as follows:

$$\begin{aligned}HT(G) = \{&\{e_1, e_2\}, \{e_1, e_4\}, \{e_1, e_7\}, \{e_2, e_3\}, \{e_2, e_5\},\\&\{e_2, e_6\}, \{e_3, e_4\}, \{e_3, e_7\}, \{e_4, e_5\},\\&\{e_4, e_6\}, \{e_5, e_7\}, \{e_6, e_7\}, \{\phi\}\}\end{aligned} \tag{5.1.4.9}$$

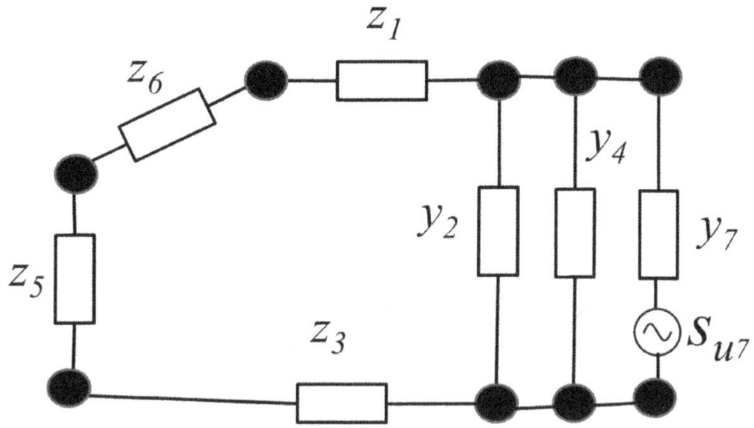

Fig. 5.1.4.3. A network with impedance and admittance elements.

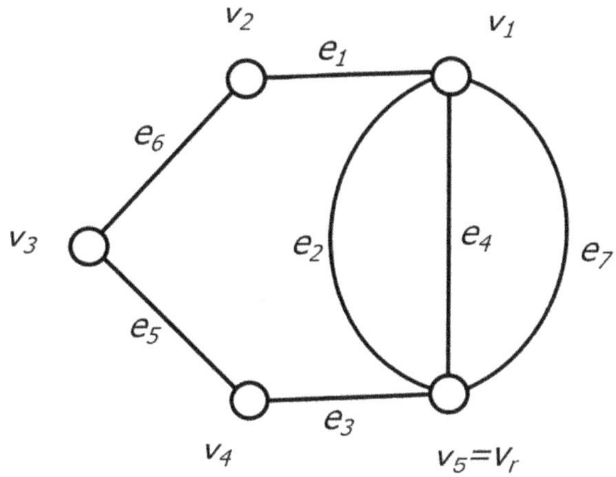

Fig. 5.1.4.4. The graph G of Fig. 5.1.4.3.

$$\begin{aligned} T(G) &= \{t | t = ht \oplus E_z,\ ht \in HT(G)\}, \\ &= \{\{e_2, e_3, e_5, e_6\}, \{e_3, e_4, e_5, e_6\}, \{e_3, e_5, e_6, e_7\}, \\ &\quad \{e_1, e_2, e_5, e_6\}, \{e_1, e_2, e_3, e_6\}, \{e_1, e_2, e_3, e_5\}, \\ &\quad \{e_1, e_4, e_5, e_6\}, \{e_1, e_5, e_6, e_7\}, \{e_1, e_3, e_4, e_6\}, \\ &\quad \{e_1, e_3, e_4, e_5\}, \{e_1, e_3, e_6, e_7\}, \{e_1, e_3, e_5, e_7\}, \\ &\quad \{e_1, e_3, e_5, e_6\}\}, \end{aligned} \tag{5.1.4.10}$$

$$
\begin{aligned}
C(G) &= \{c|c = ht \oplus E_y,\ ht \in HT(G)\} \\
&= \{\{e_1, e_4, e_7\}, \{e_1, e_2, e_7\}, \{e_1, e_2, e_4\}, \\
&\quad \{e_3, e_4, e_7\}, \{e_4, e_5, e_7\}, \{e_4, e_6, e_7\}, \\
&\quad \{e_2, e_3, e_7\}, \{e_2, e_3, e_4\}, \{e_2, e_5, e_7\}, \{e_2, e_6, e_7\}, \\
&\quad \{e_2, e_4, e_5\}, \{e_2, e_4, e_6\}, \{e_2, e_4, e_7\}\}. \qquad (5.1.4.11)
\end{aligned}
$$

Let us obtain the set of hybrid 2-tree $HT_{11,r}$, the set of 2-tree $T_{11,r}$ and the set of co2-trees $C_{11,r}$ to calculate the input impedance between v_1 and $v_5(v_r)$:

$$HT_{11,r} = \{\{e_1\}, \{e_3\}, \{e_5\}, \{e_6\}\}, \qquad (5.1.4.12)$$

$$
\begin{aligned}
T_{11,r} &= \{\{e_3, e_5, e_6\}, \{e_1, e_5, e_6\}, \{e_1, e_3, e_6\}, \\
&\quad \{e_1, e_3, e_5\}\}, \qquad (5.1.4.13)
\end{aligned}
$$

$$
\begin{aligned}
C_{11,r} &= \{\{e_1, e_2, e_4, e_7\}, \{e_2, e_3, e_4, e_7\}, \\
&\quad \{e_2, e_4, e_5, e_7\}, \{e_2, e_4, e_6, e_7\}\}. \qquad (5.1.4.14)
\end{aligned}
$$

The input impedance between v_1 and $v_5(v_r)$ is obtained by $a_{11} = \Delta_{h,11}/\Delta_{h,n}$, $a_{11} = \Delta_{11}/\Delta_n$ or $a_{11} = \Delta_{c,11}/\Delta_{c,n}$, where $\Delta_{h,n}$, Δ_n and $\Delta_{c,n}$ are the sum of hybrid tree immittance product, the sum of tree admittance product and the sum of cotree impedance product respectively, and $\Delta_{h,11}$, Δ_{11} and $\Delta_{c,11}$ are the sum of hybrid 2-tree immittance product, the sum of 2-tree admittance product and the sum of co2-tree impedance product, respectively. Note that a tree or a cotree is a hybrid tree for $E_y \neq \emptyset$ and $E_z = \emptyset$, or $E_y = \emptyset$ and $E_z \neq \emptyset$, respectively. From (5.1.4.9)–(5.1.4.14),

$$
\begin{aligned}
\Delta_{h,n} &= z_1y_2 + z_1y_4 + z_1y_7 + y_2z_3 \\
&\quad + y_2z_5 + y_2z_6 + z_3y_4 + z_3y_7 + y_4z_5 \\
&\quad + y_4z_6 + z_5y_7 + z_6y_7 + 1, \qquad (5.1.4.15)
\end{aligned}
$$

$$\Delta_{h,11} = z_1 + z_3 + z_5 + z_6, \qquad (5.1.4.16)$$

$$\begin{aligned}
a_{11} &= \Delta_{h,11}/\Delta_{h,n} = (z_1 + z_3 + z_5 + z_6)/(z_1 y_2 + z_1 y_4 \\
&\quad + z_1 y_7 + y_2 z_3 + y_2 z_5 + y_2 z_6 + z_3 y_4 + z_3 y_7 + y_4 z_5 \\
&\quad + y_4 z_6 + z_5 y_7 + z_6 y_7 + 1), \qquad (5.1.4.17)\\
\Delta_n &= y_2 y_3 y_5 y_6 + y_3 y_4 y_5 y_6 + y_3 y_5 y_6 y_7 + y_1 y_2 y_5 y_6 \\
&\quad + y_1 y_2 y_3 y_6 + y_1 y_2 y_3 y_5 + y_1 y_4 y_5 y_6 + y_1 y_5 y_6 y_7 \\
&\quad + y_1 y_3 y_4 y_6 + y_1 y_3 y_4 y_5 + y_1 y_3 y_6 y_7 + y_1 y_3 y_5 y_7 \\
&\quad + y_1 y_3 y_5 y_6, \qquad (5.1.4.18)\\
\Delta_{11} &= y_3 y_5 y_6 + y_1 y_5 y_6 + y_1 y_3 y_6 + y_1 y_3 y_5, \qquad (5.1.4.19)\\
a_{11} &= \Delta_{11}/\Delta_n = (y_3 y_5 y_6 + y_1 y_5 y_6 + y_1 y_3 y_6 \\
&\quad + y_1 y_3 y_5)/(y_2 y_3 y_5 y_6 + y_3 y_4 y_5 y_6 + y_3 y_5 y_6 y_7 \\
&\quad + y_1 y_2 y_5 y_6 + y_1 y_2 y_3 y_6 + y_1 y_2 y_3 y_5 \\
&\quad + y_1 y_4 y_5 y_6 + y_1 y_5 y_6 y_7 + y_1 y_3 y_4 y_6 + y_1 y_3 y_4 y_5 \\
&\quad + y_1 y_3 y_6 y_7 + y_1 y_3 y_5 y_7 + y_1 y_3 y_5 y_6), \qquad (5.1.4.20)\\
\Delta_{c,n} &= z_1 z_4 z_7 + z_1 z_2 z_7 + z_1 z_2 z_4 + z_3 z_4 z_7 \\
&\quad + z_4 z_5 z_7 + z_4 z_6 z_7 + z_2 z_3 z_7 + z_2 z_3 z_4 \\
&\quad + z_2 z_5 z_7 + z_2 z_6 z_7 + z_2 z_4 z_5 + z_2 z_4 z_6 \\
&\quad + z_2 z_4 z_7, \qquad (5.1.4.21)\\
\Delta_{c,11} &= z_1 z_2 z_4 z_7 + z_2 z_3 z_4 z_7 + z_2 z_4 z_5 z_7 + z_2 z_4 z_6 z_7, \qquad (5.1.4.22)\\
a_{11} &= \Delta_{c,11}/\Delta_{c,n} = (z_1 z_2 z_4 z_7 + z_2 z_3 z_4 z_7 \\
&\quad + z_2 z_4 z_5 z_7 + z_2 z_4 z_6 z_7)/(z_1 z_4 z_7 + z_1 z_2 z_7 \\
&\quad + z_1 z_2 z_4 + z_3 z_4 z_7 + z_4 z_5 z_7 + z_4 z_6 z_7 \\
&\quad + z_2 z_3 z_7 + z_2 z_3 z_4 + z_2 z_5 z_7 + z_2 z_6 z_7 + z_2 z_4 z_5 \\
&\quad + z_2 z_4 z_6 + z_2 z_4 z_7). \qquad (5.1.4.23)
\end{aligned}$$

Similarly, we can get the transfer impedance a_{ij}. **(End of Example)**

Comparing the methods of hybrid tree (5.1.4.17), tree (5.1.4.20) and cotree (5.1.4.23), it is clear that the expression by hybrid tree (5.1.4.17) is the simplest. This is one of the great advantages of using hybrid tree. Since the network of Fig. 5.1.4.3 is a simple example, it may be easily obtained by the series-parallel method without using hybrid tree, tree or cotree, but it is a great advantage for analysis in the case of a large and complex network.

5.1.5. *A network with controlled source*

An electrical network in Section 5.1.1 is composed of electrical components (Figs. 4.2.1 and 4.2.2) and the connected directed graph G. Since the current and voltage satisfy (5.1.1.4) and (5.1.1.5), the matrix $\boldsymbol{Y}$ of (5.1.1.12) is *diagonal*. If some edge current depends on other edge voltage, the component is called a *controlled source*. It is known that the components called active elements, for example transistor, IC, Operational Amplifier etc. are modeled by the controlled sources. There are four kinds of controlled sources, that is, *voltage controlled voltage source* ($VCVS$), *voltage controlled current source* ($VCCS$), *current controlled voltage source* ($CCVS$) and *current controlled current source* ($CCCS$). Figure 5.1.5.1 shows voltage controlled current source ($VCCS$) in which a current proportional to the voltage across the left edge e_j in the figure (proportional coefficient Y_m) is added as a current source in parallel with the edge e_k. Since the edges e_j and e_k represent the admittance elements y_j and y_k respectively, the voltage edge and current edge of the controlled source are described as e_{jmu} and e_{kmi} respectively, in order to avoid confusion between the controlled source and the admittance

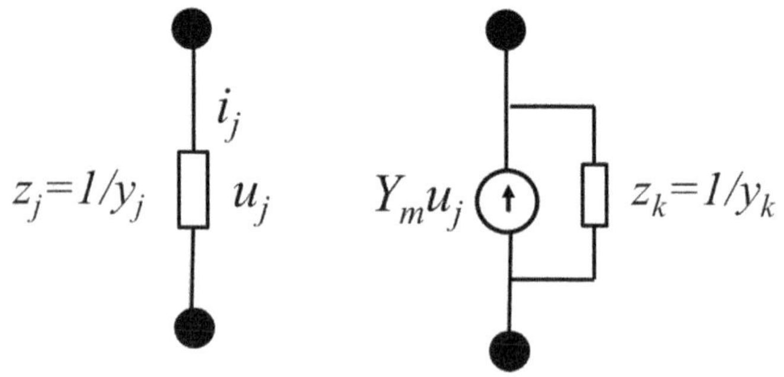

Fig. 5.1.5.1. A voltage controlled current source (VCCS).

elements. The edges e_{jmu} and e_{kmi} are called voltage edge and current edge of the voltage controlled current source, respectively. Generally, in an actual network, they can be converted to each other using (4.2.8) and (4.2.11) (see Fig. 4.2.3), so only *VCCS* will be described here.

Let us consider of a network with a voltage controlled current source of Fig. 5.1.5.1.

Then in $\boldsymbol{I} = \boldsymbol{YU} + \boldsymbol{S}_i$ of (5.1.1.11), $\boldsymbol{Y}$ is as follows:

$$\boldsymbol{Y} = \begin{bmatrix} y_1 & 0 & \cdots & \cdots & \cdots & \cdots & 0 \\ 0 & y_2 & 0 & \cdots & \cdots & \cdots & 0 \\ \vdots & 0 & \ddots & 0 & \cdots & \cdots & \vdots \\ \vdots & \vdots & 0 & y_k & 0 & -Y_m & \vdots \\ \vdots & \vdots & \vdots & 0 & \ddots & 0 & \vdots \\ \vdots & \vdots & \vdots & \vdots & 0 & y_j & 0 \\ 0 & 0 & \cdots & \cdots & \cdots & 0 & y_{|E|} \end{bmatrix}. \quad (5.1.5.1)$$

This matrix has diagonal elements $y_1, y_2, \ldots$ and a nondiagonal element Y_m. If the network includes many voltage controlled current sources, the matrix $\boldsymbol{Y}$ has many nondiagonal elements.

Let

$$\boldsymbol{I_a} = \begin{bmatrix} i_1 \\ \vdots \\ i_{|E|} \\ i_{km} \end{bmatrix}, \quad (5.1.5.2)$$

$$\boldsymbol{U_a} = \begin{bmatrix} u_1 \\ \vdots \\ u_{|E|} \\ u_{jm} \end{bmatrix} \quad (5.1.5.3)$$

and

$$\boldsymbol{Y}_a = \begin{bmatrix} y_1 & 0 & \cdots & 0 & 0 \\ 0 & y_2 & \cdots & 0 & 0 \\ \vdots & \vdots & \ddots & \vdots & \vdots \\ 0 & 0 & \cdots & y_{|E|} & 0 \\ 0 & 0 & \cdots & 0 & Y_m \end{bmatrix}, \tag{5.1.5.4}$$

where

$$i_{km} = Y_m u_{jm}. \tag{5.1.5.5}$$

Let the incidence matrix $\boldsymbol{A_{ru}}$ be a reduced incidence matrix with voltage edge of the controlled source as a column of the matrix.

$$\boldsymbol{A}_{rq} = \begin{matrix} \\ v_1 \\ v_2 \\ \vdots \\ \vdots \\ v_{|V|-1} \end{matrix} \begin{matrix} e_1 & e_2 & \cdots & \cdots & e_{|E|} & e_{jmu} \\ \left[\begin{matrix} 1 \\ 0 \\ -1 \\ \vdots \\ \vdots \end{matrix}\right. & \begin{matrix} \vdots \\ 1 \\ \vdots \\ \vdots \\ \vdots \end{matrix} & \begin{matrix} \cdots \\ \cdots \\ \cdots \\ \cdots \\ \cdots \end{matrix} & \begin{matrix} \cdots \\ \cdots \\ \cdots \\ \cdots \\ \cdots \end{matrix} & \begin{matrix} -1 \\ 0 \\ \vdots \\ \vdots \\ \vdots \end{matrix} & \left.\begin{matrix} 0 \\ 1 \\ \vdots \\ -1 \\ \vdots \end{matrix}\right] \end{matrix}. \tag{5.1.5.6}$$

Let the incidence matrix $\boldsymbol{A_{rq}}$ be a reduced incidence matrix with current edge of the controlled source as a column of the matrix.

$$\boldsymbol{A_{rq}} = \begin{matrix} \\ v_1 \\ v_2 \\ \vdots \\ \vdots \\ v_{|V|-1} \end{matrix} \begin{matrix} e_1 & e_2 & \cdots & \cdots & & e_{kmi} \\ \left[\begin{matrix} 1 \\ 0 \\ -1 \\ \vdots \\ \vdots \end{matrix}\right. & \begin{matrix} \vdots \\ 1 \\ \vdots \\ \vdots \\ \vdots \end{matrix} & \begin{matrix} \cdots \\ \cdots \\ \cdots \\ \cdots \\ \cdots \end{matrix} & \begin{matrix} \cdots \\ \cdots \\ \cdots \\ \cdots \\ \cdots \end{matrix} & \begin{matrix} -1 \\ 0 \\ \vdots \\ \vdots \\ \vdots \end{matrix} & \left.\begin{matrix} -1 \\ \vdots \\ 1 \\ \vdots \\ \vdots \end{matrix}\right] \end{matrix}. \tag{5.1.5.7}$$

Let G_u and G_q be the *graph corresponding to incidence matrix* $\boldsymbol{A_{ru}}$ *regarding voltage* and the *graph corresponding to incidence matrix* $\boldsymbol{A_{rq}}$

regarding current in which the graph G corresponding to an incidence matrix $\boldsymbol{A}_r$ means that the incidence matrix of G is $\boldsymbol{A}_r$. It is obvious that $\rho(G) = \rho(G_u) = \rho(G_q)$.

Let

$$\boldsymbol{P}_a = \begin{bmatrix} p_1 \\ \vdots \\ p_{|V|-1} \end{bmatrix}, \tag{5.1.5.8}$$

where $p_{|V|} = 0$ called reference and $\rho(G) = |V| - 1$.

Then,

$$\boldsymbol{U}_a = \boldsymbol{A}_{ru}^T \boldsymbol{P}_a. \tag{5.1.5.9}$$

$$\text{From (5.1.1.11)},\quad \boldsymbol{I}_a = \boldsymbol{Y}_a \boldsymbol{U}_a + \boldsymbol{S}_i = \boldsymbol{Y}_a \boldsymbol{A}_{ru}^T \boldsymbol{P}_a + \boldsymbol{S}_i. \tag{5.1.5.10}$$

From *KCL*, $\boldsymbol{A}_{rq}\boldsymbol{I}_a = \boldsymbol{0}$. Therefore, $\boldsymbol{A}_{rq}\boldsymbol{Y}_a\boldsymbol{A}_{ru}^T\boldsymbol{P}_a + \boldsymbol{A}_{rq}\boldsymbol{S}_i = \boldsymbol{0}$, then

$$\boldsymbol{A}_{rq}\boldsymbol{Y}_a\boldsymbol{A}_{ru}^T\boldsymbol{P}_a = -\boldsymbol{A}_{rq}\boldsymbol{S}_i.$$

Let

$$\boldsymbol{Y}_{an} = \boldsymbol{A}_{rq}\boldsymbol{Y}_a\boldsymbol{A}_{ru}^T \text{ and } \boldsymbol{J}_{an} = -\boldsymbol{A}_{rq}\boldsymbol{S}_i,$$

$$\boldsymbol{Y}_{an}\boldsymbol{P}_a = \boldsymbol{A}_{rq}\boldsymbol{Y}_a\boldsymbol{A}_{ru}^T\boldsymbol{P}_a = \boldsymbol{J}_{an}. \tag{5.1.5.11}$$

First, solve for $\boldsymbol{P}_a$ in (5.1.5.11) where $\boldsymbol{Y}_{an}$ and $\boldsymbol{J}_{an}$ are given by the electrical components (Figs. 4.2.1 and 4.2.2), *VCCS* and the connected directed graphs G_q and G_u.

$$\boldsymbol{P}_a = \boldsymbol{Y}_{an}^{-1}\boldsymbol{J}_{an}, \tag{5.1.5.12}$$

where $\boldsymbol{Y}_{an}^{-1}$ is the inverse matrix of $\boldsymbol{Y}_{an}$

Then all edge voltages $\boldsymbol{U}_a$ are obtained from (5.1.93) using $\boldsymbol{P}_a$ and all edge currents $\boldsymbol{I}_a$ are obtained from (5.1.5.10) using $\boldsymbol{U}_a$.

$$\text{Let } \boldsymbol{Y}_{an}^{-1} = [a_{aij}], \tag{5.1.5.13}$$

$$a_{aij} = \Delta_{aji}/\Delta_{an}, \tag{5.1.5.14}$$

$$\Delta_{a_n} = \det/\boldsymbol{Y}_{a_n}, \tag{5.1.5.15}$$

$$\Delta_{a_{ij}} = (-1)^{i+j}(\det \boldsymbol{N}_{a_{ij}}), \tag{5.1.5.16}$$

where $\boldsymbol{N}_{aij}$ is a matrix obtained by deleting the ith row and jth column from $\boldsymbol{Y}_{an}$.

We obtain the following theorems as Theorems 5.1.2.1 and 5.1.2.2.

Theorem 5.1.5.1. *The determinant Δ_{an} of a matrix $\boldsymbol{Y}_{an}$ is*

$$\Delta_{an} = \sum_{\text{all common trees}} (\text{tree-admittance product of common tree of } G_u \text{ and } G_q), \tag{5.1.5.17}$$

where a tree-admittance product is the product of the admittance of the edges of a common tree in G_u and G_q, and if $\{e_{p1}, e_{p2}, \ldots, e_{pp(G)}\}$ is a tree of G_u and $\{e_{o1}e_{o2}, \ldots, e_{op(G)}\}$ is a tree of G_q where $e_{pl} = e_{ol}$ ($l = 1, \ldots, \rho(G)$) or $e_{ol} = e_{omi}$ for $e_{pl} = e_{pmu}$, $\{e_{p1}, e_{p2}, \ldots, e_{pp}(G)\}$ and $\{e_{o1}, e_{o2}, \ldots, e_{op}(G)\}$ are called a common tree or a complete tree.

Proof. From (5.1.5.11),

$$\Delta_{an} = \det \boldsymbol{Y}_{an} = \det(\boldsymbol{A}_{rq}\boldsymbol{Y}_a\boldsymbol{A}_{ru}^T) = \det((\boldsymbol{A}_{qi}\boldsymbol{Y}_a)(\boldsymbol{A}_{ru}^T)). \tag{5.1.5.18}$$

In (5.1.5.18), $\boldsymbol{A}_{rq}\boldsymbol{Y}_a$ and $\boldsymbol{A}_{ru}^T$ are $\rho(G) \times (|E|+1)$ and $(|E|+1) \times \rho(G)$ matrices, respectively. Let us apply *Binet–Cauchy theorem* to Equation (5.1.5.18). Let $\boldsymbol{A}_{rqs}$ be a $\rho(G) \times \rho(G)$ submatrix of $\boldsymbol{A}_{rq}$ with column set S and $\boldsymbol{A}_{rus}^T$ be the $\rho(G) \times \rho(G)$ submatrix of $\boldsymbol{A}_{ru}^T$ with the corresponding rows S, then $\det \boldsymbol{A}_{rqs} = \det \boldsymbol{A}_{rus}^T = \pm 1$ if and only if submatrices of $\boldsymbol{A}_{rq}$ and $\boldsymbol{A}_{ru}$ with column set S correspond a common tree of G_q and G_u. Since $\boldsymbol{Y}_a$ is the diagonal matrix consisting of the admittance of edges and a voltage controlled current source in G,

$$\det((\boldsymbol{A}_{rq}\boldsymbol{Y}_a)(\boldsymbol{A}_{ru}^T)) = \sum_{\text{all common trees}} (\text{tree-admittance product of common tree of } G_u \text{ and } G_q). \qquad \square$$

Theorem 5.1.5.2. *The determinant* Δ_{aij} *is*

$$\Delta_{aij} = \sum_{\text{all common 2-trees}} (T_{aij,r}) \text{ (2-tree-admittance product of common 2-tree of } T_{aij,r} \text{ of } G_{u-j,r} \text{ and } G_{q-i,r}), \quad (5.1.5.19)$$

where a common 2-tree of $T_{aij,r}$ *consists of a subgraph obtained by removing an edge from a common tree in* G_u *and* G_q *which has only two components and the reference vertex* v_r *must be in one of the components, both* v_i *and* v_j *vertices must be in the other and a 2-tree-admittance product is the product of the admittance of the edges of a common 2-tree of* $T_{aij,r}$ *of* G_{u-jr} *and* $G_{q-i,r}$.

Proof. From (5.1.5.16),

$$\Delta_{aij} = (-1)^{i+j}(\det N_{aij}) = (-1)^{i+j} \det(\boldsymbol{A}_{rq\text{-}i}\boldsymbol{Y}_a\boldsymbol{A}^T_{ru\text{-}j}), \quad (5.1.5.20)$$

where $\boldsymbol{A}_{rq-i}$ denotes the submatrix of $\boldsymbol{A}_{rq}$ remaining after its ith row has been deleted.

In (5.1.5.20), $\boldsymbol{A}_{rq-i}\boldsymbol{Y}_a$ and $\boldsymbol{A}^T_{ru-j}$ are $(\rho(G){-}1) \times (|E| + 1)$ and $(|E| + 1) \times (\rho(G) - 1)$ matrices, respectively. Let us apply *Binet–Cauchy theorem* to Equation (5.1.5.20). Let $\boldsymbol{A}_{rq-i,s}$ be a $(\rho(G){-}1) \times (\rho(G){-}1)$ submatrix of $\boldsymbol{A}_{rq-i}$ with column set S and $\boldsymbol{A}^T_{ru-j,s}$ be the $(\rho(G){-}1) \times (\rho(G){-}1)$ submatrix of $\boldsymbol{A}^T_{ru-j}$ with the corresponding rows S. Since from the proof of Theorem 4.1.2.1, det $\boldsymbol{A}_{rq-i,s} = \pm 1$ if and only if submatrix of $\boldsymbol{A}_{rq-i}$ with column set S corresponds a tree of $G_{q-i,r}$ obtained by coinciding v_i and v_r (reference vertex) in G_q and det$\boldsymbol{A}^T_{r-j,s} = \pm 1$ if and only if submatrix of $\boldsymbol{A}^T_{ru-j}$ with rows set S corresponds a tree of $G_{u-j,r}$ obtained by coinciding v_j and v_r (reference vertex) in G_u, a tree in both $G_{q-i,r}$ and $G_{u-j,r}$ means that det$\boldsymbol{A}_{rq-i,s}$ and det$\boldsymbol{A}^T_{ru-j,s}$ are $+1$ or -1, and only the tree contributes to det($\boldsymbol{A}_{rq-i}\boldsymbol{Y}_a\boldsymbol{A}^T_{ru-j}$). A tree in both $G_{q-i,r}$ and $G_{u-j,r}$ shows a *2-tree* of $T_{aij,r}$. Since $\boldsymbol{Y}_a$ of (5.1.5.4) is the diagonal matrix consisting of the admittance of edges and a voltage controlled current source in G, $\Delta_{aij} = \sum_{\text{all common 2-trees }(T_{ij,r})}$ (2-tree-admittance product of common 2-tree of $T_{aij,r}$ of G_u and G_q). □

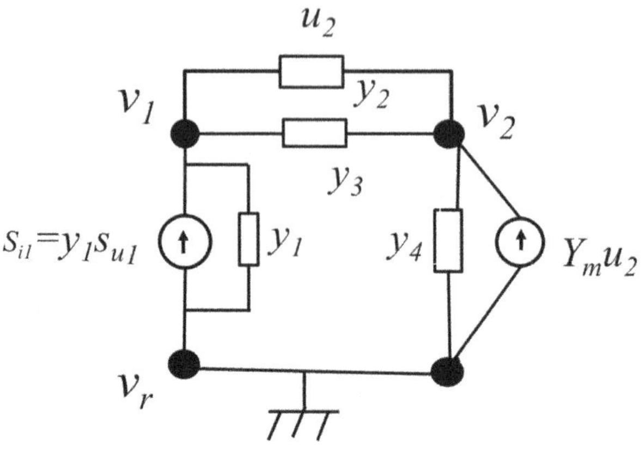

Fig. 5.1.5.1. An electrical network with a voltage controlled current source Y_m.

Theorems 5.1.5.1 and 5.1.5.2 show that the inverse matrix of $\boldsymbol{Y}_{an}$ is obtained by common trees and common 2-trees which are graphical considerations for a network with controlled sources. The formulas (5.1.5.17) and (5.1.5.19) in analysis methods are called *topological formulas* for a network with controlled sources. Similarly, formulas using cotree and hybrid trees and be obtained.

Example. Figure 5.1.5.1 shows an electrical network including a voltage controlled current source in a network of Fig. 5.1.2.2. In Fig. 5.1.5.1, the current source in parallel with the admittance y_4 is proportional (proportional coefficient Y_m) to the voltage across the admittance y_2.

The graphs of Figs. 5.1.5.2 and 5.1.5.3 are the graph G_u corresponding to incidence matrix $\boldsymbol{A}_{ru}$ regarding voltage and the graph G_q corresponding to incidence matrix $\boldsymbol{A}_{rq}$ regarding current, respectively, where

$$\boldsymbol{A}_{ru} = \begin{matrix} \\ v_1 \\ v_2 \end{matrix} \begin{matrix} \begin{matrix} e_1 & e_2 & e_3 & e_4 & e_{2mu} \end{matrix} \\ \begin{bmatrix} -1 & 1 & -1 & 0 & 1 \\ 0 & -1 & 1 & 1 & -1 \end{bmatrix} \end{matrix}, \tag{5.1.5.21}$$

$$\boldsymbol{A}_{rq} = \begin{matrix} \\ v_1 \\ v_2 \end{matrix} \begin{matrix} \begin{matrix} e_1 & e_2 & e_3 & e_4 & e_{4mi} \end{matrix} \\ \begin{bmatrix} -1 & 1 & -1 & 0 & 0 \\ 0 & -1 & 1 & 1 & -1 \end{bmatrix} \end{matrix}, \tag{5.1.5.22}$$

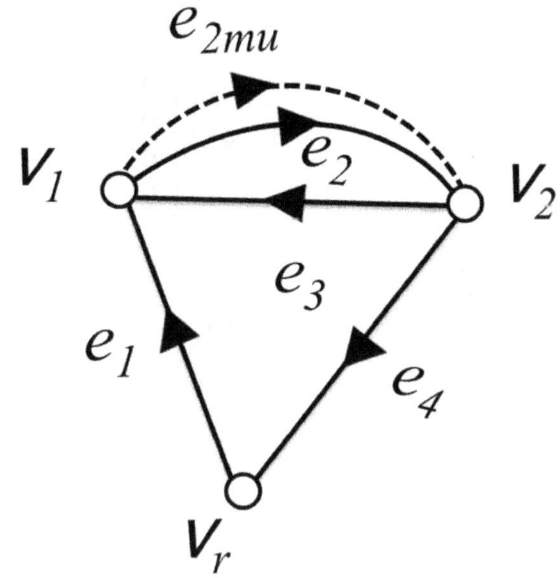

Fig. 5.1.5.2. The graph G_u of an electrical network of Fig. 5.1.5.1.

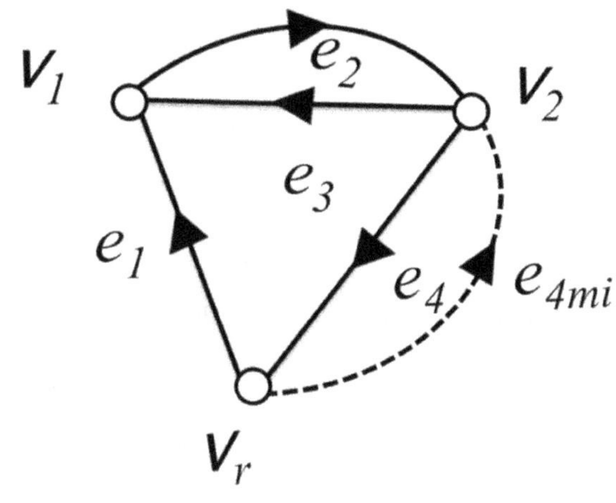

Fig. 5.1.5.3. The graph G_q of an electrical network of Fig. 5.1.5.1.

$$\boldsymbol{Y}_{an} = \boldsymbol{A}_{rq}\boldsymbol{Y}_a\boldsymbol{A}_{ru}^T = \begin{bmatrix} -1 & 1 & -1 & 0 & 0 \\ 0 & -1 & 1 & 1 & -1 \end{bmatrix} \begin{bmatrix} y_1 & 0 & 0 & 0 & 0 \\ 0 & y_2 & 0 & 0 & 0 \\ 0 & 0 & y_3 & 0 & 0 \\ 0 & 0 & 0 & y_4 & 0 \\ 0 & 0 & 0 & 0 & Y_m \end{bmatrix}$$

$$\times \begin{bmatrix} -1 & 0 \\ 1 & -1 \\ -1 & 1 \\ 0 & 1 \\ 1 & -1 \end{bmatrix} = \begin{bmatrix} y1 + y2 + y3 & -y2 - y3 \\ -y2 - y3 - Ym & y2 + y3 + y4 + Ym \end{bmatrix}. \tag{5.1.5.23}$$

From (5.1.5.11), $\boldsymbol{A}_{rq}\boldsymbol{Y}_a\boldsymbol{A}_{ru}^T\boldsymbol{P}_a = -\boldsymbol{A}_{rq}\boldsymbol{S}_i$, that is,

$$\begin{bmatrix} y1 + y2 + y3 & -y2 - y3 \\ -y2 - y3 - Ym & y2 + y3 + y4 + Ym \end{bmatrix} \begin{bmatrix} p_1 \\ p_2 \end{bmatrix} = \begin{bmatrix} s_{i1} \\ 0 \end{bmatrix}. \tag{5.1.5.24}$$

From (5.1.5.24), $\boldsymbol{P}_a = \boldsymbol{Y}_{an}^{-1}\boldsymbol{J}_{an}$ that is,

$$\boldsymbol{P}_a = \boldsymbol{Y}_{an}^{-1}\boldsymbol{J}_{an}.$$

Let

$$\boldsymbol{Y}_{an}^{-1} = [a_{aij}], \tag{5.1.5.25}$$

then using the inverse matrix calculation method for matrix theory,

$$a_{a11} = (y_2 + y_3 + y_4 + Y_m)/(y_1y_2 + y_1y_3 + y_1y_4 + y_2y_4 + y_3y_4 + y_1Y_m),$$

$$a_{a12} = (y_2 + y_3)/(y_1y_2 + y_1y_3 + y_1y_4 + y_2y_4 + y_3y_4 + y_1Y_m),$$

$$a_{a21} = (y_2 + y_3 + Y_m)/(y_1y_2 + y_1y_3 + y_1y_4 + y_2y_4 + y_3y_4 + y_1Y_m)$$

and

$$a_{a22} = (y_1 + y_2 + y_3)/(y_1y_2 + y_1y_3 + y_1y_4 + y_2y_4 + y_3y_4 + y_1Y_m).$$

Next, let us obtain a_{aij} using Theorems 5.1.5.1 and 5.1.5.2.

Let $T(G_q)$ and $T(G_u)$ be the set of trees of G_q and G_u of Figs. 5.1.5.3 and 5.1.5.2, respectively.

$$T(G_q) = \{\{e_1, e_2\}, \{e_1, e_3\}, \{e_1, e_4\}, \{e_2, e_4\}, \{e_3, e_4\}, \{e_1, e_m\}, \{e_2, e_m\}, \{e_3, e_m\}\}, \quad (5.1.5.26)$$

where $e_m = e_{4mi}$.

$$T(G_u) = \{\{e_1, e_2\}, \{e_1, e_3\}, \{e_1, e_4\}, \{e_2, e_4\}, \{e_3, e_4\}, \{e_1, e_m\}, \{e_4, e_m\}\}, \quad (5.1.5.27)$$

where $e_m = e_{2mu}$.

Since e_{4mi} and e_{2mu} are one controlled source, they are expressed as an element e_m.

Then, the common trees of $T(G_q)$ and $T(G_u)$ are as follows:

$$T(G_q) \bigcap T(G_u) = \{\{e_1, e_2\}, \{e_1, e_3\}, \{e_1, e_4\}, \{e_2, e_4\}, \{e_3, e_4\}, \{e_1, e_m\}\}. \quad (5.1.5.28)$$

Let us obtain the sets of trees $T(G_{u-j,r})$ and $T(G_{q-i,r})$:

$$T(G_{u-1,r}) = \{\{e_2\}, \{e_3\}, \{e_4\}, \{e_m\}\},$$
$$T(G_{u-2,r}) = \{\{e_1\}, \{e_2\}, \{e_3\}, \{e_m\}\},$$
$$T(G_{q-1,r}) = \{\{e_2\}, \{e_3\}, \{e_4\}, \{e_m\}\},$$
$$T(G_{q-2,r}) = \{\{e_1\}, \{e_2\}, \{e_3\}\}.$$

Let $T_{a\,ij,r}$ be the set of 2-tree:

$$T_{a11,r} = T(G_{u-1,r}) \bigcap T(G_{q-1,r}) = \{\{e_2\}, \{e_3\}, \{e_4\}, \{e_m\}\}, \quad (5.1.5.29)$$

$$T_{a12,r} = T(G_{u-2,r}) \bigcap T(G_{q-1,r}) = \{\{e_2\}, \{e_3\}, \{e_m\}\}, \quad (5.1.5.30)$$

$$T_{a21,r} = T(G_{u-1,r}) \bigcap T(G_{q-2,r}) = \{\{e_2\}, \{e_3\}\}, \quad (5.1.5.31)$$

$$T_{a22,r} = T(G_{u-2,r}) \bigcap T(G_{q-2,r}) = \{\{e_1\}, \{e_2\}, \{e_3\}\}. \quad (5.1.5.32)$$

Using (5.1.5.17) and (5.1.5.28),

$$\Delta_{an} = \sum_{\text{all common trees}} (\text{tree-admittance product of common tree of } G_u \text{ and } G_q)$$
$$= (y_1y_2 + y_1y_3 + y_1y_4 + y_2y_4 + y_3y_4 + y_1Y_m). \qquad (5.1.5.33)$$

Using (5.1.5.17), (5.1.113), (5.1.5.29), (5.1.5.31) and (5.1.5.32),

$$\Delta_{a11} = \sum_{\text{all common 2-trees } (T_{a11,r})} (\text{2-tree-admittance product of common 2-tree of } T_{a11,r} \text{ of } G_{u\text{-}1,r} \text{ and } G_{q\text{-}1,r})$$
$$= (y_2 + y_3 + y_4 + Y_m), \qquad (5.1.5.34)$$

$$\Delta_{a12} = \sum_{\text{all common 2-trees } (T_{a12,r})} (\text{2-tree-admittance product of common 2-tree of } T_{a12,r} \text{ of } G_{u\text{-}2,r} \text{ and } G_{q\text{-}1,r})$$
$$= (y_2 + y_3 + Y_m), \qquad (5.1.5.35)$$

$$\Delta_{a21} = \sum_{\text{all common 2-trees } (T_{a21,r})} (\text{2-tree-admittance product of common 2-tree of } T_{a21,r} \text{ of } G_{u-1,r} \text{ and } G_{q\text{-}2,r})$$
$$= (y_2 + y_3), \qquad (5.1.5.36)$$

$$\Delta_{a22} = \sum_{\text{all common 2-trees } (T_{a22,r})} (\text{2-tree-admittance product of common 2-tree of } T_{a22,r} \text{ of } G_{u-2,r} \text{ and } G_{q-2,r})$$
$$= (y_1 + y_2 + y_3). \qquad (5.1.5.37)$$

Therefore, from $a_{aij} = \Delta_{aji}/\Delta_{an}$ in (5.1.5.14),

$$a_{a11} = (y_2 + y_3 + y_4 + Y_m)/(y_1y_2 + y_1y_3 + y_1y_4 + y_2y_4 + y_3y_4 + y_1Y_m),$$

$$a_{a12} = (y_2 + y_3)/(y_1y_2 + y_1y_3 + y_1y_4 + y_2y_4 + y_3y_4 + y_1Y_m),$$

$$a_{a21} = (y_2 + y_3 + Y_m)/(y_1y_2 + y_1y_3 + y_1y_4 + y_2y_4 + y_3y_4 + y_1Y_m)$$

and

$$a_{a22} = (y_1 + y_2 + y_3)/(y_1y_2 + y_1y_3 + y_1y_4 + y_2y_4 + y_3y_4 + y_1Y_m). \qquad (5.1.5.38)$$

The results (5.1.5.38) by graphical calculations can be verified by matrix computing (5.1.5.25). **(End of Example)**

5.2. Algebraic Considerations of Hybrid Trees and Hybrid k-Trees

In Section 5.1, the graphical considerations of electrical and electronics networks analysis were discussed. The electrical and electronics networks form the basis of the most important hardware in mobile communication systems, etc. In order to analyze, design and improve the electrical and electronics networks, it is essential to study which part of the circuit affects its overall performance. A graphical view and considerations of the networks is indispensable for conducting research for that purpose. In Section 5.1, it was shown that the trees, cotrees, hybrid trees, 2-trees, co2-trees and hybrid 2-trees of the graph of the network play a major role in the considerations. In this section, only we discuss algebraic considerations of hybrid trees, because hybrid trees include trees and cotrees as a special case.

5.2.1. *Definitions and preliminary considerations*

Let $G = (V, E)$ be a connected graph and a subgraph of $G = (V, E)$ induced by the edge set $E_i \subseteq E$ be $G_i = (V_i, E_i)$, then for convenience sake the subgraph is represented by the "product" of the edges, for example, if $E_i = \{e_1, e_2, e_3\}$, the subgraph $s_i = e_1e_2e_3$. Let A and B be the sets of subgraphs in G.

Let us define the operators $*$, $\times$, $\frac{\partial}{\partial e_i}$ and $\int de_i$:

$$A \times B = \left\{s_i \bigcup s_j | s_i \in A, s_j \in B\right\}, \qquad (5.2.1.1)$$

$$A * B = \{s_i \oplus s_j | s_i \in A, s_j \in B\}, \qquad (5.2.1.2)$$

$$\frac{\partial s_i}{\partial e_i} = \begin{cases} s_i \oplus e_i, & \text{if } e_i \in s_i, \\ \emptyset & \text{if } e_i \notin s_i, \end{cases} \qquad (5.2.1.3)$$

$$\int s_i de_i = \begin{cases} s_i \bigcup e_i, & \text{if } e_i \notin s_i, \\ \emptyset & \text{if } e_l \in s_i, \end{cases} \qquad (5.2.1.4)$$

$$\frac{\partial A}{\partial e_i} = \left\{ \frac{\partial s_i}{\partial e_i} \mid s_i \in A \right\}, \tag{5.2.1.5}$$

$$\int A\, de_i = \left\{ \int s_i de_i | s_i \in A \right\}. \tag{5.2.1.6}$$

If $s_i = e_{i1}e_{i2}\ldots e_{im}$, we define

$$\frac{\partial A}{\partial s_i} = \frac{\partial A}{\partial e_{i1}} \oplus \frac{\partial A}{\partial e_{i2}} \oplus \cdots \oplus \frac{\partial A}{\partial e_{im}}, \tag{5.2.1.7}$$

$$\int A\, ds_i = \int A\, de_{i1} \oplus \int A\, de_{i2} \oplus \cdots \oplus \int A\, de_{im}. \tag{5.2.1.8}$$

Now $\frac{\partial 2A}{\partial s_1 \partial s_2}$ and $\int ds_1 \int A\, ds_2$ mean $\frac{\partial}{\partial s_1}(\frac{\partial A}{\partial s_2})$ and $\int(\int A\, ds_2)ds_1$, respectively. And the order of the partial differentiations or integrations is immaterial. Furthermore, it can be shown that

$$\frac{\partial(A \oplus B)}{\partial s_i} = \frac{\partial A}{\partial s_i} \oplus \frac{\partial B}{\partial s_i}, \tag{5.2.1.9}$$

$$\int (A \oplus B) ds_i = \int A\, ds_i \oplus \int B\, ds_i. \tag{5.2.1.10}$$

If no subgraph in A contains an element (edge) of a subgraph in B,

$$\frac{\partial(A \times B)}{\partial s_i} = A \times \frac{\partial B}{\partial s_i} \bigcup B \times \frac{\partial A}{\partial s_i}, \tag{5.2.1.11}$$

$$\frac{\partial(A * B)}{\partial s_i} = A * \frac{\partial B}{\partial s_i} \bigcup B * \frac{\partial A}{\partial s_i}. \tag{5.2.1.12}$$

In particular, if no subgraph in A contains an element (edge) in s_i,

$$\frac{\partial(A \times B)}{\partial s_i} = A \times \frac{\partial B}{\partial s_i}, \tag{5.2.1.13}$$

$$\frac{\partial(A * B)}{\partial s_i} = A * \frac{\partial B}{\partial s_i}, \tag{5.2.1.14}$$

$$\int (A \times B) ds_i = A \times \int B\, ds_i, \tag{5.2.1.15}$$

$$\int (A * B) ds_i = A * \int B\, ds_i. \tag{5.2.1.16}$$

5.2.2. *Hybrid trees and hybrid k-trees*

Trees, cotrees and hybrid trees are defined in Section 3 and 2-trees, co 2-trees and hybrid 2-trees are defined in Section 5.1.2. In this section, we re-explain trees, cotrees and hybrid trees and we define k-trees, co k-trees and hybrid k-trees based on 2-trees, co 2-trees and hybrid 2-trees trees.

$G = (V, E)$ is a connected graph. Let E be the set of edges of G and let the partition of E be (E_y, E_z) into two subsets E_y and E_z such that $E_y \bigcup E_z = E$ and $E_y \bigcap E_z = \emptyset$.

Let ε_z and $\bar{\varepsilon_z}$ be cutsetfree and tiesetfree in G respectively where $\varepsilon_z \bigcup \bar{\varepsilon_z} = E_z, \varepsilon_z \bigcap \bar{\varepsilon_z} = \emptyset$. Let ε_y and $\bar{\varepsilon_y}$ where $\varepsilon_y \bigcup \bar{\varepsilon_y} = E_z, \varepsilon_y \bigcap \bar{\varepsilon_y} = \emptyset$ be tiesetfree and cutsetfree respectively in $G_y = G_{os}[\varepsilon_z; \bar{\varepsilon_z}]$ which is obtained by open-circuiting of ε_z and short-circuiting of $\bar{\varepsilon_z}$ in G. Then the subgraph $\varepsilon_y \bigcup \varepsilon_z$ is called a hybrid tree in G with respect to the partition (E_y, E_z) of E.

In the definition, if $E_z = \emptyset$, a hybrid tree is a tree and if $E_y = \emptyset$, a hybrid tree is a cotree. Let $HT(G)$, $T(G)$, and $C(G)$ be the sets of hybrid trees, trees, and cotrees in G, respectively:

$$T(G) = \{t | t = ht \oplus E_z,\ htHT(G)\}, \tag{5.2.2.1}$$

$$C(G) = \{c | c = ht \oplus E_y,\ htHT(G)\}. \tag{5.2.2.2}$$

Equations (5.2.2.1) and (5.2.2.2) show the relations among $HT(G), T(G)$, and $C(G)$, and when $E_y \neq \emptyset$ and $E_z = \emptyset$, $T(G) = HT(G)$, and when $E_y = \emptyset$ and $E_z \neq \emptyset$, $C(G) = HT(G)$.

The number of edges in a tree and a cotree of G is identical with the rank $\rho(G)$ and the nullity $\mu(G)$ of G, respectively. The number of edges in a hybrid tree of G, however, is not always equal to $\rho(G)$ or $\mu(G)$ and every hybrid tree of G does not always contain the same number of edges. Let hT_k, be a subset of $HT(G)$ such that

$$hT_k = \{ht || ht| = k,\ ht \in HT(G)\}. \tag{5.2.2.3}$$

Then,

$$HT(G) = \{hT_m, hT_{m-2}, \ldots, hT_{l+2}, hT_l\}, \tag{5.2.2.4}$$

where $hT_k \neq \emptyset$, $m \geq k \geq l \geq 0$ and m, k, and l are non-negative integers.

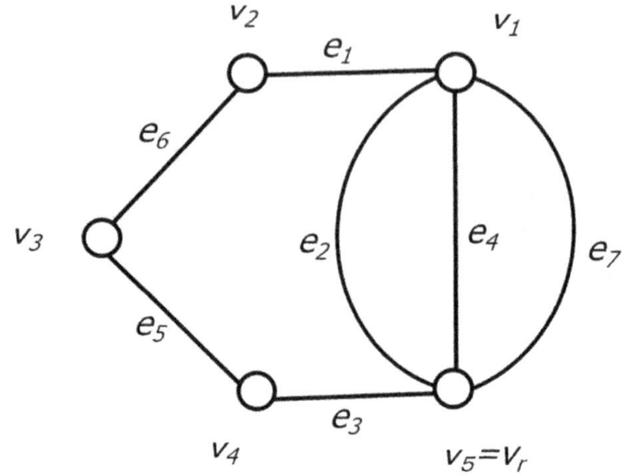

Fig. 5.2.2.1. The same graph G as Fig. 5.1.4.4.

From (5.2.2.1) and (5.2.2.2),

$$HT(G) = \{ht | ht = t \oplus E_z, t \in T(G)\} = E_z * T(G) = E_y * CT(G). \tag{5.2.2.5}$$

Let $V_1, V_2, \ldots, V_k$ be nonempty disjoint subsets of the vertex set V of G. Let us define k-trees, co k-trees and hybrid k-trees. Let $T_{V1,V2,\ldots,Vk}$ be the set of k-trees. A *k-tree* of $T_{V1,V2,\ldots,Vk}$ is a subgraph of G which is maximal tiesetfree, and also there is a path between two vertices in V_i and there is no path between a vertex v_p of V_i and a vertex v_q in V_j where $i \neq j$ and $i, j = 1, 2, \ldots, k$. Let us define co *k-trees* and *hybrid k-trees* as in (5.2.2.5).

$$HT_{V1,V2,\ldots,Vk} = E_z * T_{V1,V2,\ldots,Vk} = E_y * CT_{V1,V2,\ldots,Vk}. \tag{5.2.2.6}$$

Example. In Fig. 5.2.2.1, let us show some k-trees, co k-trees and hybrid k-trees. The graph G of Fig. 5.2.2.1 is the same of as Fig. 5.1.12, where $G = (V, E)$, $E = E_y \bigcup E_z$, $E_y \bigcap E_z = \emptyset$, $V = \{v_1, v_2, v_3, v_4, v_5\}$, $E = \{e_1, e_2, e_3, e_4, e_5, e_6, e_7\}$, $E_y = \{e_2, e_4, e_7\}$ and $E_z = \{e_1, e_3, e_5, e_6\}$.

In Fig. 5.2.2.1,

$$T_{v1,v5} = \{e_3e_5e_6, e_1e_5e_6, e_1e_3e_6, e_1e_3e_5\}, \tag{5.2.2.7}$$

$$C_{v1,v5} = \{e_1e_2e_4e_7, e_2e_3e_4e_7,$$
$$e_2e_4e_5e_7, e_2e_4e_6e_7\}, \quad (5.2.2.8)$$
$$HT_{v1,v5} = \{e_1, e_3, e_5, e_6\}. \quad (5.2.2.9)$$

Note that $HT_{v1,v5}$, $T_{v1,v5}$ and $C_{v1,v5}$ are the same as $HT_{11,r}$, $T_{11,r}$ and $C_{11,r}$ of (5.1.4.12), (5.1.4.13) and (5.1.4.14), respectively.

Let us show an cxample of 3-trees, co 3-trees and hybrid 3-trees in G.

$$T_{v1v2,v3,v4} = \{e_1e_2, e_1e_4, e_1e_7, e_1e_3\}, \quad (5.2.2.10)$$
$$C_{v1v2,v3,v4} = \{e_3e_4e_5e_6e_7, e_2e_3e_5e_6e_7, e_2e_3e_4e_5e_6,$$
$$e_2e_4e_5e_6e_7\}, \quad (5.2.2.11)$$
$$HT_{v1v2,v3,v4} = \{e_2e_3e_5e_6, e_3e_4e_5e_6, e_3e_5e_6e_7, e_5e_6\}. \quad (5.2.2.12)$$

(End of Example)

5.2.3. *Some properties of hybrid trees and hybrid k-trees*

In this section, we examine the relationships between hybrid trees (hybrid k-trees) and paths (or tiesets) or cutsets of a graph G where $G = (V, E)$ is a connected graph and $E_y \bigcup E_z = E$ and $E_y \bigcap E_z = \emptyset$. We abbreviate $HT(G)$, $T(G)$, and $C(G)$ to HT, T, and C, respectively.

Lemma 5.2.3.1. *Let P_{zij} be a path between vertices v_i and v_j in G and P_{zij} consists of only edges of E_z. Then,*

$$HT_{v_i,v_j} = \int HT dP_{zij}. \quad (5.2.3.1)$$

Proof. Let $P_{zij} = e_1e_2 \ldots e_l$, where e_i $(1 \le i \le l) \in E_z$.

First, let us prove the following (5.2.3.2) by induction on the number of edges of $P_{zi,j}$.

$$T_{v_i,v_j} = \frac{\partial T}{\partial P_{zij}} \quad (5.2.3.2)$$

If $l = 1$, (5.2.3.2) holds from Corollary 3.4.1. Suppose that (5.2.3.2) holds for $l = m-1$. Let us consider a path P_{zij} where $l = m$. Let $P_{zij} = e_1e_2...e_m$

and $e_m = (v_q, v_j)$.

$$\frac{\partial T}{\partial P_{zij}} = \left(\frac{\partial T}{\partial e_1} \oplus \frac{\partial T}{\partial e_2} \oplus \cdots \oplus \frac{\partial T}{\partial e_{m-1}}\right) \oplus \frac{\partial T}{\partial e_m}.$$

From the induction hypothesis, the first $m-1$ terms in the above equation is equal to T_{v_i,v_q} and $\frac{\partial T}{\partial e_m} = T_{v_q,v_j}$. Therefore,

$$\frac{\partial T}{\partial P_{zij}} = T_{v_i,v_q} \oplus T_{vq,vj}.$$

On the other hand, $T_{v_i,v_q} = T_{v_iv_j,v_q} \bigcup T_{v_i,v_qv_j}$ and $T_{v_qv_j} = T_{v_qv_i,v_j} \bigcup T_{v_q,v_jv_i}$ from the definition of 2-tree. Therefore,

$$\frac{\partial T}{\partial P_{zij}} = T_{v_i,v_q} \oplus T_{v_q,v_j} = T_{v_i,v_qv_j} \bigcup T_{v_qv_i,vj} = T_{v_i,v_j},$$

which is (5.2.3.2).

From (5.2.2.6),

$$\begin{aligned} HT_{v_i,v_j} &= E_z * T_{v_i,v_j} = E_z * \frac{\partial T}{\partial P_{zij}} \\ &= E_z * \left(\frac{\partial T}{\partial e_1} \oplus \frac{\partial T}{\partial e_2} \oplus \cdots \oplus \frac{\partial T}{\partial e_l}\right) \\ &= E_z * \frac{\partial T}{\partial e_1} \oplus E_z * \frac{\partial T}{\partial e_2} \oplus \cdots \oplus E_z * \frac{\partial T}{\partial e_l}. \end{aligned}$$

Since e_i $(1 \leq i \leq l) E_z$,

$$E_z * \frac{\partial T}{\partial e_i} = \int E_z * T\, de_i = \int HT\, de_i.$$

Therefore.

$$\begin{aligned} HT_{v_i,v_j} &= \int HT\, de_1 \oplus \int HT\, de_2 \oplus \cdots \oplus \int HT\, de_l \\ &= \int HT\, de_1e_2 \cdots e_l = \int HT\, dP_{zij}. \end{aligned}$$

□

Theorem 5.2.3.1. *Let P_{ij} be a path between vertices v_i and v_j in G. Then,*

$$HT_{v_i,v_j} = \frac{\partial HT}{\partial P_{yij}} \oplus \int HT\, dP_{zij} = \left(\frac{\partial}{\partial P_{yij}} \oplus \int dP_{zij}\right) HT, \tag{5.2.3.3}$$

where

$$P_{ij} = P_{yij} \bigcup P_{zij},\, P_{yij} \bigcap P_{zij} = \emptyset,\, P_{yij} \subset E_y \text{ and } P_{zij} \subset E_z.$$

Proof. From (5.2.2.6), (5.2.3.2) and (5.2.1.7),

$$HT_{v_i,v_j} = E_z * T_{v_i,v_j} = E_z * \frac{\partial T}{\partial P_{ij}} = E_z * \frac{\partial T}{\partial P_{yij}} \oplus E_z * \frac{\partial T}{\partial P_{zij}}.$$

From (5.2.1.14),

$$HT_{v_i,v_j} = \frac{\partial E_z * T}{\partial P_{yij}} \oplus E_z * \frac{\partial T}{\partial P_{zij}}.$$

Therefore using (5.2.2.5) and Lemma 5.2.3.1,

$$HT_{v_i,v_j} = \frac{\partial HT}{\partial P_{yij}} \oplus \int HT dP_{zij} = \left(\frac{\partial}{\partial P_{yij}} \oplus \int dP_{zij}\right) HT. \qquad \square$$

Let us return to the basic definition of trees, cotrees and hybrid trees, and consider about 2-trees, co 2-trees and hybrid 2-trees. There are some definitions on trees and cotrees shown in Theorems 3.1.3 and 3.2.1. Let us show a definition on a tree and a cotree below.

(T1) A tree is tiesetfree and has $|V| - 1(= \rho(G))$ edges.
(T2) A tree shares edges with every cutset and minimal.
(T3) A tree is maximal tiesetfree.
(C1) A cotree is cutsetfree and has $|E| - |V| + 1(= \mu(G))$ edges.
(C2) A cotree shares edges with every tietset and minimal.
(C3) A cotree is maximal cutsetfree.

Lemma 5.2.3.2 (1). *In a connected graph $G = (V, E)$ where $E = E_y \bigcup E_z$ and $E_y = \emptyset$, let $F = e_1e_2 \ldots e_l$ be a tieset in G. Then,*

$$\int C dF = \emptyset. \tag{5.2.3.4}$$

Proof. Let $c_i \in C, e_j \in F$ and $\int c_i de_j = c_i' \neq \emptyset$. Since c_i is maximal cutsetfree, c_i' has a cutset H_j including e_j and $H_j \bigcap F = x$ where $|x| =$ even number from Theorem 2.7.1. If two edges e_j, e_k are included in x for $e_j, e_k \in F$, then $\int c_i de_k = c_i'$. That is, $\int CdF$ consists of the term of zero and the even number of terms of the same set connected by operation $\oplus$. Therefore, $\int CdF = \emptyset$. □

Lemma 5.2.3.2 (2). *In a connected graph $G = (V, E)$ where $E = E_y \bigcup E_z$ and $E_z = \emptyset$, let $H = e_1e_2 \cdots e_l$ be a cutset in G. Then,*

$$\int TdH = \emptyset. \tag{5.2.3.5}$$

Proof. Let $t_i \in T, e_j \in H$ and $\int t_i de_j = t_i' \neq \emptyset$. Since t_i is maximal tiesetfree, t_i' has a tietset F_j including e_j and $F_j \bigcap H = \text{x}$ where $|x| =$ even number from Theorem 2.7.1. If two edges e_j, e_k are included in x for $e_j, e_k \in H$, then $\int t_i de_k = t_i'$. That is, $\int TdH$ consists of the term of zero and the even number of terms of the same set connected by operation $\oplus$. Therefore, $\int TdH = \emptyset$. □

Lemma 5.2.3.3. *In a connected graph $G = (V, E)$ where $E = E_y \bigcup E_z$ and $E_y = \emptyset$, let $P_{ij} = e_1e_2 \cdots e_l$ be a path between vertices v_i and v_j in G. Then, P_{ij} is a tieset $F_{ij} = e_1e_2 \ldots e_l$ in G_{v_i,v_j} where G_{v_i,v_j} is a graph coinciding v_i and v_j in G.*

$$C_{v_i,v_j} = \int CdP_{ij} = \int CdF_{ij}. \tag{5.2.3.6}$$

Proof. Since G_{v_i,v_j} is a graph coinciding v_i and v_j in G, the tiesets of G_{v_i,v_j} maintain every tieset of G or they are new tiesets by coinciding v_i and v_j. This means that the tiesets of G_{v_i,v_j} consist of all of tiesets of G and new tiesets by coinciding v_i and v_j. Generally, a cotree shares edges with every tietset and minimal, and a cotree is cutsetfree and has $|E|-|V|+1(= \mu(G))$ edges from (C1) and (C2). The number of vertices of G_{v_i,v_j} is one less than number of vertices of G, that is, $\mu(G_{v_i,v_j}) = \mu(G) + 1$. Since a cotree in G_{v_i,v_j} must shares edges with every tietset and minimal, each cotree of $C_{v_i,vj}$ has at least one edge of the new tiesets. The new tiesets may be one or more. If two tiesets F_1 and F_2 are included in the new tiesets, $F_1 \oplus F_2$ includes a tieset F_3 of G and the tieset F_3 shares edges with every cotree of G. Since $\int CdF_3$ is empty from Lemma 5.2.2-1, in order to obtain cotrees of

G_{v_i,v_j}, we can find the cotrees by adding a part of a new tieset to a cotree of G which does not include a part of the new tieset. Thus, since the new tiesets include $F_{ij} = e_1e_2 \cdots e_l$ as a new tieset in G_{v_i,v_j}, an element of $\int CdF_{ij}$ has an edge $e_i \in F_{ij}$ as an shared element with the new tieset and is a cotree of G_{v_i,v_j}. If $e_i, e_k \in F_{ij}$ and $e_k \in c_p, e_i \notin c_p$ for $c_p \in C$, then $c_p \cup e_i$ is not a cotree in G_{v_i,v_j} since a cotree shares edges with every tietset and minimal. In this case, the same two terms with e_i and e_k appear twice in $\int CdF_{ij}$ which are erased by the operation $\oplus$.

Note that this Lemma 5.2.3.3 can be also used to prove Lemma 5.2.3.1 and an another approach. Then, Theorem 5.2.3.1 is expressed in another expression as follows. □

Theorem 5.2.3.1′. *Let $F_{ij} = e_1e_2 \cdots e_l$ be is a tieset in G_{v_i,v_j} where G_{v_i,v_j} is a graph coinciding v_i and v_j in G and $P_{ij} = F_{ij} = e_1e_2 \cdots e_l$ be a path between vertices v_i and v_j in G. Then,*

$$HT_{v_i,v_j} = \frac{\partial HT}{\partial F_{yij}} \oplus \int HTdF_{zij} = \left(\frac{\partial}{\partial F_{yij}} \oplus \int dF_{zij}\right) HT, \tag{5.2.3.7}$$

where

$$F_{ij} = F_{yij} \bigcup F_{zij}, F_{yij} \bigcap F_{zij} = \emptyset, \quad F_{yij} \subset E_y \quad \text{and} \quad F_{zij} \subset E_z.$$

Lemma 5.2.3.4. *In a connected graph $G = (V, E)$ where $E = E_y \bigcup E_z$ and $E_z = \emptyset$, let $H_{ij} = e_1e_2 \cdots e_l$ be a cutset of $E\langle V_1, V_2\rangle$ where $v_i \in V_1$ and $v_j \in V_2$ in G. Let T and $T_{v_i,vj}$ be the set of trees of G and the set of trees of G_{v_i,v_j} which is the graph coinciding $v_i$and v_j in G, respectively. Then,*

$$T = \int T_{v_i,\, vj} dH_{ij}. \tag{5.2.3.8}$$

Proof. Since G is a graph separating one point in G_{v_i,v_j} into two points v_i and v_j, the cutsets of G maintain every cutset of G_{v_i,v_j} or they are new cutsets by separating into v_i and v_j. This means that the cutsets of G consist of all of cutsets of G_{v_i,v_j} and new cutsets by separating into v_i and v_j. Generally, a tree shares edges with every cutset and minimal, and a tree is tiesetfree and has $|V| - 1(= \rho(G))$ edges from (T1) and (T2). The number of vertices of G is one more than number of vertices of G_{v_i,v_j}, that is,

$\rho(G) = \rho(G_{v_i,v_j}) + 1$. Since a tree in G must shares edges with every cutset and minimal, each tree of T has at least one edge of the new cutsets. The new cutsets may be one or more. If two cutsets H_1 and H_2 are included in the new cutsets, $H_1 \oplus H_2$ includes a cutset H_3 of G_{v_i,v_j} and the cutset H_3 shares edges with every tree of G. Since $\int T_{v_i,vj}\, dH_3$ is empty from Lemma 5.2.2.2, in order to obtain trees of G, we can find the trees by adding a part of a new cutset to a tree of G_{v_i,v_j} which does not include a part of the new cutset. Thus, since the new cutsets include $H_{ij} = e_1e_2\cdots e_l$ as a new cutset in G, an element of $\int T_{v_i,vj} dH_{ij}$ has an edge $e_i \in H_{ij}$ as an shared element with the new cutset and is a tree of G. If $e_i, e_k \in H_{ij}$ and $e_k \in t_p, e_i \notin t_p$ for t$_p \in T_{v_i,vj}$, then $t_p \cup e_i$ is not a tree in G since a tree shares edges with every cutset and minimal. In this case, the same two terms with e_i and e_k appear twice in $\int T_{v_i,vj\, dHij}$ which are erased by the operation $\oplus$. □

Theorem 5.2.3.2. *In a connected graph* $G = (V, E)$ *where* $E = E_y \bigcup E_z$ *and* $E_y \cap E_z = \emptyset$, *let* $H_{ij} = e_1e_2\cdots e_l$ *be a cutset of* $E\langle V_1, V_2\rangle$ *where* $v_i \in V_1$ *and* $v_j \in V_2$*in G. Let HT and* HT_{v_i,v_j} *be the set of hybrid trees of G and the set of hybrid trees of* $G_{v_i,vj}$ *which is the graph coinciding* v_i *and* v_j *in G, respectively. Then,*

Let P_{ij} be a path between vertices v_i and v_j in G. Then,

$$HT = \frac{\partial HT\ v_i, vj}{\partial H_{zij}} \oplus \int HTv_i,\ vjdH_{yij}$$
$$= \left(\frac{\partial}{\partial H_{zij}} \oplus \int dH_{yij}\right) HT_{v_i,v_j}, \tag{5.2.3.9}$$

where

$$H_{ij} = H_{yij} \bigcup H_{zij}, \quad H_{yij} \bigcap H_{zij} = \emptyset, \quad H_{yij} \subset E_y \quad \text{and} \quad H_{zij} \subset E_z.$$

Proof. From (5.2.2.6), (5.2.3.7) and (5.2.1.8),

$$HT = E_z * T = E_z * \int T_{v_i,v_j}\, dH_{ij} == E_z * \int T_{v_i,v_j}\, dH_{zij} \oplus E_z$$
$$* \int T_{v_i,\,v_j}\, dH_{yij}.$$

Since

$$E_z * \int T_{v_i,v_j}\, dH_{zij} = \frac{\partial HTv_i, vj}{\partial H_{zij}}$$

and using (5.2.1.16), we have

$$HT = \frac{\partial HT\, v_i,\, vj}{\partial H_{zij}} \oplus \int HTv_i, v_j dH_{yij} = \left(\frac{\partial}{\partial H_{zij}} \oplus \int dH_{yij}\right) HT_{v_i,v_j}.$$

□

The following theorem is obtained by repeatedly using Theorems 5.2.3.1 and 5.2.3.2.

Theorem 5.2.3.3. *In a connected graph $G = (V, E)$ where $E = E_y \bigcup E_z$ and $E_y \cap E_z = \emptyset$, let $P_{i_m i_n}$ and $H_{i_m i_n}$ be a path between $v_{i_m} \in V$ and $v_{i_n} \in V$ and a cutset separating into $v_{i_m} \in V$ and $v_{i_n} \in V$ in G. Then,*

$$HT_{v_{i1},v_{i2},\ldots,v_{ik}} = \left(\frac{\partial}{\partial P_{yi_1i_2}} \oplus \int dP_{zi1i2}\right)\left(\frac{\partial}{\partial P_{yi_2i_3} \oplus \int dP_{zi2i3}}\right) \cdots \left(\frac{\partial}{\partial P_{yik-1ik}} \oplus \int dP_{zik-1ik}\right) HT, \tag{5.2.3.10}$$

$$HT = \left(\frac{\partial}{\partial H_{zi_1i_2}} \oplus \int dH_{yi_1i_2}\right)\left(\frac{\partial}{\partial H_{zi1i2}} \oplus \int dH_{yi_1i_2}\right) \cdots \left(\frac{\partial}{\partial H_{zi_{k-1}i_k}} \oplus \int dH_{zik-1ik}\right) HT_{v_{i1},v_{i2},\ldots,v_{ik}}. \tag{5.2.3.11}$$

In Theorem 5.2.3.3, (5.2.3.10) shows a method for generation of hybrid k-tree from hybrid tree and (5.2.3.11) shows a method for generation of hybrid tree from hybrid k-tree. In general the set of hybrid k-trees is small size, thus (5.2.3.11) shows a method for realization of large size hybrid trees. We also obtain a formula for generation of hT_n from hT_{n+2} or hT_{n-2} in (5.2.2.4). In these algebraic considerations, we also obtain a method for the realization of a graph with E_y and E_z from the set of hybrid trees [11].

Example.

(1) Let us consider a network of Fig. 5.1.3.6 again which is a network with impedance and admittance elements. Figure 5.2.3.1 shows the graph G of Fig. 5.1.4.1, where $G = (V, E)$, $E = E_y \bigcup E_z$, $E_y \bigcap E_z = \emptyset$,

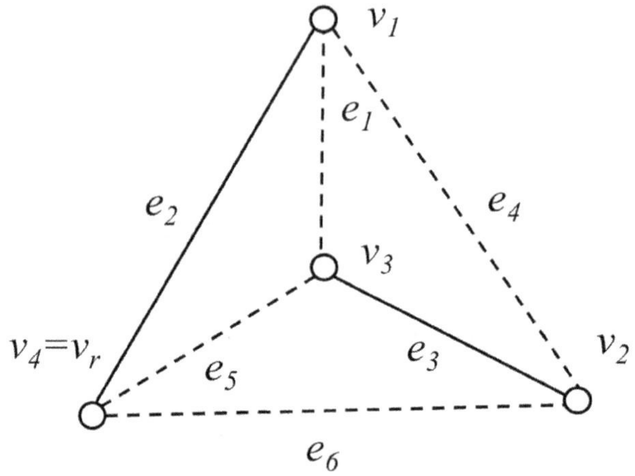

Fig. 5.2.3.1. The graph G of Fig. 5.1.4.1, where $E_y = \{e_2, e_3\}$ and $E_z = \{e_1, e_4, e_5, e_6\}$.

$V = \{v_1, v_2, v_3, v_4\}$, $E = \{e_1, e_2, e_3, e_4, e_5, e_6\}$, $E_y = \{e_2, e_3\}$ and $E_z = \{e_1, e_4, e_5, e_6\}$. Edges of E_y and E_z are shown by solid lines and broken lines respectively in Fig. 5.2.3.1.

From (5.1.4.4), the set of hybrid trees $HT(G) = HT$ is represented as the "product" of the edges as follows.

$$\begin{aligned} HT = \{&e_1e_2e_3e_4e_5, e_1e_2e_3e_4e_6, e_1e_2e_3e_5e_6, e_2e_3e_4e_5e_6, e_1e_2e_4, \\ &e_1e_2e_6, e_1e_3e_5, e_1e_3e_6, e_2e_4e_5, e_2e_5e_6, e_3e_4e_5, e_3e_4e_6, e_1, \\ &e_4, e_5, e_6\}. \end{aligned} \tag{5.2.3.12}$$

There are several paths between v_2 and v_4, for example, $\{e_6, e_2e_4, e_3e_5, e_1e_4e_5\}$. Let $P_{24} = e_3e_5$, where $e_3 \in E_y$, $e_5 \in E_z$. Let us obtain the set of hybrid 2-trees HT_{v_2,v_4} from HT using Theorem 5.2.3.1. From (5.2.3.3),

$$\begin{aligned} HT_{v_2,v_4} &= \frac{\partial HT}{\partial P_{y24}} \oplus \int HT\, dP_{z24} = \frac{\partial HT}{\partial e_3} \oplus \int HT\, de_5 \\ &= \{e_1e_2e_4e_5, e_1e_2e_4e_6, e_1e_2e_5e_6, e_2e_4e_5e_6, \\ &\quad e_1e_5, e_1e_6, e_4e_5, e_4e_6\} \oplus \{e_1e_2e_3e_4e_5e_6, e_1e_2e_4e_5, \\ &\quad e_1e_2e_5e_6, e_1e_3e_5e_6, e_3e_4e_5e_6, e_1e_5, e_4e_5, e_5e_6\} \end{aligned}$$

$$= \{e_1e_2e_3e_4e_5e_6, e_1e_2e_4e_6, e_1e_3e_5e_6,$$
$$e_2e_4e_5e_6, e_3e_4e_5e_6, e_1e_6, e_4e_6, e_5e_6\}. \qquad (5.2.3.13)$$

Next, let us obtain *HT* from HT_{v_2,v_4} using Theorem 5.2.3.2. There are several cutsets separating v_2 and v_4, for example, $\{e_2e_5e_6, e_3e_4e_6, e_1e_2e_3e_6, e_1e_4e_5e_6\}$. Let $H_{24} = e_3e_4e_6$, where $e_3 \in E_y$, $e_4e_6 \in E_z$.

$$\begin{aligned}
HT &= \frac{\partial HTv2, v4}{\partial H_{z24}} \oplus \int HTv2,\ v4\, dH_{y24} \\
&= \frac{\partial HTv2, v4}{\partial e_4e_6} \oplus \int HTv2,\ v4\, de_3 \\
&= \{e_1e_2e_3e_5e_6, e_1e_2e_6, e_2e_5e_6, e_3e_5e_6, e_6\} \\
&\quad \oplus \{e_1e_2e_3e_4e_5, e_1e_2e_4, e_1e_3e_5, \\
&\quad e_2e_4e_5, e_3e_4e_5, e_1, e_4, e_5\} \\
&\quad \oplus \{e_1e_2e_3e_4e_6, e_2e_3e_4e_5e_6, e_1e_3e_6, e_3e_4e_6, e_3e_5e_6\} \\
&= \{e_1e_2e_3e_4e_5, e_1e_2e_3e_4e_6, \\
&\quad e_1e_2e_3e_5e_6, e_2e_3e_4e_5e_6, e_1e_2e_4, e_1e_2e_6, \\
&\quad e_1e_3e_5, e_1e_3e_6, e_2e_4e_5, e_2e_5e_6, e_3e_4e_5, \\
&\quad e_3e_4e_6, e_1, e_4, e_5, e_6\}. \qquad (5.2.3.14)
\end{aligned}$$

(2) Wheatstone bridge: A bridge circuit is a circuit in which a bridge is attached to the shape of a rhombus as shown below. Above all, when no current flows through the bridge between them, three known immittance (admittance or impedance) can be prepared to measure the remaining one. This mechanism is called the *Wheatstone bridge*.

One of the features is that the potential is the same at the end of the bridge where current does not flow. If no current flows through the bridge, the input impedance is determined using the series-parallel method because the bridge is the same as removed. When some current flows through the bridge, the input impedance becomes a typical network that cannot be obtained using the series-parallel method. In this example, only the graphical method is used to determine the input impedance and the conditions under which the bridge current does not flow.

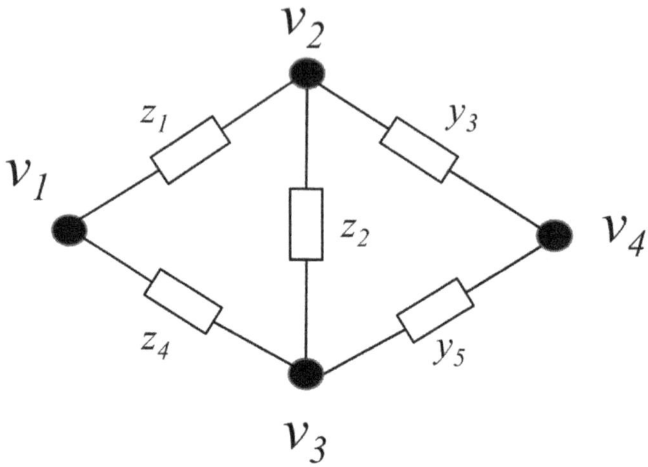

Fig. 5.2.3.2. A network called the *Wheatstone bridge*, where the element z_2 is bridge.

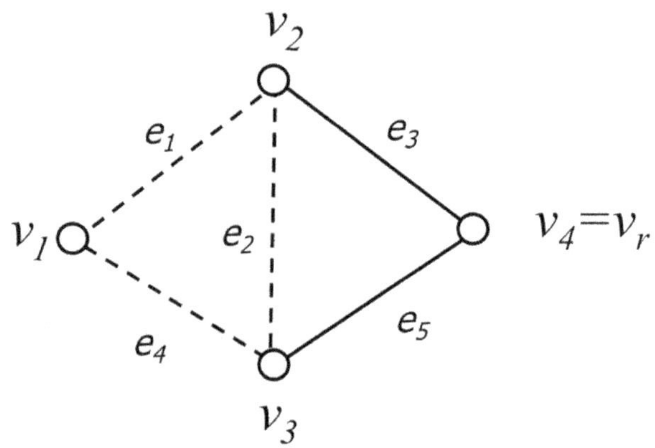

Fig. 5.2.3.3. The graph of the network of the *Wheatstone bridge* in Fig. 5.2.3.2.

Figure 5.2.3.2 shows a network called the *Wheatstone bridge*, where the element z_2 is bridge, and Fig. 5.2.3.3 shows the graph of the network of the *Wheatstone bridge*, where $G = (V, E)$, $E = E_y \bigcup E_z$, $E_y \bigcap E_z = \emptyset$, $V = \{v_1, v_2, v_3, v_4\}$, $E = \{e_1, e_2, e_3\}, e_4, e_5\}$, $E_y = \{e_1, e_2, e_4\}$ and $E_z = \{e_3, e_5\}$. Edges of E_y and E_z are shown by solid lines and broken lines respectively in Fig. 5.2.3.3.

In the graph of Fig. 5.2.3.3,

$$HT = \{e_1e_2e_3e_5, e_2e_3e_4e_5, e_1e_3, e_1e_5, e_2e_3, e_2e_5, e_3e_4, e_4e_5\}. \tag{5.2.3.15}$$

There are several paths between v_1 and v_4, for example, $\{e_1e_3, e_4e_5, e_1e_2e_5, e_2e_3e_4\}$. Let $P_{14} = e_1e_3$, where $e_3 \in E_y$, $e_1 \in E_z$. Let us obtain the set of hybrid 2-trees HT_{v_1,v_4} from HT using Theorem 5.2.3.1. From (5.2.3.3),

$$\begin{aligned} HT_{v_1,v_4} &= \frac{\partial HT}{\partial P_{y14}} \oplus \int HT\, dP_{z14} = \frac{\partial HT}{\partial e_3}, \oplus \int HT\, de_1 \\ &= \{e_1e_2e_5, e_2e_4e_5, e_1, e_2, e_4\} \oplus \{e_1e_2e_3e_4e_5, e_1e_2e_3, \\ &\quad e_1e_2e_5, e_1e_3e_4, e_1e_4e_5\} \\ &= \{e_1e_2e_3e_4e_5, e_1e_2e_3, \\ &\quad e_1e_3e_4, e_1e_4e_5, e_2e_4e_5, e_1, e_2, e_4\}. \end{aligned} \tag{5.2.3.16}$$

From (5.2.3.15) and (5.2.3.16), the sum of hybrid tree immittance product $\Delta_{h,n}$ and the sum of hybrid 2-tree immittance product $\Delta_{h,11}$ are as follows:

$$\begin{aligned} \Delta_{h,n} &= z_1z_2y_3y_5 + z_2y_3z_4y_5 + z_1y_3 + z_1y_5 \\ &\quad + z_2y_3 + z_2y_5 + y_3z_4 + z_4y_5, \end{aligned} \tag{5.2.3.17}$$

$$\begin{aligned} \Delta_{h,11} &= z_1z_2y_3z_4y_5 + z_1z_2y_3 + z_1y_3z_4 \\ &\quad + z_1z_4y_5 + z_2z_4y_5 + z_1 + z_2 + z_4. \end{aligned} \tag{5.2.3.18}$$

Then, the input impedance between v_1 and v_4,

$$\begin{aligned} \Delta_{h,11}/\Delta_{h,n} &= (z_1z_2y_3z_4y_5 + z_1z_2y_3 \\ &\quad + z_1y_3z_4 + z_1z_4y_5 + z_2z_4y_5 + z_1 \\ &\quad + z_2 + z_4)/(z_1z_2y_3y_5 + z_2y_3z_4y_5 \\ &\quad + z_1y_3 + z_1y_5 + z_2y_3 + z_2y_5 + y_3z_4 + z_4y_5). \end{aligned} \tag{5.2.3.19}$$

Note that this is driven by graphical method since this is not able to driven by the series-parallel method.

Next, we consider the conditions under which the bridge current does not flow. Since $a_{ij} = \Delta_{h,ji}/\Delta_{h,n} = \sum_{\text{all hybrid 2-trees } (HT_{ij,r})}$ (hybrid 2-tree immittance product of hybrid 2-tree of $HT_{ij,r}$of G)/ $\sum_{\text{all hybrid trees}}$ (hybrid tree immittance product of hybrid tree of G) where if $i = j$ or $i \neq j$, a_{ij} is input impedance or transfer impedance respectively, from (5.1.4.3), the condition is driven by the difference of potentials v_2 and v_3 that means the difference $a_{12} - a_{13}$ of transfer impedance.

In order to obtain the difference $a_{12} - a_{13}$, let us consider hybrid 2-trees $HT_{12,r} = HT_{v_1v_2,v_4}$ and $HT_{13,r} = HT_{v_1v_3,v_4}$.

From Theorem 5.1.2.2,

$$HT_{v_1v_2,v_4} = HT_{v_1,v_4} \cap HT_{v_2,v_4}, \tag{5.2.3.20}$$

$$HT_{v_1v_3,v_4} = HT_{v_1,v_4} \cap HT_{v_3,v_4}. \tag{5.2.3.21}$$

Let $P_{24} = e_3$, where $e_3 \in E_y$, $E_z = \emptyset$ to obtain HT_{v_2,v_4}. From (5.2.3.3),

$$\begin{aligned} HT_{v_2,v_4} &= \frac{\partial HT}{\partial P_{y24}} \oplus \int HT\, dP_{z24} \\ &= \frac{\partial HT}{\partial e_3} = \{e_1e_2e_5, e_2e_4e_5, e_1, e_2, e_4\}. \end{aligned} \tag{5.2.3.22}$$

Let $P_{34} = e_5$, where $e_5 \in E_y$, $E_z = \emptyset$ to obtain HT_{v_3,v_4}. From (5.2.3.3),

$$HT_{v_3,v_4} = \frac{\partial HT}{\partial P_{y34}} \oplus \int HT\, dP_{z34} = \frac{\partial HT}{\partial e_5} = \{e_1e_2e_3, e_2e_3e_4, e_1, e_2, e_4\}. \tag{5.2.3.23}$$

From (5.2.3.16), (5.2.3.22) and (5.2.3.20),

$$\begin{aligned} HT_{v_1v_2,v_4} &= HT_{v_1,v_4} \cap HT_{v_2,v_4} \\ &= \{e_1e_2e_3e_4e_5,\ e_1e_2e_3,\ e_1e_3e_4, \\ &\quad e_1e_4e_5, e_2e_4e_5, e_1, e_2, e_4\} \cap \{e_1e_2e_5, \\ &\quad e_2e_4e_5, e_1, e_2, e_4\} = \{e_2e_4e_5, e_1, e_2, e_4\}. \end{aligned} \tag{5.2.3.24}$$

From (5.2.3.16), (5.2.3.23) and (5.2.3.21),

$$\begin{aligned} HT_{v_1v_3,v_4} &= HT_{v_1,v_4} \cap HT_{v_3,v_4} \\ &= \{e_1e_2e_3e_4e_5, e_1e_2e_3, e_1e_3e_4, \\ &\quad e_1e_4e_5, e_2e_4e_5, e_1, e_2, e_4\} \cap \{e_1e_2e_3, \\ &\quad e_2e_3e_4, e_1, e_2, e_4\} \\ &= \{e_1e_2e_3, e_1, e_2, e_4\}. \end{aligned} \tag{5.2.3.25}$$

Since the difference of potentials v_2 and v_3 depends on the $HT_{v_1v_2,v_4} - HT_{v_1v_3,v_4}$,

$$HT_{v_1v_2,v_4} - HT_{v_1v_3,v_4} = \{e_2e_4e_5, e_1, e_2, e_4\} - \{e_1e_2e_3, e_1, e_2, e_4\}. \tag{5.2.3.26}$$

Let us consider the immittance product of (5.2.3.26), then the difference of potentials v_2 and v_3 is related to the difference $z_2z_4y_5 - z_1z_2y_3$. To be zero of the difference, $z_2z_4y_5 = z_1z_2y_3$, that is, $z_4y_5 = z_1y_3$.

Therefore, the condition under which the bridge current does not flow is as follows.

$$z_1/z_4 = y_5/y_3 = z_3/z_5. \tag{5.2.3.27}$$

This is known as the conditions for Wheatstone bridge.

(End of Example)

5.3. Sensitivity Analysis in Signal Flow Graphs

In the base part of this information and communication technology, there are electrical networks including electronic device, new components, and electronic circuits. In Sections 5.1 and 5.2, the electrical networks including electronic device, new components, and electronic circuits are discussed from a graphical point of view. We have also the digital circuits including digital filters which can be represented by signal flow graphs in the base part of this information and communication technology. There are many research results about graphical view of signal flow graphs [2,3]. Since it is essential to study which part of the circuit affects its overall performance

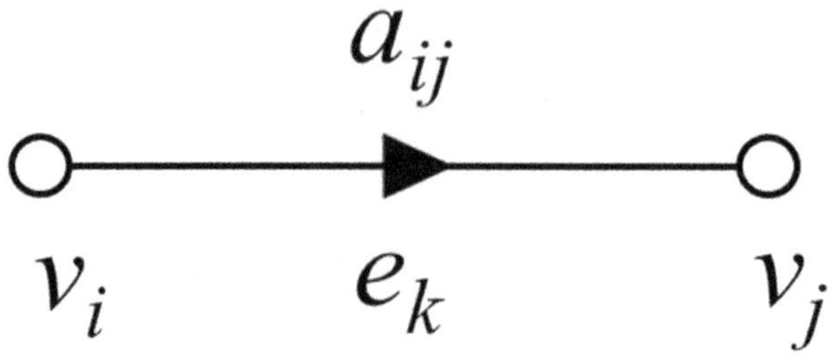

Fig. 5.3.1.1. An edge of signal flow graph where $x_k = a_{ij} z_k$.

to analyze, design and improve the digital systems including digital filters etc., a generalization of signal flow graphs and the part of the sensitivity in signal flow graphs is only discussed in this section.

5.3.1. *Tellegen's theorem in signal flow graphs*

A usual signal flow graph is used to represent a set of simultaneous linear algebraic equations. It is composed of vertices and directed edges, where each vertex represents a variable and the coefficients in the equations are assigned as the weights of directed edges [3,16]. For an edge $e_k = (v_i, v_j)$, let the *input signal* and the *output signal* of e_k be z_k and x_k, respectively. If the weight of e_k is a_{ij}, this means $x_k = a_{ij} z_k$ (see Fig. 5.3.1.1). This edge weight can be generalized as in Fig. 5.3.1.2. In this case, $x_k = f_{ij}(z_k)$, where $f_{ij}(\bullet)$ is a function. If $f_{ij}(z_k) = a_{ij} z_k$, this coincides with Fig. 5.3.1.1. Furthermore, an edge may be generalized to a block (multiport edge) as in Fig. 5.3.1.3. In this generalization, the signal flow graph is called a *signal flow network* since now we cannot see the graph structure of the signal flow system. In this section, we consider generalized signal flow graph of Fig. 5.3.1.2, and call it a *signal flow graph* for simplicity and use the arrow as in Fig. 5.3.1.1 instead of Fig. 5.3.1.2 for a directed edge.

First, we define a signal flow graph. Let G be a directed graph or a digraph for short:

$$G = (V, E), \tag{5.3.1.1}$$

where $V = \{v_1, v_2, \ldots\}$ is a set of vertices and $E = \{e_1, e_2, \ldots\}$ is a set of directed edges. For a directed edge $e_k = (v_i, v_j)$, where $v_i \in V$ is the initial vertex of e_k and $v_j \in V$ the terminal vertex of e_k, if $e_k = (v_i, v_j)$ and

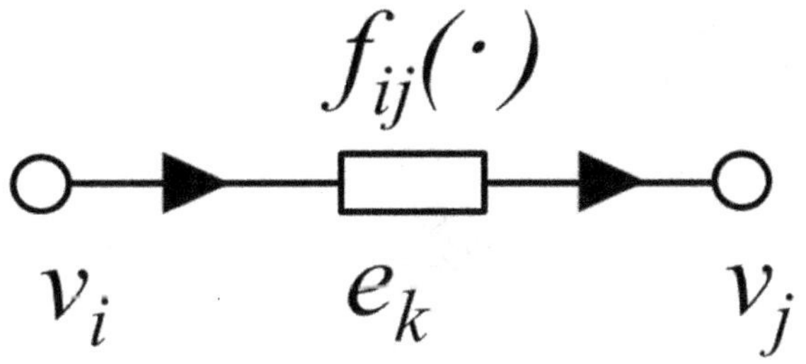

Fig. 5.3.1.2. An edge of signal flow graph where the weight is generalized as a function.

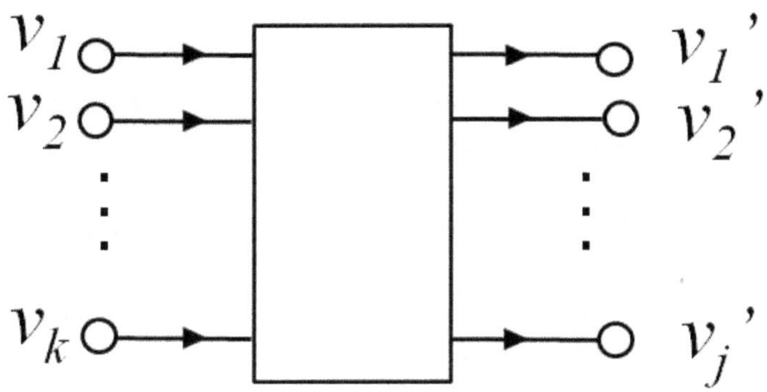

Fig. 5.3.1.3. A generalized of signal flow graphs.

$e_m = (v_i, v_j)$, then e_k and e_m are said to be *parallel*. An edge for which the initial and terminal vertices are the same is called a *self-loop* or a *self-tie*. Let

$$\delta^+ v_i = \{e_k | e_k = (v_i, v_j),\ e_k \in E\}, \tag{5.3.1.2}$$

$$\delta^- v_i = \{e_k | e_k = (v_j, v_i), e_k \in E\}. \tag{5.3.1.3}$$

To each edge, a weight is assigned, $f_k(\cdot)$ denotes the weight of an edge e_k. We consider three kinds of signals in G: signal on a vertex, *input signal* of an edge and *output signal* of an edge. We call G a *signal flow graph* if the signals of G obey the following rules (a)–(d):

(a) For an edge $e_k = (v_i, v_j)$, the input signal z_k of e_k is equal to y_i where y_i is the signal of $v_i \in V$.

(b) $$x_k = f_k(z_k), \tag{5.3.1.4}$$

where x_k is the output signal of $e_k = (v_i, v_j)$ and z_k is the input signal of e_k. From (a), (5.3.1.4) is rewritten as

$$x_k = f_k(y_i). \tag{5.3.1.5}$$

(c)

$$y_j = \sum_{e_k \in \delta^- v_j} x_k, \tag{5.3.1.6}$$

where y_j is the signal of v_j and x_k is the output signal of e_k.

(d) The signal of a source is an arbitrary constant. If $\delta^- v_i = \emptyset$, v_i is called a *source*, and if $\delta^+ v_i = \emptyset$, v_i is a *sink*.

Let us consider two different signal flow graphs $G = (V, E)$ and $G' = (V', E')$ of the same topology, that is, the digraphs G and G' are isomorphic but signals of vertices end edges of G and G' may be different. Let $V = \{v_1, v_2, \cdots\}$, $E = \{e_1, e_2, \cdots\}$, $V' = \{v'_1, v'_2, \cdots\}$ and $E' = \{e'_1, e'_2, \cdots\}$. Let $v_i \in V$ correspond to $v'_i \in V'$ and let $e_i \in E$ correspond $e'_k \in E'$. Since G and G' are isomorphic, $e'_k = (v'_i, v'_j)$ if $e_k = (v_i, v_j)$ and $e_k = (v_i, v_j)$ if $e'_k = (v'_i, v'_j)$. Let x_k and x'_k be the output signals of the corresponding edges e_k and e'_k of G and G', respectively. Let y_i and y'_i be the output signals of the corresponding edges v_i and v'_i of G and G', respectively. Let $|E| = M$, where $|X|$ denotes the number of elements in the set X. Then, we obtain the following theorem.

Theorem 5.3.1.1 (Tellegen's theorem for signal flow graphs). *Let* $e_1 = (v_{i_1}, v_{j_1})$, $e_2 = (v_{i_2}, v_{j_2})$, $\ldots$, $e_k = (v_{i_k}, v_{jk})$, $\ldots$ *and let* $e'_1 = (v_{i'_1}, v_{j'_1})$, $e'_2 = (v_{i'_2}, v_{j'_2})$, $\ldots$, $e'_k = (v_{i'_k}, v_{j'_k})$, $\ldots$. *Then,*

$$\sum_{k=1}^{M} y_{j_k} x'_k = \sum_{k=1}^{M} y'_{jk} x_k. \tag{5.3.1.7}$$

In Theorem 5.3.1.1, (5.3.1.7) is rewritten as

$$\sum_{\substack{e'_k = (v'_{ik}, v'_{jk}) \in E' \\ e_k = (v_{ik}, v_{jk}) \in E}} y_{jk} x'_k = \sum_{\substack{e_k = (v_{ik}, v_{jk}) \in E \\ e'_k = (v'_{ik}, v'_{jk}) \in E'}} y'_{jk} x_k. \tag{5.3.1.8}$$

Since $v_i \in V$ corresponds to $v'_i \in V'$, that is, vertices of G and G' are in a one-to-one correspondence, we use the following symbol instead of (5.3.1.8), for simplicity:

$$\sum_{e'_k = (v'_i, v'_j) \in E'} y_j x'_k = \sum_{e_k = (v_i, v_j) \in E} y'_j x_k. \qquad (5.3.1.9)$$

We also use the following symbol instead of (5.3.1.9) when G and G' are isomorphic:

$$\sum_{e_k=(v_i,v_j)\in E} y_j x'_k = \sum_{e_k = (v_i,v_j) \in E} y'_j x_k. \qquad (5.3.1.10)$$

Since both vertices and edges of G and G' are in a one-to-one correspondence when G and G' are isomorphic, $\sum_{e'_k = (v'_i,v'_j)\in E'} y_j x'_k$ means $\sum_{e_k = (v_i,v_j)\in E} y'_j x_k$ and (5.3.1.9) and (5.3.1.10) are equivalent. Hereafter, when $v_i \in V$ corresponds to $v'_i \in V'$ or $e_k \in E$ corresponds to $e'_k \in E'$, we use the symbol $\sum_{v_i \in V} y'_i$ instead of $\sum_{v'_i \in V'} y'_i$ or $\sum_{e_k \in E} x'_k$ instead of $\sum_{e'_k \in E'} x'_k$ for simplicity.

Theorem 5.3.1.1 is called the Tellegen's theorem for signal flow graphs and digital filters since (5.3.1.7) is similar to that of Tellegen's theorem for electrical networks (see Section 4.3). This theorem has been applied to the *sensitivity analysis* of digital filters [15–17]. We note that the weight (function) of an edge is not always linear, that is, signal flow graphs G and G' may be nonlinear.

Note that for two different electrical networks N and N' with the same topology, Tellegen's theorem is

$$\sum_{k=1}^{M} u_k i'_k = \sum_{k=1}^{M} u'_k i_k, \qquad (5.3.1.11)$$

where u_k and i_k are voltage and current of edge e_k in N respectively and u'_k and i'_k are voltage and current of edge e'_k in N' respectively, and M is the number of edges. Equation (5.3.1.11) is based on Kirchhoff's law shown in Theorem 4.3.1.

5.3.2. *Operators in signal flow graphs*

Tellegen's theorem for signal flow graphs has been extended to generalized signal flow graphs or signal flow networks. In this section, we show a different version of generalization of Tellegen's theorem for signal flow graphs.

Definition 1. Let $G = (V, E)$ and $G' = (V', E')$ be two signal flow graphs of the same topology. Let $v_i \in V$ correspond to $v'_i \in V'$ and $e_k \in E$ correspond to $e'_k \in E'$. A function $F_G(y, x)$ is called an *operator* for G and G' (*SFG operator*), if it satisfies the relation

$$\sum_{e_k=(v_i,vj)\in E} F_G(y_j, x'_k) = \sum_{e_k=(v_i,v_j)\in E} F_G(y'_j\, x_k), \tag{5.3.2.1}$$

where $y_j(x_k)$ and $y'_j(x'_k)$ are signals of the corresponding vertices v_j and v'_j (output signals of the corresponding edges e_k and e'_k) of G and G', respectively.

Before examining the SFG operator, let us first define a few terms. Let $S_1 = \{p_i\}$ and $S_2 = \{q_i\}$ be two signal spaces where p_i and q_i are complex functions of the complex variables $t_1, t_2, \ldots$. Let $p_i \in S_1$ and $q_i \in S_2$ be the signals of the vertices and the output signals of the edges, respectively. For $y, y_1, y_2 \in S_1$ and $x, x_1, x_2 \in S_2$:

(a) An operator $F(y, x)$ is said to be *symmetric*, if

$$F(y, x) = F(x, y). \tag{5.3.2.2}$$

An operator $F(y, x)$ is said to be *additive* with respect to y or x, if

$$F(y_1 + y_2, x) = F(y_1, x) + F(y_2, x) \tag{5.3.2.3}$$

or

$$F(y, x_1 + x_2) = F(y, x_1) + F(y, x_2). \tag{5.3.2.4}$$

(b) An operator $F(y, x)$ is said to be *biadditive*, if both (5.3.2.3) and (5.3.2.4) are satisfied.

(c) An operator $F(y, x)$ is said to be *bilinear*, if $F(y, x)$ is biadditive and

$$F(ax, y) = F(x, ay) = aF(x, y) \tag{5.3.2.5}$$

for any constant a.

For example, an operator

$$F_1(y, x) = yx. \tag{5.3.2.6}$$

is symmetric, additive and biadditive (bilinear). The operator

$$F_2(y, x) = yx^* \tag{5.3.2.7}$$

is biadditive but is not symmetric, where x^* is the complex conjugate of x. The operator

$$F_3(y, x) = yx^* - y^*x \tag{5.3.2.8}$$

is also biadditive but is not symmetric. The operators

$$F_4(y, x) = yx - y^*x^*, \tag{5.3.2.9}$$

$$F_5(y, x) = yx + y^*x^*, \tag{5.3.2.10}$$

$$F_6(y, x) = yx^* + y^*x \tag{5.3.2.11}$$

are symmetric and biadditive. Let y and x be differentiable with respect to t_1. The operator

$$F_7(y, x) = y\frac{\partial x}{\partial t_1} - x\frac{\partial y}{\partial t_1} \tag{5.3.2.12}$$

is biadditive but is not symmetric. The operator

$$F_8(y, x) = y\frac{\partial x}{\partial t_1} + x\frac{\partial y}{\partial t_1} \tag{5.3.2.13}$$

is symmetric and biadditive.

Theorem 5.3.2.1. *$F_G(y, x)$ is an SFG operator if and only if $F_G(y, x)$ is symmetric and*

$$F_G(x + y, z) + c = F_G(x, z) + F_G(y, z), \tag{5.3.2.14}$$

where c is a constant such that $F_G(x, 0) = c$.

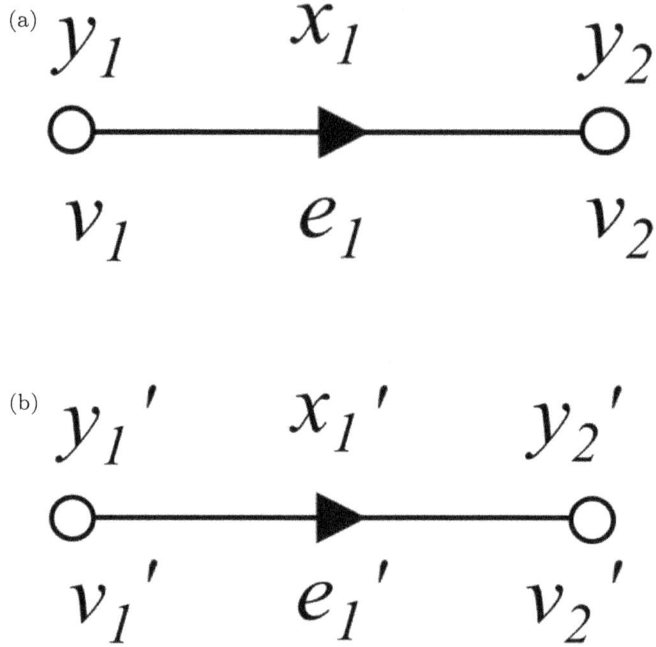

Fig. 5.3.2.1. Graphs G and G'. (a) A graph G and (b) A graph G'.

Proof. *Necessity*: Since $F_G(y, x)$ must satisfy (5.3.2.1) for any signal flow graph G, let us consider the graphs G and G' of Fig. 5.3.2.1 which are isomorphic. In G and G', using (5.3.2.1)

$$F_G(y_2, x_1') = F_G(y_2', x_1). \tag{5.3.2.15}$$

From the definition of a signal flow graph,

$$y_2 = x_1, \quad y_2' = x_1'. \tag{5.3.2.16}$$

Hence,

$$F_G(x_1, x_1') = F_G(x_1', x_1), \tag{5.3.2.17}$$

which implies that $F_G(y, x)$ is symmetric.

Next, in the graphs G and G' of Fig. 5.3.2.2 which are isomorphic, using (5.3.2.1) we obtain

$$F_G(y_2, x_1') + F_G(y_2, x_2') = F_G(y_2', x_1) + F_G(y_2', x_2). \tag{5.3.2.18}$$

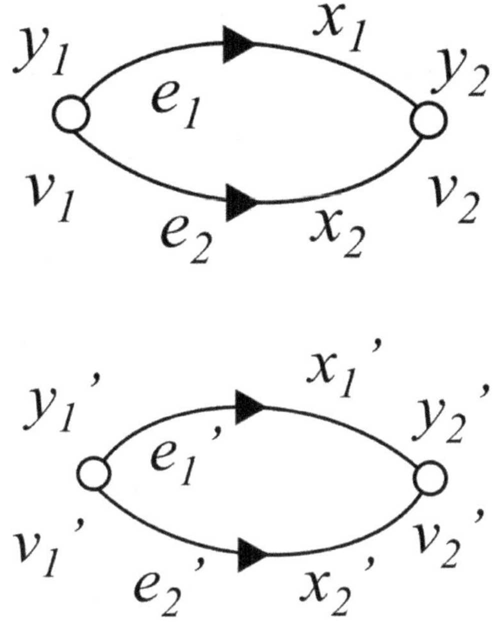

Fig. 5.3.2.2. Graphs G and G'. (a) A graph G and (b) A graph G'.

Since $y_2 = x_1 + x_2$ and $y_2' = x_1' + x_2'$,

$$F_G(x_1 + x_2, x_1') + F_G(x_1 + x_2, x_2') = F_G(x_1' + x_2', x_1) + F_G(x_1' + x_2', x_2). \tag{5.3.2.19}$$

Let $x_1' = x_2'$, then

$$2F_G(x_1 + x_2, x_1') = F_G(2x_1', x_1) + F_G(2x_1', x_2). \tag{5.3.2.20}$$

Letting $x_1 = x_2$ and $x_1' = 0$ in (5.3.2.20) and using the property (5.3.2.17), we get

$$F_G(2x_1, 0) = F_G(0, x_1) = F_G(x_1, 0). \tag{5.3.2.21}$$

Setting $x_2 = 0$ in (5.3.2.20) and using the property of (5.3.2.21) yields

$$\begin{aligned} 2F_G(x_1, x_1') &= F_G(2x_1', x_1) + F_G(2x_1', 0) \\ &= F_G(2x_1', x_1) + F_G(x_1', 0). \end{aligned}$$

Using (5.3.2.20) and the property of (5.3.2.22),

$$F_G(x_1 + x_2, x_1') = F_G(x_1, x_1') + F_G(x_2, x_1') - F_G(x_1', 0). \tag{5.3.2.23}$$

Using (5.3.2.16) and the properties of (5.3.2.15) and (5.3.2.23),

$$\begin{aligned} F_G(y_2, x_1') + F_G(y_2, x_2') &= F_G(y_2, x_1' + x_2') + F_G(y_2, 0) \\ &= F_G(y_2', x_1) + F_G(y_2', x_2) \\ &= F_G(y_2', x_1 + x_2) + F_G(y_2', 0). \end{aligned} \tag{5.3.2.24}$$

Since $y_2 = x_1 + x_2$ and $y_2' = x_1' + x_2'$,

$$F_G(y_2, y_2') + F_G(y_2, 0) = F_G(y_2', y_2) + F_G(y_2', 0). \tag{5.3.2.25}$$

Hence,

$$F_G(y_2, 0) = F_G(y_2', 0) \equiv c \tag{5.3.2.26}$$

since y_2 is independent of y_2'. Therefore, from (5.3.2.23) and (5.3.2.26)

$$F_G(x_1 + x_2, x_1') + c = F_G(x_1, x_1') + F_G(x_2, x_1'). \tag{5.3.2.27}$$

Thus $F_G(y, x)$ is symmetric from (5.3.2.17) and satisfies (5.3.2.14).

Sufficiency: Let $S \subset P$ be the set of source vertices of G. Since G and G' are isomorphic, $v_i \in V$ corresponds to $v_i' \in V'$ and $e_k \in E$ corresponds to $e_k' \in E'$.

$$\sum_{e_k=(v_i,v_j)\in E} F_G(y_j, x_k') = \sum_{\substack{v_j \notin S \\ v_j \in V}} \sum_{e_k \in \delta^- v_j} F_G(y_j, x_k') \tag{5.3.2.28}$$

Using (5.3.2.14),

$$\sum_{e_k \in \delta^- v_j} F_G(y_j x_k') = F_G\left(y_j, \sum_{e_k \in \delta^- v_j} x_k'\right) + c\left(\left|\delta^- v_j\right| - 1\right). \tag{5.3.2.29}$$

From (5.3.1.6),

$$y'_j = \sum_{e_k \in \delta^{-v_j}} x'_k \tag{5.3.2.30}$$

or

$$\sum_{e_k \in \delta^- v_j} F_G(y_j, x_{k'}), = F_G(y_j, y_{j'}) + c(|\delta^- v_j| - 1). \tag{5.3.2.31}$$

Hence, from (5.3.2.28)

$$\sum_{e_k=(v_i,v_j)\in E} F_G(y_j, x'_k) = \sum_{\substack{v_j \notin S \\ v_j \in V}} \{F_G(y_j\, y'_j) + c(|\delta^- v_j| - 1)\}. \tag{5.3.2.32}$$

Similarly,

$$\sum_{e_k=(v_i,v_j)\in E} F_G(y'_j, x_k) = \sum_{\substack{v_j \notin S \\ v_j \in V}} \{F_G(y'_j y_j) + c(|\delta^- v_j| - 1)\}. \tag{5.3.2.33}$$

Since $F_G(y, x)$ is symmetric, from (5.3.2.32) and (5.3.2.33),

$$\sum_{e_k=(v_i,\, v_j)\in E} F_G(y_j, x'_k) = \sum_{e_k=(v_i,\, v_j)\in E} F_G(y'_j, x_k). \tag{5.3.2.34}$$

□

In this theorem, if $F_G(y, x)$ is continuous, we have the following corollary.

Corollary 5.3.2.1. *A continuous real function $F_G(y, x)$ with respect to real variables y and x is an SFG operator if and only if*

$$F_G(y, x) = H(y, x) + c, \tag{5.3.2.35}$$

where $H(y, x)$ is continuous symmetric and bilinear, and c is a real constant such that $F_G(x, 0) = c$.

Proof. Let $F_G(y, x) = H(y, x) + c$. From Theorem 5.3.2.1,

$$H(y, x) = H(x, y), \tag{5.3.2.36}$$

$$H(y + x, z) = H(y, z) + H(x, z) \tag{5.3.2.37}$$

and

$$H(z, y + x) = H(z, y) + H(z, x). \tag{5.3.2.38}$$

In order to prove that $H(y, x)$ is bilinear (see (5.3.2.5)), we have to prove the following (5.3.2.39).

$$H(ay, x) = H(y, ax) = aH(y, x) \tag{5.3.2.39}$$

using (5.3.2.36)~(5.3.2.38) and the continuity of $H(x, y)$.

Let $x = 0$ in (5.3.2.37), then

$$H(y, z) = H(y, z) + H(0, z). \tag{5.3.2.40}$$

Therefore,

$$H(0, z) = H(z, 0) = 0. \tag{5.3.2.41}$$

Let $y = -x$ in (5.3.46), then

$$H(0, z) = H(-x, z) + H(x, z). \tag{5.3.2.42}$$

Hence,

$$H(-x, z) = -H(x, z). \tag{5.3.2.43}$$

From (5.3.2.43), we only must show that (5.3.2.39) holds under the condition that $a > 0$. Since $H(y, x)$ is continuous, it is sufficient to show that (5.3.2.39) holds under the condition that $a = l/m$, where l and m are positive integers. Because, for a positive real number a, a rational number or sequence $\gamma_n > 0$ converges to a, that is

$$\gamma_n \underset{n \to \infty}{\longrightarrow} a. \tag{5.3.2.44}$$

It follows that if $H(\gamma_n y, x) = \gamma_n H(y, x)$, then from the continuity of $H(y, x)$, $H(ay, x) = aH(y, x)$ for $n \to \infty$.

$$H(ly, x) = H(y + y + \cdots + y, x) = H(y, x) + H(y, x) + \cdots + H(y, x) = lH(y, x). \tag{5.3.2.45}$$

If $y = mz$, then

$$H(lmz, x) = lH(mz, x) = mH(lz, x). \tag{5.3.2.46}$$

If $z = u/m$, then

$$lH(u, x) = mH\left(\frac{l}{m}u, x\right). \tag{5.3.2.47}$$

Therefore,

$$H\left(\frac{l}{m}u, x\right) = \frac{l}{m}H(u, x) \tag{5.3.2.48}$$

and we obtain (5.3.2.39).

Conversely, it is obvious that if $H(y, x)$ is symmetric and bilinear, $F_G(y, x)$ is an SFG operator. □

Theorem 5.3.2.1 and Corollary 5.3.2.1 contain Theorem 5.3.1.1 because if $F_G(y, x) = yx$, which satisfies the condition of Theorem 5.3.2.1 or Corollary 5.3.2.1, then Theorem 5.3.2.1 can be obtained by using the SFG operator $F_G(y, x) = yx$. In other words, Theorem 5.3.1.1 can be considered as a simple example of Theorem 5.3.2.1.

5.3.3. *Operators in signal flow graphs with nonisomorphic graphs*

In the previous section, we discussed an operator for two signal flow graphs of the same topology. In this section, we consider an operator for two signal flow graphs which may not have the same topology. Thus, it is an extension of the previous result.

Let us consider two signal flow graphs $G = (V, E)$ and $G' = (V', E')$ which are not necessarily isomorphic. We assume that G and G' satisfy

Condition A: Vertices of V and V' except for the sources are in a onc-to-one correspondence.

Let $S \subset V$ and $S' \subset V'$ be the sets of sources of V and V', respectively. Let $v_i \in V - S$ correspond to $v_i' \in V' - S'$. Condition A implies that if $v_i \in V - S$, then the corresponding vertex v_i' of v_i is included in S' or v_i has no corresponding vertex in V', and if $v_i' \in S'$, then the corresponding vertex v_i of v_i' is included in S or v_i' has no corresponding vertex in V.

Definition 2. An operator $F(y, x)$ is called an *SFG operator* if for any two signal flow graphs G and G' satisfying Condition A, it satisfies the relation

$$\sum_{e_{k'} = \left(v_i', v_j'\right) \in E'} F(y_j, x_k') = \sum_{e_k = \left(v_i, v_j\right) \in E} F(y_j', x_k), \tag{5.3.3.1}$$

where $y_j(x_k)$ and $y_j'(x_k')$ are signals of the corresponding vertices v_j and v_j' (output signals of the edges e_k and e_k'), respectively.

Here, we use the same term "SFG operator" as in Definition 1, for simplicity. When $|E| = |E'|$, we obtain the following theorem.

Theorem 5.3.3.1. *Equation (5.3.3.1) holds for an operator $F(y, x)$ for any two signal flow graphs G and G' satisfying Condition A and $|E| = |E'|$ if and only if $F(y, x)$ is symmetric and*

$$F(x + y, z) + c = F(x, z) + F(y, z), \tag{5.3.3.2}$$

where c is a constant such that $F(x, 0) = c$.

Proof. *Necessity*: Since $F(y, x)$ must satisfy (5.3.3.1) for any two signal flow graphs G and G' satisfying Condition A and $|E| = |E'|$, let G and G' be isomorphic. Thus G and G' satisfy Condition A and $|E| = |E'|$. The signal flow graphs of Figs. 5.3.2.1 and 5.3.2.2 are such examples. The necessity follows from a proof similar to that of Theorem 5.3.2.1.

Sufficiency:

$$\sum_{e_k' = \left(v_i', v_j'\right) \in E'} F(y_j, x_k') = \sum_{\substack{v_j' \notin S' \\ v_j' \in V'}} \sum_{e_k' \in \delta^- v_j'} F(y_j\, x_k'). \tag{5.3.3.3}$$

Using (5.3.3.2),

$$\sum_{e_{k'} \in \delta^- v_j'} F(y_j, x_k') = F\left(y_j \sum_{e_k' \in \delta^- v_j'} x_k'\right) + c(|\delta - v_j'| - 1). \tag{5.3.3.4}$$

From (5.3.1.6),

$$y_j' = \sum_{e_k' \in \delta^- v_j'} x_k'. \tag{5.3.3.5}$$

Hence,

$$\sum_{\substack{v_j' \notin S' \\ v_j' \in V'}} \sum_{e_k' \in \delta^- v_j'} F(y_j, x_k') = \sum_{\substack{v_j' \notin S' \\ v_j' \in V'}} \{F(y_j, y_j') + c(|\delta^- v_j'| - 1)\}. \tag{5.3.3.6}$$

Similarly,

$$\sum_{\substack{v_j \notin S \\ v_j \in V}} \sum_{e_k \in \delta^- v_j} F(y_j', x_k) = \sum_{\substack{v_j \notin S \\ v_j \in V}} \{F(y_j', y_j) + c(|\delta^- v_j| - 1)\}. \tag{5.3.3.7}$$

In (5.3.3.6) and (5.3.3.7), since V and V' satisfy Condition A, $|E| = |E'|$ and $F(y, x)$ is symmetric,

$$\sum_{e_k' = (v_i', v_j') \in E'} F(y_j, x_k') = \sum_{e_k = (v_i, v_j) \in E} F(y_j', x_k) \tag{5.3.3.8}$$

□

Following a proof similar to that of Corollary 5.3.2.1, we obtain the following corollary.

Corollary 5.3.3.1. *Equation (5.3.3.1) holds for a continuous real function $F(y, x)$ with respect to real variables y and x for any two signal flow graph*

G and G' satisfying Condition A and $|E| = |E'|$ if and only if,

$$F(y, x) = H(y, x) + c \tag{5.3.3.9}$$

where $H(y, x)$ is symmetric and bilinear, and c is a real constant such that $F(x, 0) = c$.

In general, we have the following theorem that is applicable to the general situation where $|E| \neq |E'|$.

Theorem 5.3.3.2. *$F(y, x)$ is an SFG operator if and only if $F(y, x)$ is symmetric and additive.*

Proof. *Necessity*: $F(y, x)$ is an operator satisfying (5.3.3.1) for any signal flow graphs G and G' with Condition A. First, we consider the graphs G and G' satisfying Condition A and $|E| = |E'|$. From Theorem 5.3.3.1, $F(y, x)$ must satisfy (5.3.3.2). Next we consider the graphs G end G' satisfying Condition A and having a different number of edges. Since $F(y, x)$ satisfies (5.3.3.1) for G and G', we have $c = 0$ in (5.3.3.2). Thus $F(y, x)$ is symmetric and additive.

Sufficiency: Sufficiency can be proved in a manner similar to that of Theorem 5.3.3.1 except for $c = 0$. □

Corollary 5.3.3.2. *A continuous real function $F(y, x)$ is an SFG operator if and only if $F(y, x)$ is symmetric and bilinear.*

From Theorem 5.3.3.2, the operators of (5.3.2.6), (5.3.2.9)–(5.3.2.11), and (5.3.2.13) are SFG operators. For an SFG operator $F(y \cdot x)$, y and x are signals of the vertices and edges respectively. Let ψ be an operator which transforms from one set of signals of vertices and edges to another. Let

$$\bar{y} = \psi(\mathrm{y}), \tag{5.3.3.10}$$

$$\bar{x} = \psi(\mathrm{x}) \tag{5.3.3.11}$$

and let ψ be an additive operator. Consider an operator $F_\psi(\psi(\mathrm{y}) \cdot \psi(x))$, where $F_\psi(y, x)$ is symmetric and additive, that is, $F_\psi(y, x)$ is an SFG operator. Since ψ is additive, $F_\psi(\psi(\mathrm{y}) \cdot \psi(x))$ is also symmetric and additive with respect to y and x. This means that $F_\psi(\psi(y)\cdot\psi(x))$ is considered to be an SFG operator $F(y,x)$.

$$F(y,x) \triangleq F_\psi(\psi(y),\psi(x)) = F_\psi(\bar{y},\bar{x}) \tag{5.3.3.12}$$

Hence, (5.3.3.1) can be rewritten as follows:

$$\sum_{e'_k=\left(v'_i,v'_j\right)\in E'} F_\psi(\bar{y}_j,\bar{x}'_k) = \sum_{e_k=\left(v_i,\,v_j\right)\in E} F_\psi(\bar{y}'_j\bar{x}'_k). \tag{5.3.3.13}$$

Since $F_\psi(y,x)$ is an SFG operator, we can say that for signals $\bar{y}$ and $\bar{x}$ derived from y and x, respectively, through an operator ψ, $F_\psi(\bar{y},\bar{x})$ also possesses the property (5.3.3.1) for an SFG operator $F(y\cdot x)$. Obviously, the operator includes the linear transformations (e.g., the Fourier transform and z-transform). For signals Y and X derived from y and x respectively through a linear transformation, it is known that [15,17]

$$\sum_{e'_k=\left(v'_i,v'_j\right)\in E'} Y_jX'_k = \sum_{e_k=\left(v_i,v_j\right)\in E} Y'_jX_k. \tag{5.3.3.14}$$

Let $F_\psi(y,x) = yx$ in (5.3.3.13). Then from (5.3.3.13), we see that Theorem 5.3.3.2 includes (5.3.3.14) as a special case.

5.3.4. *Applications of the SFG operators*

In this section, we present some applications of the SFG operators. Before we do this, we present a method of treating the source and sink as edges, and define the *transmission* T_{ij} in a signal flow graph.

The source and sink can be treated as edges by using the self-loop of Fig. 5.3.4.1, where

$$x_k = f_k(y_j) = d, \tag{5.3.4.1}$$

$$v_i:\ \text{source, if } d \neq 0, \tag{5.3.4.2}$$

$$v_i:\ \text{sink, if } d = 0, \tag{5.3.4.3}$$

and in (5.3.4.2), d may be zero if the sourcing at v_i is unnecessary.

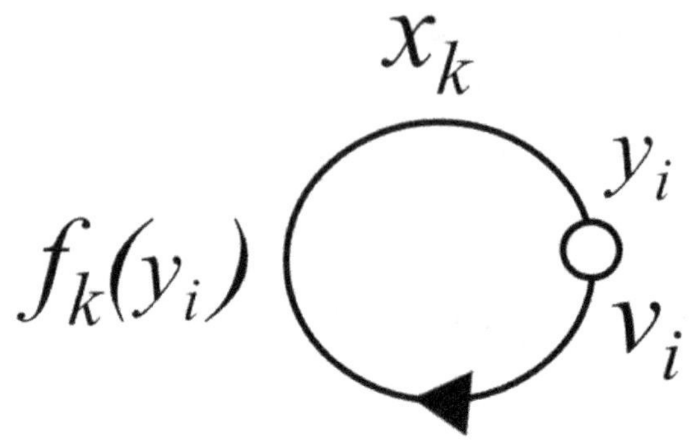

Fig. 5.3.4.1. Representation of a source or a sink by means of a self-loop (self-tie).

Next, we define the transmission (or the *gain*) T_{ij}. Let $U \subset V$ be the set of sources of G. Self-loops of (5.3.4.1) are attached to every source and sink in G. For vertices $v_i, v_j \in V$, if v_i is a source, attach a self-loop of (5.3.4.1) and (5.3.4.2) to v_i and if v_j is a sink, attach a self-loop of (5.3.4.1) and (5.3.4.3) to v_j. For simplicity, let self-loops attached to $v_m \in V$ be e_m. Then, the output signal of e_m is x_m. We define the transmission (or the gain) T_{ij} as follows:

$$T_{ij} = \frac{y_j}{x_i}\bigg|x_i = 0 \quad \text{and} \quad x_k = 0 \quad \text{for } v_k \in U - \{v_i\} \tag{5.3.4.4}$$

if v_i is a source.

$$T_{ij} = \frac{y_j}{x_i}\bigg|x_i \neq 0 \quad \text{and} \quad x_k = 0 \quad \text{for } v_k \in U \tag{5.3.4.5}$$

if v_i is not a source.

We sometimes call T_{ij} the transmission from v_i to v_j.

Observe that in the definition if v_i is a source,

$$T_{ij} = \frac{y_j}{y_i}\bigg|x_i \neq 0 \quad \text{and} \quad x_k = 0 \quad \text{for } v_k \in U - \{v_i\} \tag{5.3.4.6}$$

(because $x_i = y_i$), but if v_i is not a source, T_{ij} is not always equal to

$$\frac{y_j}{y_i}\bigg|x_i \neq 0 \quad \text{and} \quad x_k = 0 \quad \text{for } v_k \in U - \{v_i\}. \tag{5.3.4.7}$$

Note that the value of T_{ij} remains unaltered whether or not v_i is a sink. The transmission T_{ij} from v_i to v_j equals the ratio of the signal of v_j to the signal of v_i if v_i is a source, and it is not always equal to this ratio if v_i is not a source.

(1) *Application I*

A signal flow graph in which every edge function $f_k(\cdot)$ except for the self-loops of sources and sinks is linear, that is, $f_k(y) = a_k y$ where a_k is a constant, is called a *linear signal flow graph.* There are various graph-theoretic properties of linear signal flow graphs. For instance, it is known that the transmission from source to sink of a linear signal flow graph G with one source and one sink, and the transmission from source to sink of G' derived from G by reversing the direction of every edge in G with the same edge function (the value of the function of the self-loop of a source or a sink is changed) are identical, where G' is called the *transposed graph* of G. This property is used for the realization of digital filters and may not hold for nonlinear signal flow graphs. We state the following theorem using the SFG operator.

Theorem 5.3.4.1. *The transmission from source to sink for any signal flow graph G with one source and one sink equals the transmission from source to sink of its transposed graph G' if and only if G is a linear signal flow graph.*

Proof. *Necessity*: Consider a signal flow graph $G = (V, E)$ and its transposed graph $G' = (V', E')$ in Fig. 5.3.4.2(a) and (b). In G and G', self-loops are attached to represent the source and sink. Let the vertices.

v_1 and v_2 in V correspond to v_1' and v_2' in V', respectively. G and G' satisfy Condition A. Thus an SFG operator $F(y, y)$ can be applied to G and G'. Let $F(y, x) = yx$ which satisfies the condition of either Theorems 5.3.3.1 or 5.3.3.2. From (5.3.3.1),

$$y_1x_s' + y_1x_1' + y_2x_0' = y_1'x_s + y_2'x_1 + y_2'x_0. \qquad (5.3.4.8)$$

Let $y_1 = x_s$, $y_2' = x_0'$, $x_0 = 0$ and $x_s' = 0$. Since G' is the transposed signal flow graph of G,

$$x_1' = f_1'(y_2') = f_1(y_2') \qquad (5.3.4.9)$$

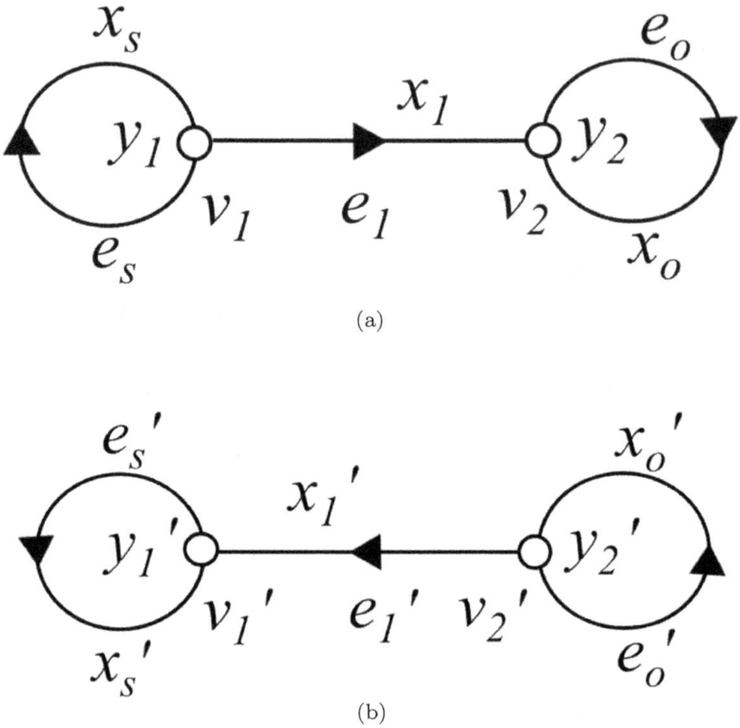

Fig. 5.3.4.2. (a) A signal flow graph G and (b) its transposed graph G'.

Hence,

$$y_1 f_1(y_2') + y_2 y_2' = y_1' y_1 + y_2' f_1(y_1). \tag{5.3.4.10}$$

Since the transmission from source to sink of G and G' are identical,

$$\frac{y_2}{y_1} = \frac{y_1'}{y_2'}. \tag{5.3.4.11}$$

Substituting (5.3.4.11) into (5.3.4.10), we obtain

$$\frac{f_1(y_1)}{y_1} = \frac{f_1(y_2')}{y_2'}. \tag{5.3.4.12}$$

Since y_1 is independent of y_2' for any y_1 and y_2',

$$\frac{f_1(y_1)}{y_1} = \frac{f_1(y_2')}{y_2'} = a_1 \tag{5.3.4.13}$$

where a_1 is a constant. Therefore,

$$f_1(y_1) = a_1 y_1 \tag{5.3.4.14}$$

Sufficiency: Let functions of edges be linear, that is, $f_i(y) = a_i y$. In $G = (V, E)$ and its transposed graph $G' = (V', E')$, let v_s, $v_o \in V$ and v'_o, $v'_s \in V'$ be the source and the sink of G and G', respectively. If self-loops are attached to the vertices of the source and sink of G and G', the resulting graphs satisfy Condition A. Let these new graphs be still represented by G and G', respectively. Let the self-loops of the source and sink of G be e_s and e_o, respectively. Then the self-loops of the source and sink of G' are e'_o and e'_s, respectively. If $F(y, x) = yx$ in (5.3.3.1), (5.3.3.1) becomes

$$\sum_{e'_k = \left(v'_i, v'_j\right) \in E'} y_j x'_k = \sum_{e_k = (v_i,\, v_j) \in E} y'_j x_k. \tag{5.3.4.15}$$

Note that e_k and e'_k are not corresponding edges in (5.3.86). Next, we apply (5.3.4.15) to G and its transposed graph G'. If $x_k = f_k(y_i) = a_k y_i$ for an edge $e_k = (v_i, v_j) \in E$ of G, there is an edge $e'_k = (v'_i, v'_j) \in E'$ with $x_{k'} = f_k(y'_j) = a'_k y'_j = a_k y'_j$ in G', where $k \neq s, o$. Then (5.3.4.15) can be rewritten as

$$\sum_{e'_k = \left(v'_j, v'_i\right) \in E' - \{e'_s, e'_o\}} y_i a_k y'_j + y_s x'_s + \{y_o x'_o\}$$
$$= \sum_{e_k = (v_i,\, v_j) \in E - \{e_s,\, e_o\}} y_{j'} a_k y_i + y'_s x_s + \{y'_o x_o\}. \tag{5.3.4.16}$$

Note that in this equation, e'_k in G' represents the edge whose direction is opposite to that of e_k in G. Hence, (5.3.4.16) becomes

$$y_s x'_s + \{y_o x'_o\} = y'_s x_s + \{y'_o x_o\}. \tag{5.3.4.17}$$

Let $x_o = 0$ and $x'_s = 0$ since v_o and v'_s are the sinks of G and G', respectively. Then

$$\frac{y_o}{x_s} = \frac{y'_s}{x'_o}. \tag{5.3.4.18}$$

Thus the transmission T_{so} of G is equal to the transmission T'_{os} (note that T'_{os} means the transmission from source to sink of its transposed graph G') of G'. □

In Theorem 5.3.4.1, a signal flow graph G has one source and one sink. Next, we consider a signal flow graph G with multiple sources and sinks.

Theorem 5.3.4.2. *In a linear signal flow graph G and its transposed graph G′,*

$$T_{pq} = T'_{qp} \tag{5.3.4.19}$$

where T_{pq} and T'_{qp} are the transmissions of G and G′, respectively.

Proof. Let $U \subset V$ and $W \subset V$ be the sets of sources and sinks of $G = (V, E)$, respectively, and let $U' \subset V'$ and $W' \subset V'$ be the sets of sources and sinks of $G' = (V', E')$, respectively, where G' is the transposed graph of G. Let a self-loop be attached to the vertex of every source and sink of G and G'. The resulting graphs satisfy Condition A. If $v_p \in U$, a self-loop of (5.3.4.1) and (5.3.4.2) is attached to v_p, and if $v_q \in W$, a self-loop of (5.3.4.1) and (5.3.4.3) is attached to v_q. Similarly, a self-loop of (5.3.4.1) and (5.3.4.2) and a self-loop of (5.3.4.1) and (5.3.4.3) are attached to v'_q and v'_p, respectively. Let the self-loops attached to $v_m \in V$ and $v'_m \in V'$ be e_m and e'_m and let the sets of self-loops attached to G and G' be L and L', respectively. For simplicity, the graphs obtained from G and G' after attaching the self-loops are again represented by G and G', respectively. If $F(y, x) = yx$ in (5.3.3.1), (5.3.3.1) becomes,

$$\sum_{e'_k = \left(v'_i, v'_j\right) \in E'} y_j x'_k = \sum_{e_k = \left(v_i, v_j\right) \in E} y'_j x_k. \tag{5.3.4.20}$$

Note that e_k and e'_k are not the corresponding edges in (5.3.4.2). Next, we apply (5.3.3.20) to G and its transposed graph G'. If $x_k = f_k(y_i) = a_k y_i$ for an edge $e_k(v_i, v_j)$ of G, there is an edge $e'_k \triangleq (v'_j, v'_i)$ with $x'_k = f_k(y'_j) = a'_k y'_j = a_k y'_j$ in G', where e_k is not an attached self-loop. Using the above result, (5.3.4.20) becomes

$$\sum_{e'_k = (v'_k, v'_k) \in L'} y_k x'_k = \sum_{e_k = (v_k, v_k) \in L} y'_k x_k \tag{5.3.4.21}$$

which can be rewritten as

$$\begin{aligned} &\sum_{v'_j \in U' - \{v'_p\}} y'_j x_j + \sum_{v'_k \in W' - \{v'_q\}} y'_k x_k + y'_p x_p + y'_q x_q \\ &= \sum_{v_j \in U - \{v_p\}} y'_j x_j + \sum_{v_k \in W - \{v_q\}} y'_k x_k + y'_p x_p + y'_q x_q \end{aligned} \tag{5.3.4.22}$$

with $x_k = 0$ for $v_k \in W$ and $x'_j = 0$ for $v'_j \in U'$ since W and U' are the sets of sinks of G and G', respectively. Let

$$x_j = 0 \quad \text{for} \quad v_j \in U - \{v_p\}, \tag{5.3.4.23}$$

$$x'_k = 0 \quad \text{for} \quad v'_k \in W' - \{v_q\}', \tag{5.3.4.24}$$

$$x_p \neq 0 \quad \text{and} \quad x'_q \neq 0, \tag{5.3.4.25}$$

$$x_q = 0 \quad \text{and} \quad x'_p = 0, \tag{5.3.4.26}$$

Then, (5.3.4.22) becomes

$$y_q x'_q = y'_p x_p. \tag{5.3.4.27}$$

From (5.3.4.23)–(5.3.4.27), and the transmission definitions (5.3.4.4) and (5.3.4.5),

$$T_{pq} = \frac{y_q}{x_p}, \tag{5.3.4.28}$$

$$T'_{qp} = \frac{y'_p}{x'_q}. \tag{5.3.4.29}$$

Therefore, we obtain

$$T_{pq} = \frac{y_q}{x_p} = \frac{y'_p}{x'_q} = T'_{qp}. \tag{5.3.4.30}$$

□

(2) *Application 2*

In a linear signal flow graph $G = (V, E)$, the following property is known:

$$\frac{\partial T_{so}}{\partial a_{nm}} = T_{sn} T_{mo}, \tag{5.3.4.31}$$

where $f_k(y_i) = a_{ij} y_i$ for $e_k = (v_i, v_j)$. This formula has been applied to system sensitivity analysis and used as a basis of a deduction method employing Tellegen's theorem for signal flow graphs [15,17]. Here, we give a new simple deduction method using an SFG operator,

$$F(y, x) = y \frac{\partial x}{\partial a_{nm}} + x \frac{\partial y}{\partial a_{nm}}, \tag{5.3.4.32}$$

which is the same as (5.3.2.13).

Consider a linear signal flow graph $G = (V, E)$ and its transposed signal flow graph $G' = (V', E')$ of Fig. 5.3.4.3(a) and (b). For simplicity, let G contain one source and one sink and let $v_s, v_o \in V$, and $v'_s, v'_o \in V'$ be the sources and the sinks of G and G', respectively. If self-loops are attached to the vertices of the sources and sinks of G and G', the resulting graphs satisfy Condition A. Let the self-loops of the source and sink of G be e_s, and e_o, respectively, and let the self-loops of the source and sink of G' be e'_o and e'_s, respectively. The graphs obtained from G and G' after attaching the self-loops are again represented by G and G', respectively. If we use (5.3.4.32) as $F(y, x)$ in (5.3.31), (5.3.31) becomes

$$\sum_{e'_k = (v'_i, v'_j) \in E'} \left(y_j \frac{\partial x'_k}{\partial a_{nm}} + x'_k \frac{\partial y_j}{\partial a_{nm}} \right) = \sum_{e_k = (v_i, v_j) \in E} \left(y'_j \frac{\partial x_k}{\partial a_{nm}} + x_k \frac{\partial y'_j}{\partial a_{nm}} \right). \tag{5.3.4.33}$$

Note that e_k and e'_k are not the corresponding edges in (5.3.4.33). Next, we apply (5.3.4.33) to G and its transposed graph G'.

Since

$$\frac{\partial x'_k}{\partial a_{nm}} = 0, \quad \frac{\partial y'_j}{\partial a_{nm}} = 0 \tag{5.3.4.34}$$

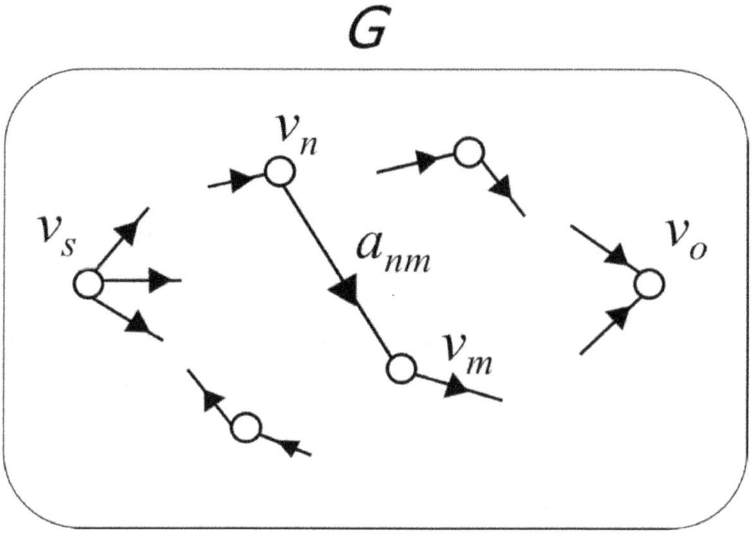

(a)

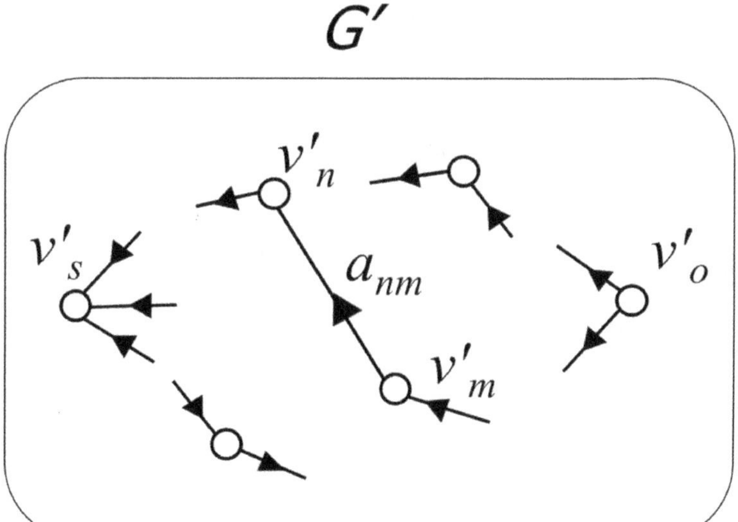

(b)

Fig. 5.3.4.3. A signal flow graph G and its transposed graph G'.

and

$$x_k = a_{ij}y_i, x'_k = a'_{ij}y'_i \tag{5.3.4.35}$$

it follows that

$$\sum_{e'_k = (v'_i, v'_j) \in E'} a'_{ij}y'_i \frac{\partial y_j}{\partial a_{nm}} = \sum_{e_k = (v_i,\, v_j) \in E} y'_j \frac{\partial a_{ij}\, y_i}{\partial a_{nm}} \tag{5.3.4.36}$$

Since G' is the transposed graph of G, if $(v_i, v_j) \in E$, then $(v'_j, v'_i) = E'$ and $a'_{ji} = a_{ij}$. Then, $\frac{\partial a_{ij}}{\partial a_{nm}} = 0$ for $i \neq n$ and $j \neq m$. Hence, (5.3.4.36) becomes

$$\begin{aligned} &a_{nm}y'_m \frac{\partial y_n}{\partial a_{nm}} + x'_o \frac{\partial y_o}{\partial a_{nm}} + x'_s \frac{\partial y_s}{\partial a_{nm}} \\ &= y'_m \frac{\partial a_{nm}\, y_n}{\partial a_{nm}} + y'_o \frac{\partial x_o}{\partial a_{nm}} + y'_s \frac{\partial x_s}{\partial a_{nm}} \end{aligned} \tag{5.3.4.37}$$

Since $x_o = 0$, $x'_s = 0$, and x_s is a constant,

$$\frac{\partial y_o}{\partial a_{nm}} = \frac{y'_m y_n}{x'_o} \tag{5.3.4.38}$$

Hence

$$\frac{\partial (y_o/x_s)}{\partial a_{nm}} = \frac{y_n}{x_s} \cdot \frac{y'_m}{x'_o} \tag{5.3.4.39}$$

Since $y_o/x_s = T_{so}$, $y_n/x_s = T_{sn}$, and from (5.3.4.18), $y'_m/x'_o = T_{mo}$, we obtain

$$\frac{\partial T_{so}}{\partial a_{nm}} = T_{sn}T_{mo}. \tag{5.3.4.40}$$

In the case where G has many sources and sinks, and where v_s and v_o are not the source and sink, (5.3.4.31) can be obtained in a similar manner by using Theorem 5.3.4.2.

The deduction method of (5.3.4.31) is simpler than those used in [15] and [17], because, in the latter, G and its transposed graph G' were used twice, while G and G' were employed only once here.

(3) *Application 3*

Let us apply the following SFG operator to a linear Signal flow graph $G = (V, E)$ and its transposed signal flow graph G' in Application 2 (see Fig. 5.3.4.3):

$$F(y, x) = y(a_{nm}) \frac{x(a_{nm} + \Delta a_{nm}) - x(a_{nm})}{\Delta a_{nm}} + x(a_{nm}) \frac{y(a_{nm} + \Delta a_{nm}) - y(a_{nm})}{\Delta a_{nm}} \tag{5.3.4.41}$$

where y and x are functions of a_{nm} which is the coefficient of the function $f_k(z) = a_{nm}z$ of the edge $e_k = (v_n, v_m)$, and $y(a_1)$ and $x(a_1)$ are the values of y and x in the case where $a_{nm} = a_1$, respectively. This operator is symmetric and bilinear. If we use (5.3.4.41) as $F(y, x)$ in (5.3.31), then (5.3.31) becomes

$$\begin{aligned} &\sum_{e'_k = (v'_i, v'_j) \in E'} \left\{ y_j(a_{nm}) \frac{x'_k(a_{nm} + \Delta a_{nm}) - x'_k(a_{nm})}{\Delta a_{nm}} + x'_k(a_{nm}) \frac{y_j(a_{nm} + \Delta a_{nm}) - y_j(a_{nm})}{\Delta a_{nm}} \right\} \\ &= \sum_{e_k = (v_i, v_j) \in E} \left\{ y'_j(a_{nm}) \frac{x_k(a_{nm} + \Delta a_{nm}) - x_k(a_{nm})}{\Delta a_{nm}} + x_k(a_{nm}) \frac{y'_j(a_{nm} + \Delta a_{nm}) - y'_j(a_{nm})}{\Delta a_{nm}} \right\}. \end{aligned} \tag{5.3.4.42}$$

Note that e_k and e'_k are not the corresponding edges in (5.3.4.42). Next, we apply (5.3.4.42) to G and its transposed graph G'.

Since y'_i and x'_i of G' do not depend on a_{nm},

$$x'_k(a_{nm} + \Delta a_{nm}) = x'_k(a_{nm}), \tag{5.3.4.43}$$

$$y'_j(a_{nm} + \Delta a_{nm}) = y'_j(a_{nm}). \tag{5.3.4.44}$$

Therefore,

$$\sum_{e'_k = \left(v'_i, v'_j\right) \in E'} x'_k(a_{nm}) \frac{y_j(a_{nm} + \Delta a_{nm}) - y_j(a_{nm})}{\Delta a_{nm}}$$

$$= \sum_{e_k = \left(v_i, v_j\right) \in E} y'_j(a_{nm}) \frac{x_k(a_{nm} + \Delta a_{nm}) - x_k(a_{nm})}{\Delta a_{nm}}. \quad (5.3.4.45)$$

Since

$$x_k(a_{nm}) = a_{ij} y_i(a_{nm}), \; x'_k(a_{nm}) = a'_{ij} y'_i(a_{nm}), \quad (5.3.4.46)$$

$$x_k(a_{nm} + \Delta a_{nm}) = a_{ij} y_j(a_{nm} + \Delta a_{nm}), \quad i \neq n; \quad j \neq m, \quad (5.3.4.47)$$

and since $(v'_i, v'_j) = E'$ for $(v_j, v_i) \in E$ and $a'_{ji} = a_{ij}$, (5.3.4.45) becomes

$$\begin{aligned} a_{nm} y'_m(a_{nm}) & \frac{y_n(a_{nm} + \Delta a_{nm}) - y_n(a_{nm})}{\Delta a_{nm}} \\ & + x'_o(a_{nm}) \frac{y_o(a_{nm} + \Delta a_{nm}) - y_o(a_{nm})}{\Delta a_{nm}} \\ & + x'_s(a_{nm}) \frac{y_s(a_{nm} + \Delta a_{nm}) - y_s(a_{nm})}{\Delta a_{nm}} \\ & = y'_m(a_{nm}) \frac{x_p(a_{nm} + \Delta a_{nm}) - x_p(a_{nm})}{\Delta a_{nm}} \\ & + y'_o(a_{nm}) \frac{x_o(a_{nm} + \Delta a_{nm}) - x_o(a_{nm})}{\Delta a_{nm}} \\ & + y'_s(a_{nm}) \frac{x_s(a_{nm} + \Delta a_{nm}) - x_s(a_{nm})}{\Delta a_{nm}}, \end{aligned} \quad (5.3.4.48)$$

where $e_p = (v_n, v_m)$. Since

$$x'_s(\cdot) = 0, \quad (5.3.4.49)$$

$$x_o(a_{nm} + \Delta a_{nm}) = x_o(a_{nm}), \quad (5.3.4.50)$$

$$x_s(a_{nm} + \Delta a_{nm}) = x_s(a_{nm}), \quad (5.3.4.51)$$

$$x_p(a_{nm}) = a_{nm} y_n(a_{nm}) \quad (5.3.4.52)$$

and

$$x_p(a_{nm} + \Delta a_{nm}) = (a_{nm} + \Delta a_{nm}) y_n(a_{nm} + \Delta a_{nm}) \quad (5.3.4.53)$$

Equation (5.3.4.48) becomes

$$x_o'(a_{nm}) \frac{y_o(a_{nm} + \Delta a_{nm}) - y_o(a_{nm})}{\Delta a_{nm}} = y_m'(a_{nm})\, y_n(a_{nm} + \Delta a_{nm}) \quad (5.3.4.54)$$

Hence,

$$\begin{aligned} &\frac{\{y_o(a_{nm} + \Delta a_{nm}) - y_o(a_{nm})\}/x_s(a_{nm})}{\Delta a_{nm}} \\ &= \frac{y_m'(a_{nm})}{x_o'(a_{nm})} \cdot \frac{y_n(a_{nm} + \Delta a_{nm})}{x_s(a_{nm})} \\ &= \frac{y_m'(a_{nm})}{x_o'(a_{nm})} \cdot \left\{ \frac{y_n(a_{nm})}{x_s(a_{nm})} + \frac{y_n(a_{nm} + \Delta a_{nm}) - y_n(a_{nm})}{x_s(a_{nm})} \right\}, \end{aligned} \quad (5.3.4.55)$$

Let

$$\Delta T_{sj} = \frac{y_j(a_{nm} + \Delta a_{nm}) - y_j(a_{nm})}{x_s(a_{nm})} \quad (5.3.4.56)$$

Since from Theorem 5.3.4.2

$$\frac{y_m'(a_{nm})}{x_o'(a_{nm})} = T_{mo}, \quad (5.3.4.57)$$

Equation (5.3.4.55) becomes

$$\frac{\Delta T_{so}}{\Delta a_{nm}} = T_{mo} T_{sn} + \Delta T_{sn}. \quad (5.3.4.58)$$

Likewise,

$$\frac{\Delta T_{sn}}{\Delta a_{nm}} = T_{mn}\{T_{sn} + \Delta T_{sn}\}. \quad (5.3.4.59)$$

Note that the value of T_{sq} remains unaltered whether or not v_q is a sink. From (5.3.4.58) and (5.3.4.59),

$$\frac{\Delta T_{so}}{\Delta T_{sn}} = \frac{T_{mo}}{T_{mn}}. \quad (5.3.4.60)$$

This equation shows that the ratio of the transmission changes ΔT_{so} and ΔT_{sn} by a large change Δa_{nm} of a_{nm} can be expressed in terms of the transmission of G. Equations (5.3.4.58)–(5.3.4.60) are new expressions for the transmission changes. Using (5.3.4.58) and (5.3.4.60)

$$\Delta T_{so} = \frac{T_{sn} T_{mo} \Delta a_{nm}}{1 - T_{mn} \Delta a_{nm}}. \tag{5.3.4.61}$$

Equation (5.3.4.61) coincides with that given by Crochiere [18]. The above deduction method is different from that given in [17] or [18] in that Crochiere's method uses the Taylor Series expansion of ΔT_{so}. Clearly, the method of using an SFG operator is simpler and employs only elementary calculus.

5.3.5. *Summary of the SFG operators*

In this section, we generalized Tellegen's theorem for signal flow graphs using the concept of an SFG operator. We also presented some applications of the SFG operator. This result can be applied to the signal flow network which in turn is a generalization of the signal flow graph. The SFG operator has some resemblance to the operators [1–3] for electrical networks governed by Kirchhoff's laws. For example, the operator $F_8(y, x) = y(\partial x/\partial t_1) + x(\partial y/\partial t_1)$ is an SFG operator and it is also applicable to the edge current i and the edge voltage u of an electrical network. On the other hand, the operator $F_7(y, x) = y(\partial x/\partial t_1) - x(\partial y/\partial t_1)$ which is applicable to electrical networks, is not applicable to signal flow graphs as can be seen from Theorems 5.3.2.1–5.3.3.2. Like the generalized forms of Tellegen's theorem, which have been applied to many electrical engineering problems, an SFG operator is also applicable to a wide range of problems in signal flow graphs and signal flow networks. In this section, we discussed only the applications of the SFG operators (5.3.2.6), (5.3.2.13), and (5.3.4.41). We believe that the applications of the operators (5.3.2.9)–(5.3.2.11), (5.3.2.13) and their variants to digital filters and their related areas can lead to fruitful and useful results.

- **Summary:** Graphical consideration of electrical networks analysis including impedance elements, admittance elements and controlled sources are described. In this case, hybrid trees are useful and algebraic considerations of hybrid trees and hybrid k-trees are described. Signal flow graphs are introduced for digital networks and, sensitivity analysis, Tellegen's theorem and operators in signal flow graphs are described.

- **Interesting challenges:** (1) Consider an electrical network synthesis including impedance elements, admittance elements and controlled sources using graphs.

(2) Consider an digital network synthesis using signal flow graphs.

Bibliography

[1] L. O. Chua, C. A. Desoer and E. S. Kuh, *Linear and Nonlinear Circuits*, McGraw-Hill, 1987.
[2] S. Seshu and M. B. Reed, *Linear Graphs and Electrical Networks*, Addison-Wesley, 1961.
[3] W. K. Chen, *Applied Graph Theory*, North-Holland, Amsterdam, The Netherlands, 1976.
[4] W. Mayeda, *Graph Theory*, Wiley, 1972.
[5] N. Deo, *Graph Theory with Applications to Engineering and Computer Science*, Prentice-Hall, 1974.
[6] M. N. S. Swamy and K. Thulasiraman, *Graphs: Theory and Algorithms*, Wiley, 1992.
[7] S. L. Hakimi and D. G. Green, "Generation and realization of trees and trees," *IEEE Trans. Circuit Theory*, Vol-CT-11, pp. 247–255, 1964.
[8] W. K. Chen and S. K. Mark, "On the algebraic relationships of trees, co-trees," *IEEE Trans. Circuit Theory*, Vol-CT-16, pp. 176–184, 1969.
[9] M. Iri, I. Shirakawa, Y. Kajitani, S. Shinoda, H. Ariyoshi, K. Onaga, T. Ozawa, T. Ohtsuki, T. Nishi, N. Tomizawa, M. Sengoku, R. Ishii, T. Kashiwabara, T. Nishizeki, S. Fujishige and T. Asano, *Graph Theory with Exercise*, Corona Pub. 1983.
[10] Y. Kajitani, *Graph Theory for Circuits*, Shokodo Pub., 1979.
[11] M. Sengoku, T. Kurobe and Y. Ogawa, "On the realization of a set of hybrid trees," *IEICE*, vol. 55-A, no. 2, pp. 60–66, 1972.
[12] M. Sengoku, "On hybrid tree graphs," *IEICE*, vol. 57-A, no. 5, pp. 353–358, 1974.
[13] M. Sengoku, "Hybrid trees and hybrid tree graphs," *IEEE Trans. Circuits Syste.*, vol. CAS-22, no. 10, pp. 786–790, 1975.
[14] S. Seviora and M. Sablatash, A Tellegen's theorem for digital filters," *IEEE Trans. Circuit Theory*, vol. CT-18, pp. 201–203, Jan. 1971.
[15] A. Fettweis, A general theorem for signal-flow networks, with applications," *Arch. Electron. Ubertragung.*, vol. 25, pp. 557–561, 1971.
[16] S. J. Mason and H. J. Zimmermann, *Electronic Circuits, Signals and Systems*, Wiley, New York, 1960.

[17] A. V. Oppenheim and R. W. Schafer, *Digital Signal Processing*, Prentice-Hall, 1975.
[18] R. Crochiere, "Some network properties of digital filters," *Proc. Int. Symp. Circuit Theory*, pp. 146–148, Apr. 9–11, 1973.
[19] M. Sengoku and W. K. Chen, "An operator in signal flow graphs and its applications," *IEEE Trans. Circuits Syst.*, vol. 35, no. 11, pp. 1383–1392, 1988.

Chapter 6

Mobile Communication Systems (1)

6.1. Introduction for Mobile Communication Systems

Mobile communication is a general term for *telecommunications* in which one or both terminals can move in a broad sense whose terminal is not connected to a communication line and not a fixed radio station. Mobile communications can be used while moving on foot, on a car or on a ship, etc. They are also usually used in *disaster areas* including *satellite internet* for *securing communication function*.

Figure 6.1.1 shows a classification of mobile communications.

(1) *Underground place*

   (a) Underground shopping mall, subway, mobile phones, etc.

(2) *Ground place*

   (a) Cars, trains, ships, dogs, mobile phones, etc.

(3) *Unmanned aircraft*

   (a) Drones, unmanned helicopters, etc.

(4) *Aircraft*

   (a) Airplanes, etc.

(5) HAPS: *High-Altitude Pseudo Satellite*

   (a) A solar-powered high-altitude platform-station

(6) NGSO: *Non-geostationary Orbit Satellite*

(7) GEO: *Geostationary Orbit Satellite*

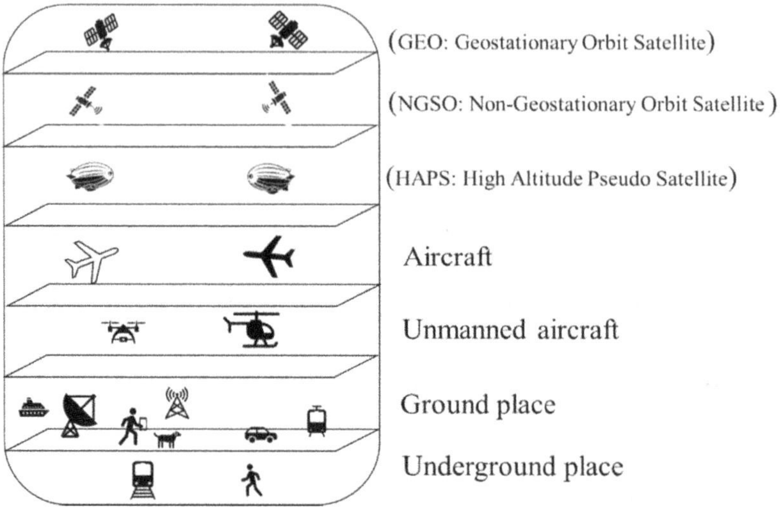

Fig. 6.1.1. A classification of mobile communications.

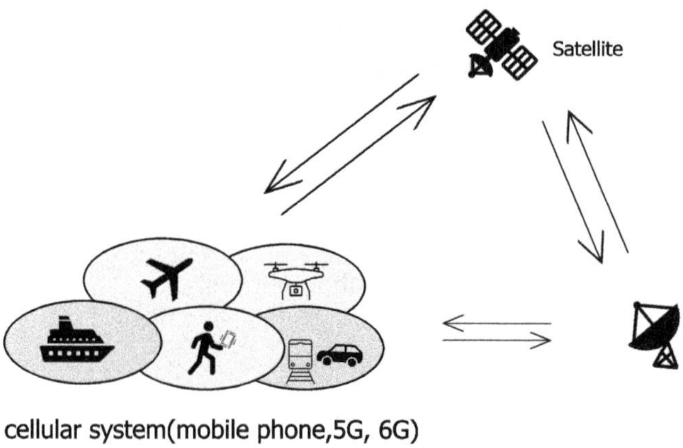

Fig. 6.1.2. A mobile communication system.

Figure 6.1.2 shows an example of mobile communication systems. In this example, a *cellular system* which is described later, is used as the communication part of the moving terminals. In Fig. 6.1.2, the cellular systems, the satellites and its earth base stations are connected by *wired communication* and *wireless communication*, where the wired communication means the communication using coaxial cable and optical fiber cable, etc. and

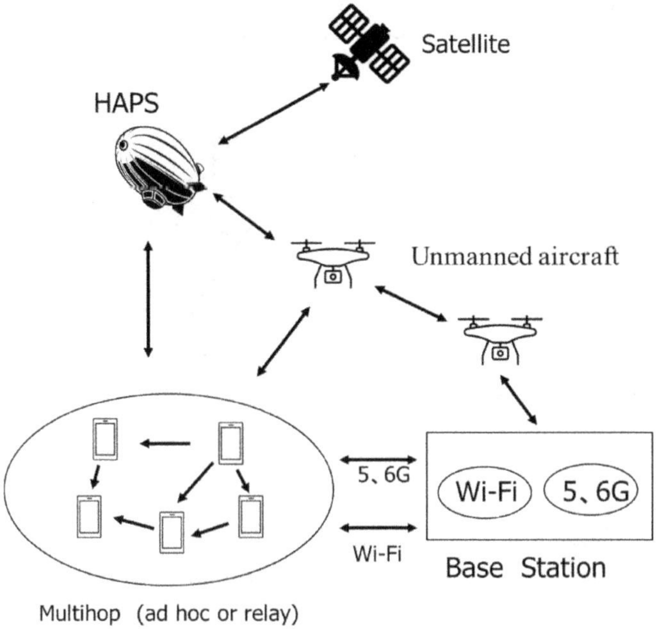

Fig. 6.1.3. A communication system in natural disasters.

the wireless communication means the communication using *radio wave*. Figure 6.1.3 shows an example of providing communication when natural disasters occur such as an earthquake. If the communication form mobile phones is lost in the disasters, an *mobile ad hoc network* (or *multihop network*) is configured among the *mobile phones*, and the network is connected wirelessly (by Wi-Fi, 5G, 6G, etc.) to a base station that is alive in the disaster. The base station is connected to unmanned aircraft, HAPS and satellite to secure communication. These communication systems have developed into a system that connects everything called IoT (*Internet of Things*) or IoE (*Internet of Everything*), and they have become the foundation that supports CPS (*Cyber-Physical System*), *Metaverse*, and so on.

Figure 6.1.4 shows a diagram explaining a *cellular system* of mobile communication. A cellular system or a cellular mobile network is a communication network where the link to and from end nodes (terminals) is wireless. The network is distributed over land areas called *cells* (or *zones*), each served by at least one fixed-location transceiver (called *base station*). These base stations provide the cell with the network coverage which can

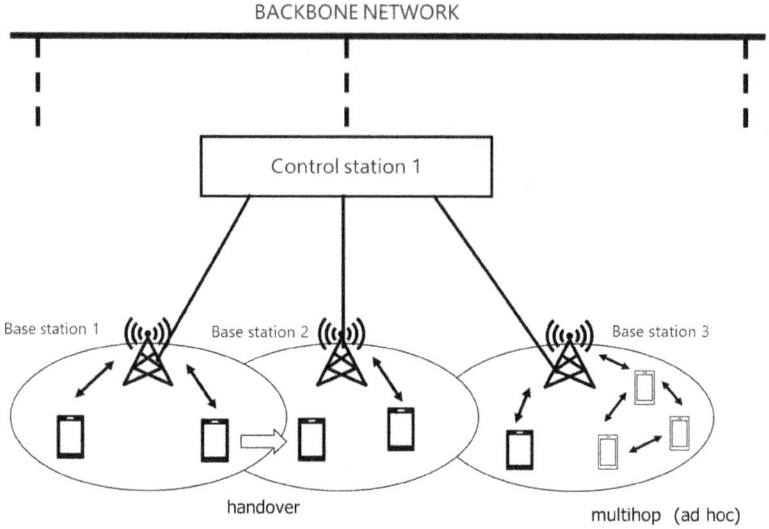

Fig. 6.1.4. A cellular communication system.

be used for transmission of voice, data, and other types of content. A cell typically uses a different set of *channels* (see the definition in Section 6.2) from neighboring cells, to avoid interference and provide guaranteed service quality within each cell. The case of moving between cells during a communication is called *handover* or *handoff*.

In Fig. 6.1.4, a cell is a land area to be supplied with radio service, where the total service area is divided into cells in a pattern dependent on the terrain and reception characteristics on a two-dimensional plane which are called *two-dimensional cells*. Figure 6.1.5 shows a cellular communication system with *multidimensional cells*. It has long been used in antennas that can actively control directivity, such as active-phased array antennas. The directivity is changed by controlling the phase of the high frequency that feeds a large number of small antennas. Using active-phased array antennas and MIMO (*Multi Input Multi Output*) systems, RIS (*Reconfigurable Intelligent Surface*) etc., the multidimensional cells can be constructed in which the upward directivity of the antenna in the figure (Fig. 6.1.5) is used for *three-dimensional cells*. Figure 6.1.6 shows cells on a road, a high way or a rail way which are considered as *one-dimensional* cells from a macro-perspective.

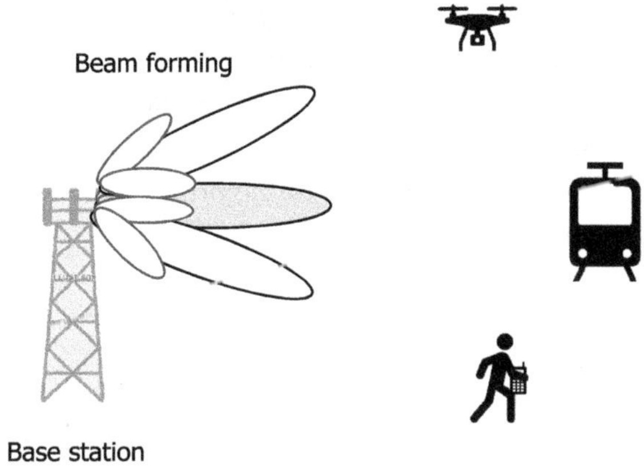

Fig. 6.1.5. A cellular communication system with multidimensional cells.

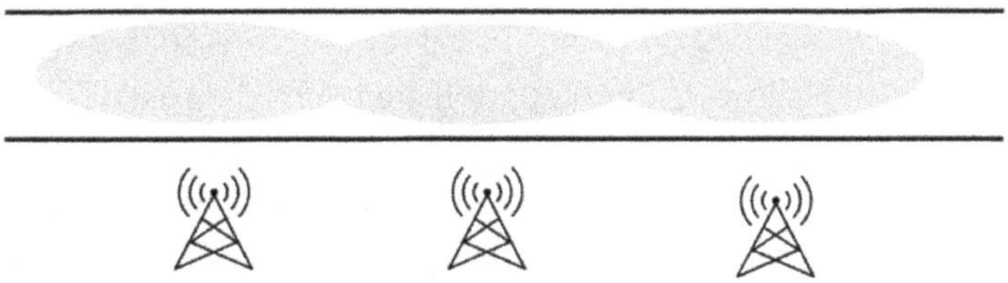

Fig. 6.1.6. A cellular communication system with one-dimensional cells.

In Fig. 6.1.4, a cellular communication system consists of a part of cells including a base station, a part of connecting between a base station and a *control station*, and *backbone networks*. This means that mobile communication systems include both a part of wired communication and a part of wireless communication. In the mobile communication systems, analog circuits and digital circuits are used in a part of hardware. The graph theoretical application to the hardware is discussed in Chapter 5. In this chapter, the location of stations, and in the next chapter, the assignment of channels etc. are discussed from the point of view of graphs.

6.2. Location Problems on Graphs and Networks

Location Problems on Graphs and Networks involve the decision problem of determining the location of one or more facilities, e.g., communication stations including mobile communication systems, to provide service to several clients located at known vertices or edges of graphs, or vertices, edges or points on edges of networks. To provide service to a client, a facility must be located "close enough" to the client. In a more general version of the problems, the purpose of the problems may be to minimize the total cost required in the system.

6.2.1. *Minimum tree problem*

Suppose we have several towns in an area, and we want to build roads or railroads that allow us to travel between all the towns. When designing the roads or railroads, it is assumed that the place where the road or railroad branches is not in the middle of the road or railroad but in a town, considering the occasional maintenance. To minimize the cost of building roads or railroads, let us find a method to build roads or railroads so that we can travel between towns by minimizing the total distance of the roads or railroads.

Let $G = (V, E)$ be a connected graph with a real number weight associated with each edge of G. The weight $l(e_i)$ of an edge $e_i \in E$ means the length of e_i. Let every vertex and every edge correspond to a town and a roads or railroad. Then the above problem comes down to the problem of finding a connected subgraph with the minimum sum of the weights of all edges in the subgraph. If the subgraph includes a tieset, it is not minimum, as we can travel between any towns even if we remove an edge to eliminate the tieset. That is, the subgraph is a tree. The problem is finding a tree with the minimum sum of the weights of the tree, which is called *minimum tree problem, minimum spanning tree problem* or *shortest spanning tree problem.*

The *distance* $d(t_i, t_j)$ between two trees t_i and t_j of G is defined as follows:

$$d(t_i, t_j) = \frac{1}{2}\left|t_i \oplus t_j\right|. \tag{6.2.1.1}$$

Since $|t_i| = |t_j| = \rho(G)$ and $|t_i \oplus t_j|$ is an even number (including 0), $d(t_i, t_j)$ is non-negative integer.

Let us consider a fundamental tieset induced by some edges of a tree t_i and an edge of the cotree and a new tree t_j is generated through addition of an edge of the cotree and deletion of an appropriate edge of t_i in the fundamental tieset. This generation of one tree from another is called an *elementary tree transformation*. Similarly, there is an elementary tree transformation using a fundamental cutset. Clearly, the distance $d(t_i, t_j)$ in which t_j is generated by an elementary tree transformation from t_i is one. Obviously, $d(t_i, t_j) = 0$ if and only if $t_i = t_j$. If $d(t_i, t_j) = 1$, then t_i and t_j are called *neighboring trees*.

It is not difficult to see that the number $d(t_i, t_j)$ is the minimum number of elementary tree transformations involved in going from t_i to t_j. The reader is encouraged to prove the following theorem.

Theorem 6.2.1.1. *The distance between the trees of a graph is a metric. That is, it satisfies*

(i) $d(t_i, t_j) \geq 0$, *and* $d(t_i, t_j) = 0$ *if and only if* $t_i = t_j$;
(ii) $d(t_i, t_j) = d(t_j, t_i)$;
(iii) $d(t_i, t_j) \leq d(t_i, t_k) + d(t_k, t_j)$.

The tree t of G is *cut minimal* if the weights of the edges of the fundamental cutset determined by the edge e_i of t are greater than or equal to $l(e_i)$ for every edge e_i of t. Similarly, the t of G is *tie minimal* if the weights of the edges of the fundamental tieset determined by the edge e_j of $c = E - t$ are greater than or equal to $l(e_j)$ for every edge e_j of c. The tree t of G is *locally minimal* if the sum of the weights of the tree t is less than or equal to the sum of the weights of the tree t' for every neighboring tree t' of t.

Lemma 6.2.1.2. *For a tree t, the following three propositions are equivalent*:

(1) *t is cut minimal*;
(2) *t is tie minimal*;
(3) *t is locally minimal.*

Proof. Let us prove (1) to (2), (2) to (3) and (3) to (1). Sometimes this is abbreviated as (1)⇒(2), (2)⇒(3) and (3)⇒(1).

(1)⇒(2): Let us assume that t is cut minimal. This means that the weights of the edges of the fundamental cutset h_i determined by the edge e_i of t are greater than or equal to $l(e_i)$ for every edge e_i of t. Let us consider a fundamental tieset f_j of G corresponding to the edge e_j of $c = E - t$, where in the edges of the fundamental tieset, the edges other than e_j are the edges of t. Every such edge e_i defines a fundamental cutset of t, which also contains e_j from Theorem 3.6. Since t is cut minimal, $l(e_i) \leq l(e_j)$. That is, t is tie minimal.

(2)⇒(3): Let us assume that t is circuit minimal and let t' be a neighboring tree of t. t' has exactly one edge e_j which is not in t, that is, the edge e_j is in $c = E - t$. Let us consider a fundamental tieset f_j defined by e_j. In the edges of the fundamental tieset f_j, the edges other than e_j are the edges of t. Let us choose an edge e_i in f_j that is not in t'. Then, e_i is an edge of t which is not in t'. Now, let us remove e_i out of t and add e_j to t. The result has to be t'. Because of tieset minimality, $l(e_j) \geq l(e_i)$, that is, the sum of the weights of the tree t is less than or equal to the sum of the weights of the neighboring tree t'. Thus, t is locally minimal.

(3)⇒(1): Let us assume that t is locally minimal. Let e_i be an arbitrary edge of t. Let us consider a fundamental cutset h_i of G corresponding to the edge e_i of t and let e_j be an edge of h_i where $e_j \neq e_i$. Then, e_i belongs to the fundamental tieset of t defined by e_j from Theorem 3.6. By removing the branch e_i from t and adding the edge e_j to t, we get the neighboring tree t' of t. Because of local minimality, the sum of the weights of the tree t is less than or equal to the sum of the weights of the neighboring tree t'. This means that $l(e_j) \geq l(e_i)$. Thus, t is cut minimal. □

Using Lemma 6.2.2, we get the following theorem.

Theorem 6.2.1.2. *If the sum of the weights of the tree t is less than or equal to the sum of the weights of the tree t' for every neighboring tree t' of t, that is, if the tree t of G is locally minimal, then the sum of the weights of the tree t is minimum in G.*

Proof. Since a minimum tree is a minimal tree, the tree is obviously locally minimal. Thus, from Lemma 6.2.1.2, it suffices to prove that a cut minimal tree is also minimum in G. We will prove by contradiction. Suppose that there is a cut minimal tree t which is not minimum. Let us consider the minimum tree t' and choose t and t' so that the distance $d(t, t')$ is as small as possible.

By Lemma 6.2.1.2 (3), $d(t_i, t_j) > 1$. t has an edge e_i which is not in t', that is, it is an edge of $c = E - t$. Let us consider the fundamental cutset h_i of t defined by e_i and the fundamental tieset f_i' of t' defined by e_i. Since e_i is included in the intersection $h_i \cap f_i'$, there are also other edges besides e_i from Theorem 2.7.1. Let us choose such an edge e_j. Then, e_j is an edge of $c = E - t$ and an edge of t'. Since t is cut minimal, $l(e_j) \geq l(e_i)$. Since t' is tieset minimal, $l(e_j) \leq l(e_i)$. Therefore, $l(e_j) = l(e_i)$. By removing e_j from t' and adding e_i to t', we get a minimum tree t'' which has the same weight as t'. However, $d(t, t'') < d(t, t')$ which is a contradiction. □

Algorithm for minimum tree: There are several methods available for actually finding a minimum tree in a given graph.

(1) Kruskal's algorithm [8, 5]: List all edges of the graph $G = (V, E)$ in order of nondecreasing weight. Next, select a smallest edge of G. Then for each successive step select (from all remaining edges of G) another smallest edge that makes no tieset with the previously selected edges. Continue until $\rho(G)$ edges have been selected, and these edges t (which is a tree) will constitute the desired minimum tree. The validity of the method follows from Theorem 6.2.1.2.

(1)′ Dual of Kruskal's algorithm: List all edges of the graph G in order of nonincreasing weight. Next, select a largest edge of G. Then for each successive step select (from all remaining edges of G) another largest edge that makes no cutset with the previously selected edges. Continue until $\mu(G)$ edges have been selected and let these edges be c. $t = E - c$ will constitute the desired minimum tree.

(2) Prim's algorithm [9, 5]: This algorithm does not require listing all edges in order of nondecreasing weight in the graph $G = (V, E)$. Select an appropriate vertex $v_1 \in V$ and let V_s be the set of vertices of V consisting of v_1. Select the edge $e_1 = (v_1, v_2)$ with the smallest weight from the edges incident to v_1, and add v_2 to V_s. Select the edge e_2 with the smallest weight from the edges incident to vertices other than V_s from V_s, where $e_1 = (v_i, v_j)$ and $v_i \in V_s, v_j \notin V_s$. Add v_j to V_s. Repeat this until all vertices join V_s. The set selecting edges is the desired minimum tree.

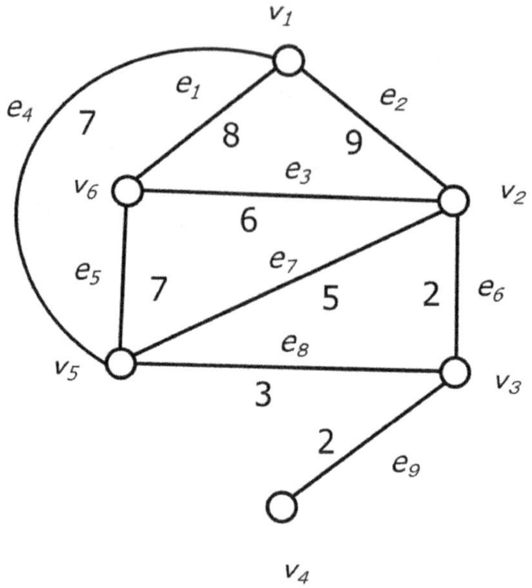

Fig. 6.2.1.1. A graph $G = (V, E)$ with edge weight of length.

Example. In Fig. 6.2.1.1, $G = (V, E)$, where $V = \{v_1, v_2, v_3, v_4, v_5, v_6\}$, $E = \{e_1, e_2, e_3, e_4, e_5, e_6, e_7, e_8, e_9\}$ and $e_1 = (v_1, v_6)$, $e_2 = (v_1, v_2)$, $e_3 = (v_2, v_6)$, $e_4 = (v_1, v_5)$, $e_5 = (v_5, v_6)$, $e_6 = (v_2, v_3)$, $e_7 = (v_2, v_5), e_8 = (v_3, v_5), e_9 = (v_3, v_4)$. Let $l(e_i)$ be the weight of an edge $e_i \in E$. $l(e_1) = 8, l(e_2) = 9, l(e_3) = 6, l(e_4) = 7, l(e_5) = 7, l(e_6) = 2, l(e_7) = 5, l(e_8) = 3, l(e_9) = 2$.

(1) **Kruskal's algorithm:**

(i) List all edges of the graph $G = (V, E)$ in order of nondecreasing weight.

$$(l(e_6) = 2, l(e_9) = 2, l(e_8) = 3, l(e_7) = 5, l(e_3) = 6, \ l(e_4) = 7,$$
$$l(e_5) = 7, \ l(e_1) = 8, l(e_2) = 9).$$

(ii) Select a smallest edge of G. Then for each successive step select (from all remaining edges of G) another smallest edge that makes no tieset with the previously selected edges.

Select e_6,e_9 and e_8. Next, if we select e_7, $\{e_6, e_8, e_7\}$ is a tieset and so we do not select e_7. Select e_3. Next, if we select e_4 and e_5, $\{e_6, e_8, e_3, e_5\}$ is a tieset and so we select e_4 and do not select e_5.

(iii) Continue until $\rho(G) = 5$ edges have been selected.
Thus, since the number of selecting edges $\{e_6, e_9, e_8, e_3, e_4\}$ is $5 = \rho(G)$, $\{e_6, e_9, e_8, e_3, e_4\}$ is a minimum tree and the sum of the weights of the tree is 20 ($= 2 + 2 + 3 + 6 + 7$).

(1)′ Dual of Kruskal's algorithm:

(i) List all edges of the graph G in order of nonincreasing weight.

$$(l(e_2) = 9, l(e_1) = 8, l(e_4) = 7,\ l(e_5) = 7,\ l(e_3) = 6,\ l(e_7) = 5,\ l(e_8) = 3, l(e_9) = 2,\ l(e_6) = 2).$$

(ii) Select a largest edge of G. Then for each successive step select (from all remaining edges of G) another largest edge that makes no cutset with the previously selected edges.
Select e_2 and e_1. Next, if we select e_4 and e_5, $\{e_2, e_1, e_4\}$ is a cutset and so we select e_5 and do not select e_4. Next, if we select e_3, $\{e_1, e_5, e_3\}$ is a cutset and so we do not select e_3. Select e_7.

(iii) Continue until $\mu(G) = 4$ edges have been selected and let these edges be c. $t = E - c$ will constitute the desired minimum tree.

Thus, since the number of selecting edges $c = \{e_2, e_1, e_5, e_7\}$ is $4 = \mu(G), t = E - c = \{e_4, e_3, e_8, e_6, e_9\}$ is a minimum tree and the sum of the weights of the tree is 20 ($= 7 + 6 + 3 + 2 + 2$).

(2) Prim's algorithm:

(i) Select an appropriate vertex $v_1 \in V$ and let V_s be the set of vertices of V consisting of v_1. Select the edge $e_1 = (v_1, v_2)$ with the smallest weight from the edges incident to v_1, and add v_2 to V_s.
Let V_s be $\{v_1\}$ and select the edge $e_4 = (v_1, v_5)$ with the smallest weight from the edges incident to v_1, and add v_2 to V_s. Then $V_s = \{v_1, v_5\}$.

(ii) Select the edge e_2 with the smallest weight from the edges incident to vertices other than V_s from V_s, where $e_1 = (v_i, v_j)$ and $v_i \in V_s, v_j \notin V_s$. Add v_j to V_s. Repeat this until all vertices join V_s.

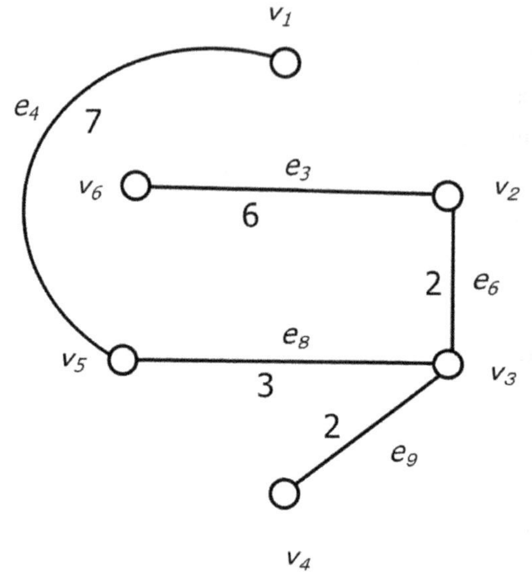

Fig. 6.2.1.2. A minimum tree of the graph $G = (V, E)$ of Fig. 6.2.1.1.

Select the edge $e_8 = (v_5, v_3)$ with the smallest weight from the edges incident to $V_s = \{v_1, v_5\}$, and add v_3 to V_s. Then $V_s = \{v_1, v_5, v_3\}$. Select the edges $e_6 = (v_2, v_3)$ and $e_9 = (v_3, v_4)$ with the smallest weight from the edges incident to $V_s = \{v_1, v_5, v_3\}$, and add v_2 and v_4 to V_s. Then, $V_s = \{v_1, v_5, v_3, v_2, v_4\}$. Select the edge $e_3 = (v_2, v_6)$ with the smallest weight from the edges incident to $V_s = \{v_1, v_5, v_3, v_2, v_4\}$, and add v_6 to V_s. Then, $V_s = \{v_1, v_5, v_3, v_2, v_4, v_6\} = V$.

(iii) The set selecting edges is a minimum tree $t = \{e_4, e_8, e_6, e_9, e_3\}$ and the sum of the weights of the tree is 20 $(= 7 + 3 + 2 + 2 + 6)$.

Figure 6.2.1.2 shows a minimum tree of the graph $G = (V, E)$ of Fig. 6.2.1.1. **(End of Example)**

The above algorithm makes the optimal choice at each step as it attempts to find the overall optimal way to solve the entire problem. Such algorithms are sometimes called *Greedy Algorithms*. However, note that in many problems, a greedy strategy does not produce an optimal solution. Therefore, if greedy algorithms are developed in many problems, they are very effective, and so the research to develop such an algorithm in various fields will be attractive.

6.2.2. *Maximum tree problem*

Let us consider the following two problems (A) and (B) where to locate files in a network including mobile communication networks.

(A) In a network including mobile communication networks, it is assume that the data is received from one server at a certain speed via the same route. For example, when we watch a video from a computer at home by streaming, the data is actually received from one server that distributes the video via basically the same route. Let $G = (V, E)$ be a connected graph with a real number weight associated with each edge of G. The weight $s(e_i)$ of an edge $e_i \in E$ means the maximum data speed of e_i. From the assumption that the routes of data are the same, the routes are tree-shaped, and finding the fastest network is a problem of finding the *maximum tree*. *Maximum tree problem* is finding a tree with the maximum sum of the weights of the tree.

Algorithm for maximum tree:

There are several methods available for actually finding a maximum tree in a given graph, which is a method similar to an algorithm for minimum tree.

(1) (Kruskal's algorithm): List all edges of the graph $G = (V, E)$ in order of nonincreasing weight. Next, select a largest edge of G. Then for each successive step select (from all remaining edges of G) another largest edge that makes no tieset with the previously selected edges. Continue until $\rho(G)$ edges have been selected, and these edges t (which is a tree) will constitute the desired maximum tree. The validity of the method follows from similar to Theorem 6.2.2.

(1)′ (Dual of Kruskal's algorithm): List all edges of the graph G in order of nondecreasing weight. Next, select a smallest edge of G. Then for each successive step select (from all remaining edges of G) another smallest edge that makes no cutset with the previously selected edges. Continue until $\mu(G)$ edges have been selected and let these edges be c. $t = E - c$ will constitute the desired maximum tree.

(2) (Prim's algorithm) [9, 5]: This algorithm does not require listing all edges in order of nonincreasing weight in the graph $G = (V, E)$. Select an appropriate vertex $v_1 \in V$ and let V_s be the set of vertices

of V consisting of v_1. Select the edge $e_1 = (v_1, v_2)$ with the largest weight from the edges incident to v_1, and add v_2 to V_s. Select the edge e_2 with the largest weight from the edges incident to vertices other than V_s from V_s, where $e_1 = (v_i, v_j)$ and $v_i \in V_s, v_j \notin V_s$. Add v_j to V_s. Repeat this until all vertices join V_s. The set selecting edges is the desired maximum tree.

Example. In Fig. 6.2.2.1, $G = (V, E)$, where $V = \{v_1, v_2, v_3, v_4, v_5, v_6, v_7, v_8, v_9, v_{10}\}$, $E = \{e_1, e_2, e_3, e_4, e_5, e_6, e_7, e_8, e_9, e_{10}, e_{11}, e_{12}, e_{13}, e_{14}, e_{15}\}$ and $e_1 = (v_1, v_2), e_2 = (v_1, v_3), e_3 = (v_1, v_7)$, $e_4 = (v_1, v_6)$, $e_5 = (v_3, v_7)$, $e_6 = (v_3, v_4)$, $e_7 = (v_3, v_6)$, $e_8 = (v_6, v_7)$, $e_9 = (v_4, v_6)$, $e_{10} = (v_4, v_8), e_{11} = (v_8, v_9), e_{12} = (v_4, v_9), e_{13} = (v_4, v_5), e_{14} = (v_5, v_6)$, $e_{15} = (v_5, v_{10})$. Let $s(e_i)$ be the weight of an edge $e_i \in E$. $s(e_1) = 7$, $s(e_2) = 7$, $s(e_3) = 8$, $s(e_4) = 7$, $s(e_5) = 7$, $s(e_6) = 5$, $s(e_7) = 6$, $s(e_8) = 6$, $s(e_9) = 4$, $s(e_{10}) = 5$, $s(e_{11}) = 5$, $s(e_{12}) = 9$, $s(e_{13}) = 1$, $s(e_{14}) = 3$, $s(e_{15}) = 6$.

(Kruskal's algorithm):

1. List all edges of the graph $G = (V, E)$ in order of nonincreasing weight. $(s(e_{12}) = 9, s(e_3) = 8, s(e_1) = 7, s(e_2) = 7, s(e_4) = 7, s(e_5) = 7, s(e_7) = 6, s(e_8) = 6, s(e_{11}) = 6, s(e_{15}) = 6, s(e_6) = 5, s(e_{10}) = 5, s(e_9) = 4, s(e_{14}) = 3, s(e_{13}) = 1)$.
2. Select a largest edge of G. Then for each successive step select (from all remaining edges of G) another smallest edge that makes no tieset with the previously selected edges.
   Select e_{12} and select e_3. Next, if we select e_1, e_2, e_4 and e_5, $\{e_2, e_3, e_5\}$ is a tieset and so we do not select e_5. Next, if we select e_7, e_8, e_{11} and e_{15}, $\{e_2, e_4, e_7\}$ and $\{e_3, e_4, e_8\}$ are tiesets and so we do not select e_7 and e_8. If select e_6 and e_{10}. $\{e_{10}, e_{11}, e_{12}\}$ is a tieset and so we do not select e_{10}. Next select e_{14}.
3. Continue until $\rho(G) = 5$ edges have been selected.
   Thus, since the number of selecting edges $\{e_1, e_2, e_3, e_4, e_6, e_{11}, e_{12}, e_{14}, e_{15}\}$ is $9 = \rho(G)$, $\{e_1, e_2, e_3, e_4, e_6, e_{11}, e_{12}, e_{14}, e_{15}\}$ is a maximum tree and the sum of the weights of the tree is 58 $(= 9 + 8 + 7 + 7 + 7 + 6 + 6 + 5 + 3)$ which is shown in Fig. 6.2.2.2.

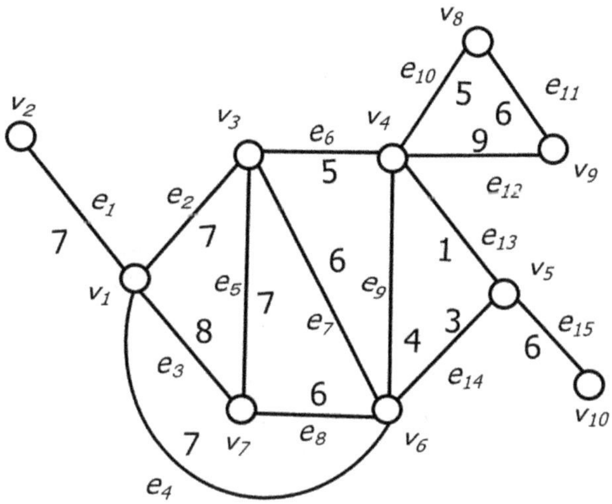

Fig. 6.2.2.1. A graph $G = (V, E)$ with edge weight of date speed.

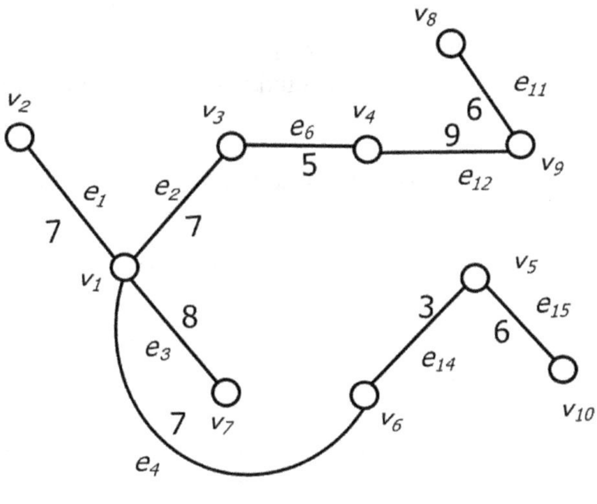

Fig. 6.2.2.2. A maximum tree of the graph $G = (V, E)$ of Fig. 6.2.2.1.

Similarly, the maximum tree is also obtained by the methods (1)′ Dual of Kruskal's algorithm and (2) Prim's algorithm, so the process of those algorithms is omitted here.

Now, let us assume that the required data speed is 6 or more on this network. Then, let us remove the edges weight less than 6 in graph of

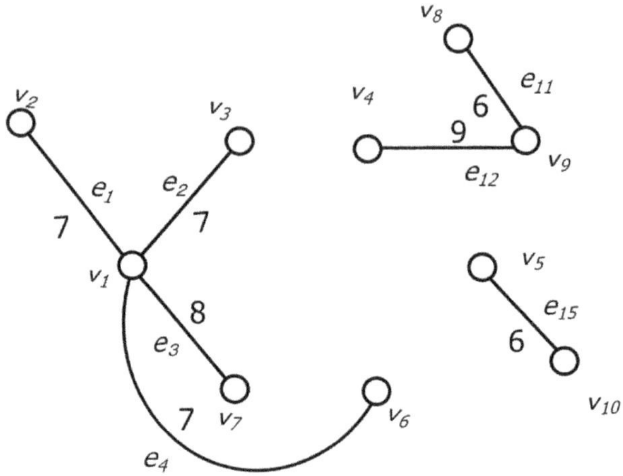

Fig. 6.2.2.3. A file location under the condition that the minimum speed of the route is 6 or more in the graph $G = (V, E)$ of Fig. 6.2.2.1.

Fig. 6.2.2.2. The graph of the result is shown in Fig. 6.2.2.3. There are three connected components in the graph of Fig. 6.2.2.3 and a file has to be located at some vertex in each connected component. For example, the files are located at v_1, v_5 and v_9. **(End of Example)**

(B) Let us consider a network like internet including mobile communications in which terminals have computers with files and users can access the file. If a file with certain contents is stored on only one computer in the network, the access from users may be concentrated and congestion may occur, so we try to distribute the files in several terminal computes to prevent congestion. This is also the case when creating a mirror site of a popular site on the internet.

In the above network, we have the issue of how to arrange files so that communication can be guaranteed even if some network links fails. The *r-covering problem* in a flow network is a problem of finding a method of arranging the minimum number of files in which access to a file is ensured even if some (less than r) edges fail in the flow network.

Here, let us consider the r-covering problem. Let $G = (V, E)$ be a connected graph with a real number weight associated with each edge of G. The weight $r(e_i)$ of an edge $e_i \in E$ means reliability or nonfailure rate of e_i. For simplicity, let us assume that the weight $r(e_i) = 1$ for every edge $e_i \in E$. Let us consider path-sets between $v_i \in V$ and $v_j \in V$ where

$v_i \neq v_j$ and each path-set does not include the same edge. These path-sets between v_i and v_j are called *edge-disjoint paths* connecting v_i and v_j in G.

A $v_i v_j$*-disconnecting set* of G is an edge set such that each path from v_i to v_j includes an edge of the edge which means that the edge set includes cutsets in G.

It is known the following theorem called *Menger's theorem.*

Theorem 6.2.2.1. *The maximum number of edge-disjoint paths connecting two distinct vertices v_i and v_j of a connected graph G is equal to the minimum number of edges in a $v_i v_j$-disconnecting set.*

Since every weight of edges is assumed to be 1, the minimum number of edges in Theorem 6.2.2.1 is the sum of edges weight. If the every weight of edges is not 1, Theorem 4.4.5.1 called max-flow min-cut theorem is applicable.

Let us show the process of solving an r-covering problem in a connected graph $G = (V, E)$.

(1) *Let $|V|$ be n. Obtain maximum flows between every pair of vertices of G. Let K_n be a complete graph in which the weight of an edge $e_k = (v_i, v_j)$ of K_n is the maximum flow between v_i and v_j.*
(2) *Obtain a maximum tree of K_n using the algorithms for maximum tree* (1), (1)′ *and* (2) *in (A). The maximum tree is sometimes called Gomory–Hu tree.*
(3) *Locate a file at some vertex in each connected component removing edges with edge weight less than r in the maximum tree.*

Example. Figure 6.2.2.4 shows a graph $G = (V, E)$ for explanation of an *r-covering problem* in a flow network. In the graph, every edge weight is *1* and so the weight is omitted in Fig. 6.2.2.4. $V = \{v_1, v_2, v_3, v_4, v_5, v_6 v_7, v_8, v_9\, v_{10}\}$, $E = \{e_1, e_2, e_3, e_4, e_5, e_6, e_7, e_8, e_9, e_{10}, e_{11}, e_{12}, e_{13}, e_{14}, e_{15}\}$ and $e_1 = (v_1, v_2)$, $e_2 = (v_1, v_3)$, $e_3 = (v_2, v_4)$, $e_4 = (v_3, v_4)$, $e_5 = (v_4, v_5)$, $e_6 = (v_3, v_6)$, $e_7 = (v_5, v_6)$, $e_8 = (v_5, v_{10})$, $e_9 = (v_6, v_7)$, $e_{10} = (v_7, v_{10})$, $e_{11} = (v_7, v_8)$, $e_{12} = (v_8, v_9)$, $e_{13} = (v_7, v_9)$, $e_{14} = (v_9, v_{10})$, $e_{15} = (v_5, v_9)$.

(1) $|V| = 10$. Obtain maximum flows between every pair of vertices of G. For example, the maximum flow between v_1 and v_2 is 2, and the maximum

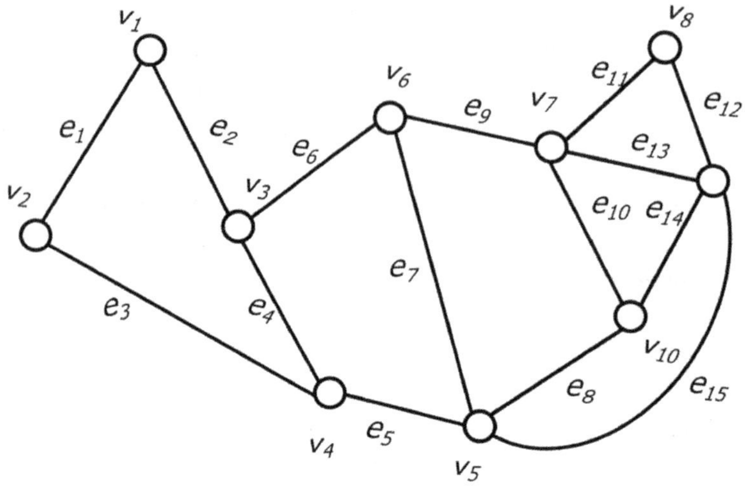

Fig. 6.2.2.4. A graph $G = (V, E)$ for explanation of an *r-covering problem* in a flow network.

flow between v_7 and v_9 is *4*. Let K_{10} be a complete graph in which the weight of an edge $e_k = (v_i, v_j)$ of K_{10} is the maximum flow between v_i and v_j. In K_{10}, the edge weights of (v_1, v_2) and (v_7, v_9) are *2* and *4*, respectively.

(2) Obtain a maximum tree of K_{10} using the algorithms for maximum tree (1), (1)′ and (2) in (A). Figure 6.2.2.5 shows a maximum tree of the complete graph K_{10} obtaining maximum flows of $G = (V, E)$ in Fig. 6.2.2.4.

(3) Let us assume that $r = 3$. Then, let us remove the edges weight less than 3 in graph of Fig. 6.2.2.5. The graph of the result is shown in Fig. 6.2.2.6. There are five connected components in the graph of Fig. 6.2.2.6 and a file has to be located at some vertex in each connected component. For example, the files are located at v_1, v_2, v_3, v_6 and v_8. **(End of Example)**

6.2.3. *Center problems*

6.2.3.1. *Eccentricity of a vertex by shortest paths*

In a connected graph $G = (V, E)$, a path and its length are defined in Section 2.4. The length of a path is defined as the number of edges in the

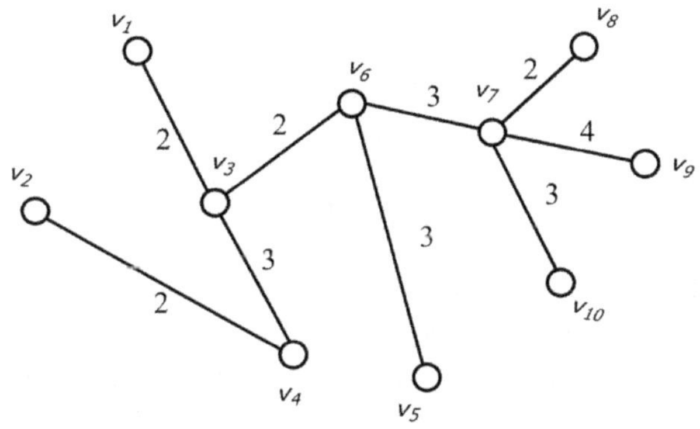

Fig. 6.2.2.5. A maximum tree of the complete graph K_{10} obtaining maximum flows of $G = (V, E)$ in Fig. 6.2.2.4.

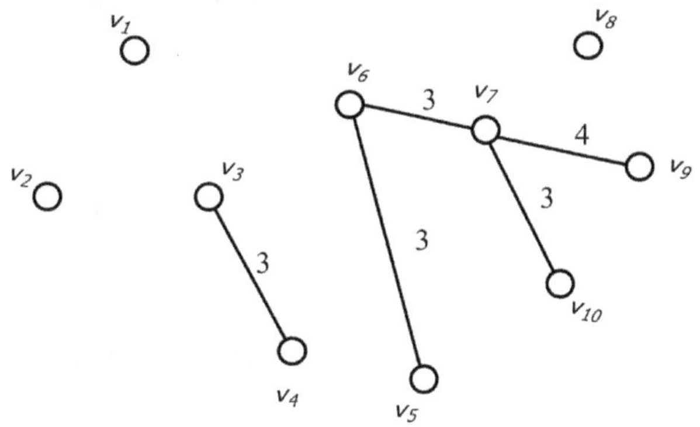

Fig. 6.2.2.6. A file location under the condition that $r = 3$ in the graph $G = (V, E)$ of Fig. 6.2.2.4.

path. If the length of a path in paths between v_i and v_j is shortest in G, the length is called the *distance* between v_i and v_j.

Next let us consider a connected graph $G = (V, E)$ with a positive real number weight associated with each edge of G. Let the length of a path define as the sum of the weights of all edges in the path. If the length of a path in paths between v_i and v_j is shortest in G, the length is called the *distance* between v_i and v_j.

Let the distance between v_i and v_j be $d(v_i, v_j)$ whether the edges of G are weighted or not, then it is clear that $d(v_i, v_j)$ is *metric*. That is, it satisfies

$$d(v_i, v_j) = 0 \quad \text{if and only if } v_i = v_j, \tag{6.2.3.1.1}$$

$$d(v_i, v_j) = d(v_j, v_i), \tag{6.2.3.1.2}$$

$$d(v_i, v_j) \leq d(v_i, v_k) + d(v_k, v_j). \tag{6.2.3.1.3}$$

The equation set (6.2.3.1.1)–(6.2.3.1.3) is called *axiom of metric for distance* and (6.2.3.1.3) is called *triangle inequality*.

Let us define the *eccentricity* $\varepsilon(v_i)$ of a vertex v_i as follows:

$$\varepsilon(v_i) = \max_{v_j \in V} d(v_i, v_j). \tag{6.2.3.1.4}$$

That is, the eccentricity $\varepsilon(v_i)$ of v_i in a graph G is the distance from v_i to the vertex farthest from v_i in G. Using the eccentricity, we define the *radius* of G:

$$r = \min_{v_i \in V} \varepsilon(v_i). \tag{6.2.3.1.5}$$

The *diameter* of G is defined as maximum of the eccentricity:

$$d = \max_{v_i \in V} \varepsilon(v_i). \tag{6.2.3.1.6}$$

A vertex with minimum eccentricity in graph G is called a *center* of G.

Let $X \subset V$,

$$d(X, v_j) = \min_{v_i \in X} d(v_i, v_j),$$

$$\varepsilon(X) = \max_{v_j \in V} d(X, v_j), \tag{6.2.3.1.7}$$

$$r_p = \min_{X \subset V} [\varepsilon(X) \mid |X| = p], \tag{6.2.3.1.8}$$

then r_p is called p-radius of G and the problem finding the set of vertices with p-radius is called *p-center problem*.

For a positive real number r, let $q(r) = \min\{|X| \mid X \subset V, \varepsilon(X) \leq r\}$. A subset X of V such that $\varepsilon(X) \leq r$ and $|X| = q(r)$ is called an r-cover of G. We simply call the problem of finding an r-cover the *r-cover problem.*

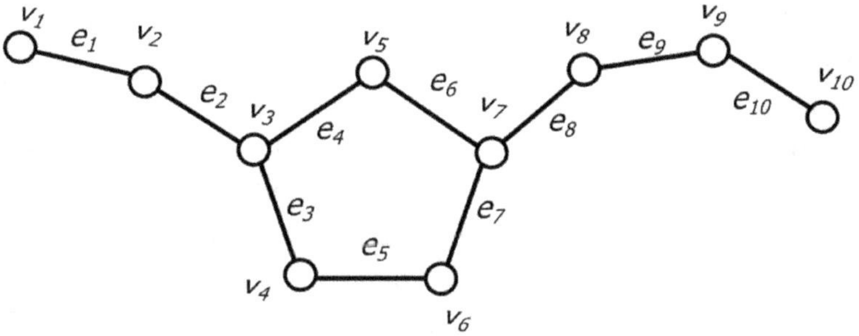

Fig. 6.2.3.1.1. A graph $G = (V, E)$ for explanation of a *p-center problem.*

Example. In Fig. 6.2.3.1.1, $G = (V, E)$, where $V = \{v_1, v_2, v_3, v_4, v_5, v_6, v_7, v_8, v_9, v_{10}\}$, $E = \{e_1, e_2, e_3, e_4, e_5, e_6, e_7, e_8, e_9, e_{10}$ and $e_1 = (v_1, v_2)$, $e_2 = (v_2, v_3)$, $e_3 = (v_3, v_4)$, $e_4 = (v_3, v_5)$, $e_5 = (v_4, v_6)$, $e_6 = (v_5, v_7)$, $e_7 = (v_6, v_7)$, $e_8 = (v_7, v_8)$, $e_9 = (v_8, v_9)$, $e_{10} = (v_9, v_{10})$.

Find the shortest path between all pairs of vertices of G and then we obtain the eccentricity of every vertex, as follows:

$\varepsilon(v_1) = 7$, $\varepsilon(v_2) = 6$, $\varepsilon(v_3) = 5$, $\varepsilon(v_4) = 5$, $\varepsilon(v_5) = 4$, $\varepsilon(v_6) = 4$, $\varepsilon(v_7) = 4$, $\varepsilon(v_8) = 5$, $\varepsilon(v_9) = 6$, $\varepsilon(v_{10}) = 7$.

Thus, the radius r of G is 4 and the center of G consists of v_5, v_6 and v_7. Since the maximum eccentricity is 7, the diameter of G is $d = 7$.

Next, let us consider 2-center problem in G.

Let us find X using $r_2 = \min_{X\subset V}[\varepsilon(X) \,|\, |X| = 2]$. If $X = \{v_2, v_8\}$ or $X = \{v_3, v_8\}$, $r_2 = \min_{X\subset V}[\varepsilon(X) \,|\, |X| = 2] = 2$. Thus, the 2-center of G consists of $\{v_2, v_8\}$ and $\{v_3, v_8\}$. **(End of Example)**

In the above example, the diameter of G is 7 and the radius is 4. Note that a radius in a graph is not necessarily half of its diameter.

Theorem 6.2.3.1.1. *In a graph G, the upper and lower bounds of the diameter d are as follows*

$$r \leq d \leq 2r, \qquad (6.2.3.1.9)$$

where r is the radius of G.

Proof. From the definitions of the radius and the diameter in (6.2.3.1.5) and (6.2.3.1.6), it is clear that $r \leq d$.

Let $d(v_i, v_j)$ be the diameter d of G and let v_k be a center of G, that is, there exists a vertex v_l in G such that $d(v_k, v_l) = \text{radius}\, r$. Since $d(v_k, v_l) = \text{radius}\, r$, $d(v_k, v_i) \geq r$ and $d(v_k, v_j) \geq r$. Using $d(v_k, v_i) = d(v_i, v_k)$ from (6.2.3.1.2), $d(v_k, v_i) + d(v_k, v_j) = d(v_i, v_k) + d(v_k, v_j) \geq d(v_i, v_j)$ by (6.2.3.1.3). Therefore, $d \leq 2r$. □

If G is a tree, we get the following theorem.

Theorem 6.2.3.1.2. *Every tree has either one or two centers.*

Proof. Let G be a tree and $\max d(v, v_j)$ be the maximum distance from a given vertex v to any other vertex v_j. Then obviously v_j is a pendant vertex. Remove all the pendant vertices from the tree G. The resulting graph G' is still a tree. The removal of all pendant vertices from G reduces uniformly the eccentricities of the remaining vertices (i.e., vertices in G') by one. Therefore, all vertices that G have as centers will still remain centers in G'. From G' we can again remove all pendant vertices and get another tree G''. We continue this process until there is left either a vertex (which is the center of G) or an edge (whose end vertices are the two centers of G'). Thus a tree G has either one or two centers. □

From the argument used in proving Theorem 6.2.3.1.2, we get he following corollary.

Corollary 6.2.3.1.2. *If a tree has two centers, the two centers must be adjacent.*

Example: In Fig. 6.2.3.1.2(a), $G = (V, E)$, where $V = \{v_1, v_2, v_3, v_4, v_5, v_6 v_7\}$, $E = \{e_1, e_2, e_3, e_4, e_5, e_6\}$ and $e_1 = (v_1, v_2)$, $e_2 = (v_2, v_3)$, $e_3 = (v_3, v_4)$, $e_4 = (v_3, v_7)$, $e_5 = (v_3, v_5)$, $e_6 = (v_4, v_6)$. The eccentricities of every vertex in G of Fig. 6.2.3.1.2(a) are $\varepsilon(v_1) = 4$, $\varepsilon(v_2) = 3$, $\varepsilon(v_3) = 2$, $\varepsilon(v_4) = 3$, $\varepsilon(v_5) = 3$, $\varepsilon(v_6) = 4$ and $\varepsilon(v_7) = 3$.

Figure 6.2.3.1.2(b) shows the graph by removing all of pendant vertices in G of Fig. 6.2.3.1.2(a). The eccentricities of every vertex in G of Fig. 6.2.3.1.2(b) are $\varepsilon(v_2) = 2$, $\varepsilon(v_3) = 1$, $\varepsilon(v_4) = 2$, that is, the removal of all pendant vertices from G of Fig. 6.2.3.1.2(a) reduces uniformly the

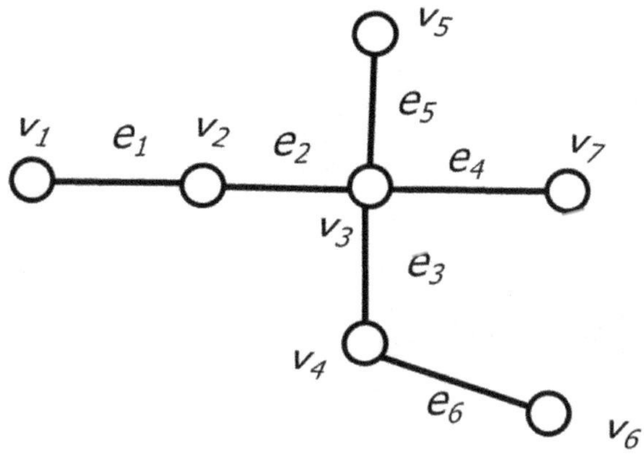

(a)

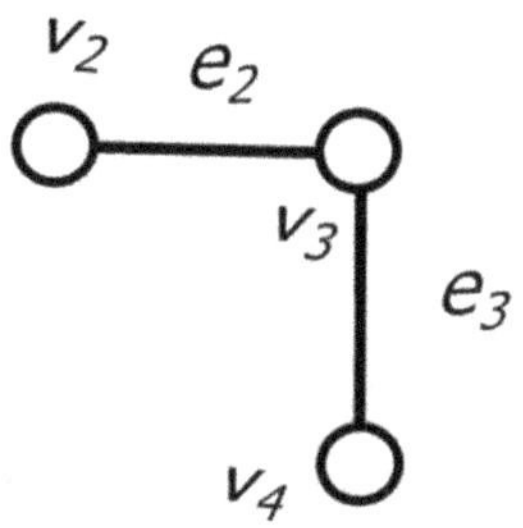

(b)

Fig. 6.2.3.1.2. A tree $G = (V, E)$ and a center. (a) A tree G. (b) Removing pendant vertices of G.

eccentricities of the remaining vertices by one. If we remove pendant vertices v_2 and v_4 from the graph of Fig. 6.2.3.1.2(b), only the vertex v_3 remains. The vertex v_3 is a center of the graph (tree) of Fig. 6.2.3.1.2(a).

In Fig. 6.2.3.1.3, $G = (V, E)$, where $V = \{v_1, v_2, v_3, v_4, v_5, v_6, v_7, v_8, v_9\}$, $E = \{e_1, e_2, e_3, e_4, e_5, e_6, e_7, e_8\}$ and $e_1 = (v_1, v_2)$, $e_2 = (v_2, v_3)$, $e_3 = (v_3, v_4)$, $e_4 = (v_3, v_7)$, $e_5 = (v_3, v_5)$, $e_6 = (v_4, v_6)$, $e_7 = (v_5, v_8)$, $e_8 = (v_8, v_9)$. The eccentricities of every vertex in G of Fig. 6.2.3.1.3 are $\varepsilon(v_1) = 5$, $\varepsilon(v_2) = 4$, $\varepsilon(v_3) = 3$, $\varepsilon(v_4) = 4$, $\varepsilon(v_5) = 3$, $\varepsilon(v_6) = 5$, $\varepsilon(v_7) = 4$, $\varepsilon(v_8) = 4$ and $\varepsilon(v_9) = 5$. The centers of the graph(tree) G of Fig. 6.2.3.1.3 are two vertices v_3 and v_5. **(End of Example)**

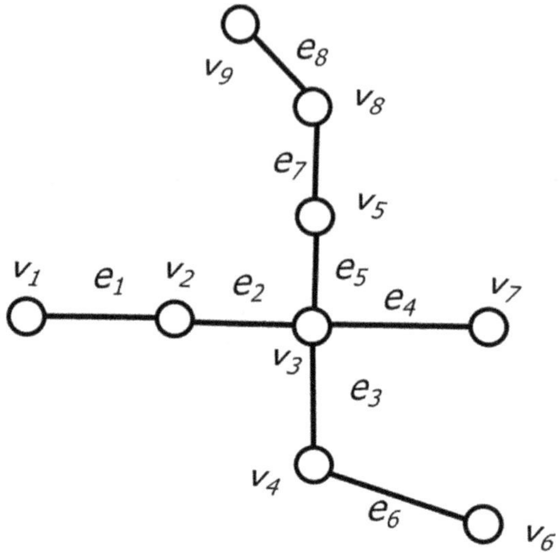

Fig. 6.2.3.1.3. A tree $G = (V, E)$ and two centers.

6.2.3.2. *Eccentricity of a vertex by maximum flows*

Let us consider a network (called a *flow network*) with a connected graph $G = (V, E)$ where each edge has a positive real number weight which is the capacity, that is, the maximum amount of flow in the edge. We discussed the problem finding the maximum flow possible through the network between a specified pair of vertices in Section 4.4.5 and explained max-flow min-cut theorem. In Section 4.4.5, the maximum flow in a digraph is discussed, and note that the discussion is applicable to a graph with nondirected edges in which each nondirected edge is replaced with two directed edges in both directions. In this chapter, let us consider a connected graph with nondirected edges which have capacity weights. From the max-flow min-cut theorem, the maximum flow possible through the network between a specified pair of vertices is the minimum of the capacities in all cutsets with respect to the pair of vertices, where the capacity of a cutset is defined as the sum of the capacity weights of all the edges in the cutset. Let the weight (capacity) of an edge e_j be $c(e_j)$. If $c(e_j) = 1$ for $e_j \in E$, the capacity of a cutset is the number of the edges in the cutset.

Let the maximum flow between v_i and v_j be $g(v_i, v_j)$, then it is known that $g(v_i, v_j)$ is sometimes also called the *gauge* or *tension* between v_i and v_j and it satisfies

$$g(v_i, v_j) = \infty, \quad \text{if and only if } v_i = v_j, \tag{6.2.3.2.1}$$

$$g(v_i, v_j) = g(v_j, v_i), \tag{6.2.3.2.2}$$

$$g(v_i, v_j) \geq \min\{g(v_i, v_k), g(v_k, v_j)\}. \tag{6.2.3.2.3}$$

The equation set (6.2.3.2.1)–(6.2.3.2.3) is sometimes called *axiom of metric for gauge*.

Let us define the *eccentricity* $\varepsilon(v_i)$ of a vertex v_i for *gauge* (or *flow network*) as follows:

$$\varepsilon(v_i) = \min_{v_j \in V} g(v_i, v_j). \tag{6.2.3.2.4}$$

That is, the eccentricity $\varepsilon(v_i)$ of v_i in a graph G is the max-flow from v_i to the vertex minimum max-flow from v_i in G. Using the eccentricity, we define the *radius* of G.

$$r = \min_{v_i \in V} \varepsilon(v_i). \tag{6.2.3.2.5}$$

The *diameter* of G is defined as maximum of the eccentricity.

$$d = \max_{v_i \in V} \varepsilon(v_i). \tag{6.2.3.2.6}$$

A vertex with minimum eccentricity in graph G is called a *center* of G.

Let $X \subset V$,

$$g(X, v_j) = \max_{v_i \in X} g(v_i, v_j),$$

$$\varepsilon(X) = \min_{v_j \in V} g(X, v_j), \tag{6.2.3.2.7}$$

$$r_p = \max_{X \subset V}[\varepsilon(X) \,|\, |X| = p], \tag{6.2.3.2.8}$$

then r_p is called *p-radius* of G and the problem finding the set X of vertices with p-radius is called *p-center problem*.

For a positive real number r, let $q(r) = \min\{|X| \,|\, X \subset V, \varepsilon(X) \geq r\}$. A subset X of V such that $\varepsilon(X) \geq r$ and $|X| = q(r)$ is called a r-cover of G. We simply call the problem of finding a r-cover the *r-cover problem*.

Theorem 6.2.3.2.1. *In a graph G with edge capacity weight, the diameter d is as follows*:

$$r = d, \qquad (6.2.3.2.8)$$

where r is the radius of G.

Proof. From the definitions of r: (6.2.3.2.5) and d: (6.2.3.2.6) of a graph with edge capacity weight, $r \leq d$.

Let $g(v_i, v_j)$ be the diameter d of G and let v_k be a center of G, that is, there exists a vertex v_l in G such that $g(v_k, v_l) =$ radius r. Since $g(v_k, v_l) =$ radius r, $g(v_l, v_i) \geq r$ and $g(v_l, v_j) \geq r$. Since $g(v_l, v_i) = g(v_i v_l)$ from (6.2.3.2.2), $\min\{g(v_l, v_i), g(v_l, v_j)\} = \min\{g(v_i v_l), g(v_l, v_j)\} \leq g(v_i, v_j)$ by (6.2.3.2.3). Let $g(v_i, v_l) \leq g(v_l, v_j)$, then $g(v_i v_l) \leq g(v_i, v_j)$. If $g(v_i v_l) < g(v_i, v_j)$, this is contrary to the assumption that $g(v_i, v_j)$ is the diameter of G. Hence, $g(v_i v_l) = g(v_i, v_j)$. Similarly, $g(v_i v_k) = g(v_i, v_j)$. Let us consider $g(v_k v_l)$, then $\min\{g(v_k v_i), g(v_i, v_l)\} \leq g(v_k, v_l)$, that is, $g(v_i, v_j) \leq g(v_k, v_l)$. Hence, $d \leq r$. Therefore, $r = d$. □

Example. In Fig. 6.2.3.2.1, $G = (V, E)$, where $V = \{v_1, v_2, v_3, v_4, v_5\}$, $E = \{e_1, e_2, e_3, e_4, e_5, e_6 e_7\}$ and $e_1 = (v_1, v_5), e_2 = (v_1, v_2), e_3 = (v_2, v_5)$, $e_4 = (v_4, v_5)$, $e_5 = (v_1, v_4)$, $e_6 = (v_1, v_3)$, $e_7 = (v_3, v_4)$. Let the weight (capacity) of an edge e_j be $c(e_j), c(e_1) = 2$, $c(e_2) = 4$, $c(e_3) = 3$, $c(e_4) = 1$, $c(e_5) = 1$, $c(e_6) = 2$, $c(e_7) = 3$.

The eccentricity of every vertex in G of Fig. 6.2.3.2.1 is $\varepsilon(v_i) = 4$ for $v_i \in V$. Thus, the radius r and diameter d are $r = d = 4$.

Let $X \subset V$ be $\{v_2, v_5\}$.

Since

$$g(X, v_1) = g(v_5, v_1) = 6,$$

$$g(X, v_3) = g(v_2, v_3) = 5,$$

and

$$g(X, v_4) = g(v_5, v_4) = 6,$$

the eccentricity of X is as follows:

$$\varepsilon(X) = \min\{6, 5, 6\} = 5.$$

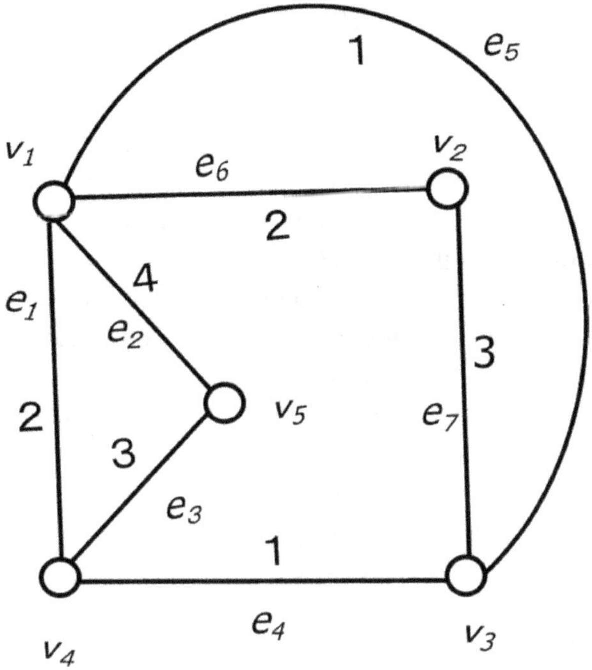

Fig. 6.2.3.2.1. A graph $G = (V, E)$ with edge capacities for explanation of a p-center problem.

(End of Example)

The algorithms for the location problems of a vertex by maximum flows are shown in [20].

6.2.3.3. *Eccentricity of an edge by tiesets and cutsets*

By a graph we mean a finite non-directed and connected graph which may include multiple edges in this section. The eccentricity of a vertex which is defined using the distance between two vertices is well known as a basic concept of graphs shown in Section 6.2.3.1. Edges are also important elements as well as vertices in the graphs. The concept of the eccentricity $\varepsilon_\tau(e_j)$ of an edge e_j by tiesets in a graph is defined using the distance between two edges in a graph, where the distance between two edges is the minimum number of edges in a tieset with the two edges. The concept of the eccentricity $\varepsilon_\gamma(e_j)$ of an edge e_j by cutsets in a graph is defined using the capacity between two edges in a graph, where the capacity between two edges is the minimum number of edges in a cutset with the two edges. We define the radius and diameter and show upper and lower bounds for diameter.

(1) Eccentricity of an edge by tiesets: In a connected graph $G = (V, E)$, a path and its length are defined in Section 2.5. The length of a path is defined as the number of edges in the path. If the length of a path in paths between v_i and v_j is shortest in G, the length is called the *distance* between v_i and v_j.

Next let us consider a connected graph $G = (V, E)$ with a positive real number weight $l(e_i)$ associated with an edge e_i of G. Let the length of a path define as the sum of the weights of all edges in the path. If the length of a path in paths between v_i and v_j is shortest in G, the length is called the *distance* between v_i and v_j.

If the weight $l(e_i) = 1$ for every edge $e_i \in E$, the results are the same as them in a graph without edge weights. Since the results are applicable to a graph with edge weights, let us consider a consider a graph without edge weights.

In a connected graph $G = (V, E)$, Let $\tau_k(e_i, e_j)$ be a tieset containing both an edge e_i and an edge e_j of G and let $R(e_i, e_j)$ be the set of these tiesets.

That is,

$$R(e_i, e_j) = \{\tau_k(e_i, e_j)\}. \tag{6.2.3.3.1}$$

The length of $\tau_k(e_i, e_j)$ is $|\tau_k(e_i, e_j)|$. We define the distance $d(e_i, e_j)$ between two edges e_i and e_j as follows:

(a) $$d(e_i, e_j) = \min_{\tau_k(e_i,e_j)\in R(e_i,e_j)} |\tau_k(e_i, e_j)|,$$

$$\text{if } R(e_i, e_j) \neq \emptyset \quad \text{for} \quad e_i, e_j \in E. \tag{6.2.3.3.2}$$

(b) $$d(e_i, e_j) = \infty, \quad \text{if } R(e_i, e_j) = \emptyset \quad \text{for } e_i, e_j \in E. \tag{6.2.3.3.3}$$

For an edge $e_i \in E$, let

(c) $$d(e_i, e_i) = 0 \tag{6.2.3.3.4}$$

In the definition (a)–(c) of the distance between edges, (c) is optional. We may define the distance $d(e_i, e_j)$ by (a) and (b) instead of by (c), where

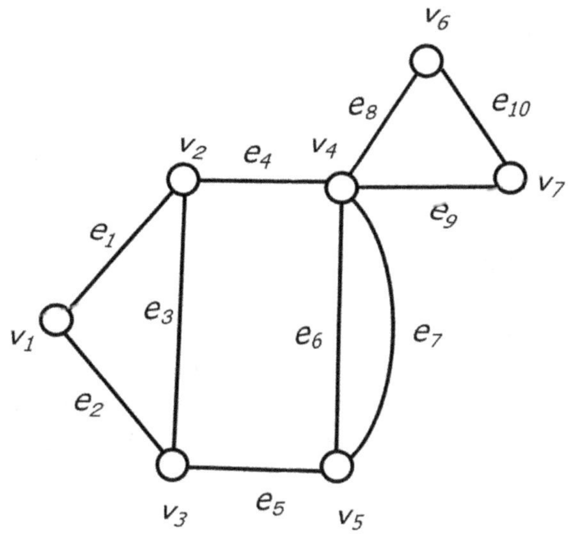

Fig. 6.2.3.3.1. A graph $G = (V, E)$ for explanation of the distance between edges.

$d(e_i, e_j) = \min_{\tau_k(e_i,e_j)\in R(e_i,e_j)} |\tau_k(e_i, e_j)|$ for $R(e_i, e_j) \neq \emptyset$ and $d(e_i, e_j) = \infty$ for $R(e_i, e_j) = \emptyset$.

Example. In Fig. 6.2.3.3.1, $G = (V, E)$, where $V = \{v_1, v_2, v_3, v_4, v_5, v_6, v_7\}$, $E = \{e_1, e_2, e_3, e_4, e_5, e_6 e_7, e_8, e_9, e_{10}\}$ and $e_1 = (v_1, v_2)$, $e_2 = (v_1, v_3)$, $e_3 = (v_2, v_3)$, $e_4 = (v_2, v_4)$, $e_5 = (v_3, v_5)$, $e_6 = (v_4, v_5)$, $e_7 = (v_4, v_5)$, $e_8 = (v_4, v_6)$, $e_9 = (v_4, v_7)$, $e_{10} = (v_6, v_7)$.

Since $\tau_1(e_1, e_2) = \{e_1, e_2, e_3\}$, $\tau_2(e_1, e_2) = \{e_1, e_2, e_4, e_5, e_6\}$ and $\tau_2(e_1, e_2) = \{e_1, e_2, e_4, e_5, e_7\}$, $R(e_1, e_2) = \{\tau_1(e_1, e_2), \tau_2(e_1, e_2), \tau_2(e_1, e_2)\}$.

Since there is no tieset with e_1 and e_8, $R(e_1, e_8) = \emptyset$.

From the definition (a)–(c), $d(e_1, e_2) = 3$ and $d(e_1, e_8) = \infty$.

(End of Example)

From Theorem 2.7.5(1), we get the following lemma.

Lemma 6.2.3.3.1. *If there are tiesets F_1 containing both e_i and e_j and F_2 containing both e_j and e_k in G, then there exists a tieset containing both e_i and e_k in a subgraph $F_1 \cup F_2$.*

Using Lemma 6.2.3.3.1, we obtain Theorem 6.2.3.3.1.

Theorem 6.2.3.3.1. *$d(e_i, e_j)$ is a metric.*

Proof. In order to prove that $d(e_i, e_j)$ satisfies the distance axiom, it suffices to show that the following equations hold:

(I) $d(e_i, e_j) = 0 \Leftrightarrow e_i = e_j$,
(II) $d(e_i, e_j) = d(e_j, e_i)$,
(III) $d(e_i, e_k) \leq d(e_i, e_j) + d(e_j, e_k)$.

From (6.2.3.3.4), $d(e_i, e_i) = 0$ and from the definition of $d(e_i, e_j)$, $d(e_i, e_j) \neq 0$ for $e_i \neq e_j$, that is, (I) holds.

In G, a tieset containing e_i and e_j is also a tieset containing e_j and e_i and thus we have $R(e_i, e_j) = R(e_j, e_i)$. Hence, from Eqs. (6.2.3.3.2) and (6.2.3.3.3), (II) holds.

Next we prove (III). From Eq. (6.2.3.3.2), $d(e_i, e_j)$ is the minimum among lengths of tiesets containing both e_i and e_j. Let $\tau_1(e_i, e_j)$ be one of tiesets of the minimum length in $R(e_i, e_j)$. Similarly, let $\tau_2(e_j, e_k)$ be one of tiesets of the minimum length in $R(e_j, e_k)$. If both of $\tau_1(e_i, e_j)$ and $\tau_2(e_j, e_k)$ exist, then from Lemma 6.2.3.3.1 there is a tieset containing both e_i and e_k in the subgraph induced by $\tau_1(e_i, e_j) \cup \tau_2(e_j, e_k)$. Let $\tau_3(e_i, e_k)$ be the set of edges in this tieset. Then,

$$\tau_3(e_i, e_k) \subseteq \tau_1(e_i, e_j) \cup \tau_2(e_j, e_k). \quad (6.2.3.3.5)$$

Hence,

$$|\tau_3(e_i, e_k)| \leq |\tau_1(e_i, e_j)| + |\tau_2(e_j, e_k)|. \quad (6.2.3.3.6)$$

From (6.2.3.3.2),

$$d(e_i, e_k) \leq \tau_3(e_i, e_k). \quad (6.2.3.3.7)$$

Since $\tau_1(e_i, e_j)$ and $\tau_2(e_j, e_k)$ are tiesets of the minimum length in $R(e_i, e_j)$ and $R(e_j, e_k)$, respectively,

$$d(e_i, e_j) = |\tau_1(e_i, e_j)| \quad \text{and} \quad d(e_j, e_k) = |\tau_2(e_j, e_k)|.$$

Therefore from (6.2.3.3.6) and (6.2.3.3.7)

$$d(e_i, e_k) \leq d(e_i, e_j) + d(e_j, e_k). \quad (6.2.3.3.8)$$

Next we consider the case where one of $\tau_1(e_i, e_j)$ and $\tau_2(e_j, e_k)$ does not exist. Without loss of generality, we assume that $\tau_2(e_j, e_k)$ does not exist. Then G is a separable graph and $e_j(e_i)$ and e_k belong to different nonseparable components. Therefore from (6.2.3.3.3), we have

$$d(e_j, e_k) = d(e_i, e_k) = \infty, \tag{6.2.3.3.9}$$

$$d(e_i, e_k) \leq d(e_i, e_j) + d(e_j, e_k). \tag{6.2.3.3.10}$$

Further in the case where neither $\tau_1(e_i, e_j)$ nor $\tau_2(e_j, e_k)$ exist, we have

$$d(e_i, e_j) = d(e_j, e_k) = \infty. \tag{6.2.3.3.11}$$

It follows from $d(e_j, e_k) \leq \infty$ that

$$d(e_i, e_k) \leq d(e_i, e_j) + d(e_j, e_k). \tag{6.2.3.3.12}$$

Equation (III) follows from (6.2.3.3.8), (6.2.3.3.10) and (6.2.3.3.12). □

Next let us consider the eccentricity of an edge of a connected nonseparable graph G since if G has a separable component, some distance between edges is infinity. We define the concept of eccentricity with respect to edges using the distance between edges.

Let

$$\varepsilon_\tau(e_i) = \max_{e_j \in E} d(e_i, e_j), \tag{6.2.3.3.13}$$

we call $\varepsilon_\tau(e_i)$ the *eccentricity* of e_i. The *radius* $\mathrm{rad}_\tau(G)$ concerning edges of G is defined as

$$\mathrm{rad}_\tau(G) = \min_{e_i \in E} \varepsilon_\tau(e_i). \tag{6.2.3.3.14}$$

The *diameter* $\mathrm{diam}_\tau(G)$ is defined as

$$\mathrm{diam}_\tau(G) = \max_{e_i \in E} \varepsilon_\tau(e_i). \tag{6.2.3.3.15}$$

Example. In Fig. 6.2.3.3.2, $G = (V, E)$, where $V = \{v_1, v_2, v_3, v_4, v_5, v_6\}$, $E = \{e_1, e_2, e_3, e_4, e_5, e_6, e_7, e_8, e_9, e_{10}\}$ and $e_1 = (v_1, v_6)$, $e_2 = (v_1, v_5)$, $e_3 = (v_1, v_2)$, $e_4 = (v_2, v_5)$, $e_5 = (v_5, v_6)$, $e_6 = (v_2, v_3)$, $e_7 = (v_3, v_5)$, $e_8 = (v_4, v_5)$, $e_9 = (v_3, v_4)$, $e_{10} = (v_3, v_4)$.

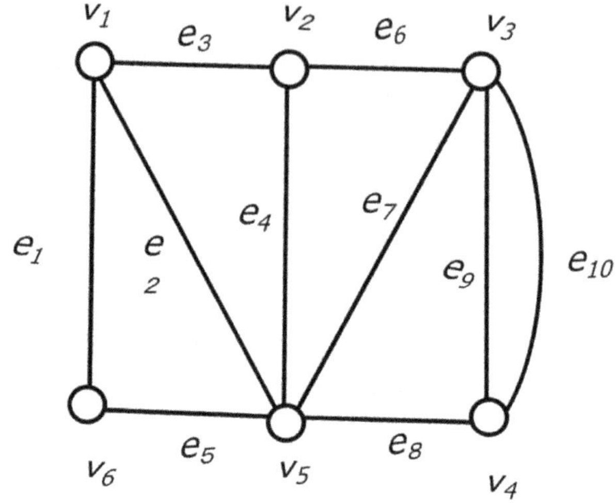

Fig. 6.2.3.3.2. A graph $G = (V, E)$ for explanation of the eccentricity of edges.

$\varepsilon_\tau(e_1) = 6$, $\varepsilon_\tau(e_2) = 5$, $\varepsilon_\tau(e_3) = 6$, $\varepsilon_\tau(e_4) = 4$, $\varepsilon_\tau(e_5) = 6$, $\varepsilon_\tau(e_6) = 6$, $\varepsilon_\tau(e_7) = 5$, $\varepsilon_\tau(e_8) = 6$, $\varepsilon_\tau(e_9) = 6$, $\varepsilon_\tau(e_{10}) = 6$. $\mathrm{rad}_\tau(G) = 4$ and $\mathrm{diam}_\tau(G) = 6$. **(End of Example)**

Let us consider the upper and lower bounds of $\mathrm{diam}_\tau(G)$.

Theorem 6.2.3.2. *For any connected and nonseparable graph $G = (V, E)$,*

$$\mathrm{rad}_\tau(G) \leq \mathrm{diam}_\tau(G) \leq 2\mathrm{rad}_\tau(G) - 2. \tag{6.2.3.3.16}$$

Proof. The inequality $\mathrm{rad}_\tau(G) \leq \mathrm{diam}_\tau(G)$ follows from the definition of radius and diameter.

Next we show $\mathrm{diam}_\tau(G) \leq 2\,\mathrm{rad}_\tau(G) - 2$.

(i) The case $E = \{e_i, e_j\}$ in which $|E| = 2$.
Since G is nonseparable, $\mathrm{diam}_\tau(G) = \mathrm{rad}_\tau(G) = 2$. Thus, we have (6.2.3.3.16).

(ii) The case where the cardinality of E is greater than $3(|E| \geq 3)$. Consider edges e_i, e_j and e_k such that $d(e_i, e_j) = \mathrm{diam}_\tau(G)$ and $\varepsilon_\tau(e_k) = \mathrm{rad}_\tau(G)$ hold. If $e_i = e_k$ or $e_j = e_k$, $\mathrm{diam}_\tau(G) = \mathrm{rad}_\tau(G) \geq 2$ holds. Thus, the inequality (6.2.3.3.16) holds. Next we consider the case $e_i \neq e_j, e_i \neq e_k$ and $e_j \neq e_k$. Let F_1 be a tieset of the minimum length among tiesets containing both e_i and e_k, and let $\tau_1(e_i, e_k)$ be the set of edges in it. Also, let F_2 be a tieset of the minimum length among tiesets containing both e_j and e_k and let $\tau_2(e_j, e_k)$ be a set of edges in it. From Lemma 6.2.3.3.1, there is a tieset containing both e_i and e_j on the subgraph of G consisting of F_1 and F_2. Let $\tau_3(e_i, e_j)$ be a set of edges in the tieset.

Then,

$$\tau_3(e_i, e_j) \subseteq \tau_1(e_i, e_k) \cup \tau_2(e_j, e_k). \tag{6.2.3.3.17}$$

We define a subset E_x of edges as

$$E_x = \tau_1(e_i, e_k) - (\tau_1(e_i, e_k) \cap \tau_3(e_i, e_j)). \tag{6.2.3.3.18}$$

(a) The case $E_x = \emptyset$.
Since $\tau_3(e_i, e_j)$ is a tieset, its subset does not contain any other tiesets. Thus we have $\tau_1(e_i, e_k) = \tau_3(e_i, e_j)$. Hence, e_j is contained in $\tau_1(e_i, e_k)$. Therefore from (6.2.3.3.17) we have

$$\begin{aligned} |\tau_3(e_i, e_j)| &\le |\tau_1(e_i, e_k) \cup \tau_2(e_j, e_k)| \\ &\le |\tau_1(e_i, e_k)| + |\tau_2(e_j, e_k)| - |\{e_k\}| - |\{e_j\}|. \end{aligned} \tag{6.2.3.3.19}$$

By the definition of distance between edges, $d(e_i, e_j) \le |\tau_3(e_i, e_j)|$. From the assumption,

$$\begin{aligned} d(e_i, e_j) = \operatorname{diam}_\tau(G), |\tau_1(e_i, e_k)| \le \operatorname{rad}_\tau(G), \quad \text{and} \\ |\tau_2(e_j, e_k)| \le \operatorname{rad}_\tau(G). \end{aligned} \tag{6.2.3.3.20}$$

Thus, from the inequality (6.2.3.3.19),

$$\operatorname{diam}_\tau(G) \le 2\operatorname{rad}_\tau(G) - 2. \tag{6.2.3.3.21}$$

(b) The case $E_x \ne \emptyset$.
Let e_a be an edge which belongs to E_x. From (6.2.3.3.17),

$$\tau_3(e_i, e_j) \subseteq \tau_1(e_i, e_k) \cup \tau_2(e_j, e_k) - \{e_a\}. \tag{6.2.3.3.22}$$

Thus, we have

$$\begin{aligned} |\tau_3(e_i, e_j)| &\le |\tau_1(e_i, e_k) \cup \tau_2(e_j, e_k)| - |\{e_a\}|, \\ |\tau_3(e_i, e_j)| &\le |\tau_1(e_i, e_k)| + |\tau_2(e_j, e_k)| - |\{e_k\}| - |\{e_a\}|. \end{aligned} \tag{6.2.3.3.23}$$

Therefore, from (6.2.3.3.20),

$$\mathrm{diam}_\tau(G) \leq 2\,\mathrm{rad}_\tau(G) - 2 \qquad (6.2.3.3.24)$$

□

(2) Eccentricity of an edge by cutsets

Let us consider the weight (capacity) of an edge e_j be $c(e_j)$ in a connected graph $G = (V, E)$. If $c(e_j) = 1$ for $e_j \in E$, the capacity of a cutset is the number of the edges in the cutset. Thus, if the weight $c(e_j) = 1$ for every edge $e_i \in E$, the results are the same as them in a graph without edge weights. Since the results are applicable to a graph with edge weights, let us consider a consider a graph without edge weights.

In a connected graph $G = (V, E)$, Let $\gamma_k(e_i, e_j)$ be a cutset containing both an edge e_i and an edge e_j of G and let $Q(e_i, e_j)$ be the set of these cutsets.

That is,

$$Q(e_i, e_j) = \{\gamma_k(e_i, e_j)\}. \qquad (6.2.3.3.25)$$

The capacity of $\gamma_k(e_i, e_j)$ is $|\gamma_k(e_i, e_j)|$. We define the distance $d(e_i, e_j)$ between two edges e_i and e_j as follows:

(a) $$d(e_i, e_j) = \min_{\gamma_k(e_i,e_j)\in Q(e_i,e_j)} |\gamma_k(e_i, e_j)| \quad \text{if } R(e_i, e_j) \neq \emptyset$$

$$\text{for } e_i, e_j \in E. \qquad (6.2.3.3.26)$$

(b) $$d(e_i, e_j) = \infty, \quad \text{if } Q(e_i, e_j) = \emptyset \quad \text{for } e_i, e_j \in E. \qquad (6.2.3.3.27)$$

For an edge $e_i \in E$, let

(c) $$d(e_i, e_i) = 0 \qquad (6.2.3.3.28)$$

In the definition (a)–(c) of the distance between edges, (c) is optional. We may define the distance $d(e_i, e_j)$ by (a) and (b) instead of by (c), where $d(e_i, e_j) = \min_{\gamma_k(e_i,e_j)\in Q(e_i,e_j)} |\gamma_k(e_i, e_j)|$ for $R(e_i, e_j) \neq \emptyset$ and $d(e_i, e_j) = \infty$ for $Q(e_i, e_j) = \emptyset$.

Example. Let us consider Fig. 6.2.3.3.1 again. In Fig. 6.2.3.3.1, $G = (V, E)$, where $V = \{v_1, v_2, v_3, v_4, v_5, v_6, v_7\}$, $E = \{e_1, e_2, e_3, e_4, e_5, e_6, e_7, e_8, e_9, e_{10}\}$ and $e_1 = (v_1, v_2)$, $e_2 = (v_1, v_3)$, $e_3 = (v_2, v_3)$, $e_4 = (v_2, v_4)$, $e_5 = (v_3, v_5)$, $e_6 = (v_4, v_5)$, $e_7 = (v_4, v_5)$, $e_8 = (v_4, v_6)$, $e_9 = (v_4, v_7)$, $e_{10} = (v_6, v_7)$.

Since $\gamma_1(e_1, e_3) = \{e_1, e_3, e_4\}$, $\gamma_2(e_1, e_3) = \{e_1, e_3, e_5\}$ and $\gamma_3(e_1, e_3) = \{e_1, e_3, e_6, e_7\}$, $Q(e_1, e_3) = \{\gamma_1(e_1, e_3), \gamma_2(e_1, e_3), \gamma_3(e_1, e_3)\}$.

Since there is no tiesets with e_2 and e_8, $R(e_2, e_8) = \emptyset$.

From the definition (a)–(c), $d(e_1, e_3) = 3$ and $d(e_2, e_8) = \infty$.

(End of Example)

The property of the eccentricity using the cutset can be obtained as well as the eccentricity using the tieset. The property of the eccentricity using the cutset is shown below.

Lemma 6.2.3.3.2. *If there are cutsets H_1 containing both e_i and e_j and H_2 containing both e_j and e_k in G, then there exists a cutset containing both e_i and e_k in a subgraph $H_1 \cup H_2$.*

Using Lemma 6.2.3.3.2, we obtain Theorem 6.2.3.3.2.

Theorem 6.2.3.3. *$d(e_i, e_j)$ is a metric.*

Next let us consider the eccentricity of an edge of a connected nonseparable graph G using cutsets since if G has a separable component, some distance between edges is infinity. We define the concept of eccentricity with respect to edges using the distance between edges.

Let

$$\varepsilon_\gamma(e_i) = \min_{e_j \in E} d(e_i, e_j), \tag{6.2.3.3.29}$$

we call $\varepsilon_\gamma(e_i)$ the *eccentricity* of e_i. The *radius* $\mathrm{rad}_\gamma(G)$ concerning edges of G is defined as

$$\mathrm{rad}_\gamma(G) = \min_{e_i \in E} \varepsilon_\gamma(e_i). \tag{6.2.3.3.30}$$

The *diameter* $\mathrm{diam}_\gamma(G)$ is defined as

$$\mathrm{diam}_\gamma(G) = \max_{e_i \in E} \varepsilon_\gamma(e_i). \tag{6.2.3.3.31}$$

Example. Let us consider Fig. 6.2.3.3.2 again. In Fig. 6.2.3.3.2, $G = (V, E)$, where $V = \{v_1, v_2, v_3, v_4, v_5, v_6\}$, $E = \{e_1, e_2, e_3, e_4, e_5, e_6, e_7, e_8, e_9, e_{10}\}$ and $e_1 = (v_1, v_6)$, $e_2 = (v_1, v_5)$, $e_3 = (v_1, v_2)$, $e_4 = (v_2, v_5)$, $e_5 = (v_5, v_6)$, $e_6 = (v_2, v_3)$, $e_7 = (v_3, v_5)$, $e_8 = (v_4, v_5)$, $e_9 = (v_3, v_4)$, $e_{10} = (v_3, v_4)$, $\varepsilon_\gamma(e_1) = 6$, $\varepsilon_\gamma(e_2) = 6$, $\varepsilon_\gamma(e_3) = 5$, $\varepsilon_\gamma(e_4) = 5$, $\varepsilon_\gamma(e_5) = 5$, $\varepsilon_\gamma(e_6) = 4$, $\varepsilon_\gamma(e_7) = 5$, $\varepsilon_\gamma(e_8) = 5$, $\varepsilon_\gamma(e_9) = 6$, $\varepsilon_\gamma(e_{10}) = 6$, $\text{rad}_\gamma(G) = 4$ and $\text{diam}_\gamma(G) = 6$. **(End of Example)**

Let us consider the upper and lower bounds of $\text{diam}_\tau(G)$.

Theorem 6.2.3.4. *For any connected and nonseparable graph $G = (V, E)$,*

$$\text{rad}_\tau(G) \le \text{diam}_\tau(G) \le 2\,\text{rad}_\tau(G) - 2 \qquad (6.2.3.3.32)$$

6.2.3.4. *Meaning of application of the concept of center of graph*

The concept of center on a graph is often used in facility location problems. As an example of a graph in which shortest paths are funded, consider the problem of the placement of a fire station or a police station. In a graph, the edges represent the road and the vertices represent the town. Consider the problem that the fire station should be placed in a town and decide the optimal location. Let us assume that the number of fire stations is 1. The fire station should be located in a place where they can be dispatched as soon as possible in any town where a fire breaks out. The optimum place corresponds a center of the graph. In communication networks, the optimum place of a base station corresponds to a center of the graph.

In a graph in which max-flows are funded, a center means that max-flow from the center to other vertices is minimum. This is also considered to be a weak point in the event of a disaster.

Let us show a simple example to explain the eccentricity of an edge by tiesets. Let us consider a road network where the length of each road is 1 and each road has a police station at the middle point of the road. A patrol car of a police station must patrol his edge (road) and other edges besides. In order to avoid the repetition point of the patrol, we assume that a patrol car of a police station have to patrol some other edges from his station to his station passing once and only once through some vertices and some edges except for his edge. If a patrol car is ordered the patrol that must make a round via a specific police station, the patrol car must include the edge of the specific police station in the patrol. $d(e_i, e_j)$ represents the shortest patrol tour length from the police station of e_i going through that of e_j which is the specific station in the patrol. Note that the shortest patrol tour from the police station of e_i going through that of e_j may not include the shortest path between the two police stations. The above assumption of the patrol tour is only used for the patrol. Of course, in case of emergency at e_j, the patrol car of e_i can use the shortest path between the two police stations. The maximum value of the shortest patrol tour length of e_i in any specific police stations to make a round via is the eccentricity of e_i. The central edge e_k of G means that a patrol car of e_k can have minimum patrol tour length in any specific stations to make a round via. In a communication network, each road has a base station at the middle point of the road, and the patrol corresponds to failure inspection.

In the eccentricity of an edge by cutsets, the eccentricity shows the a weak point in the event of a disaster and also shows the place to do failure inspection.

6.2.4. *Median problems*

In a communication network, such as a mobile communication system, a mobile phone connects usually to a base station through wireless. The base station sends the message to the proper destination via some communication stations. A model of such a system may be assumed to be a finite graph (network) G with weights attached to each of its vertices and edges. The weight attached to the edge represents the length or capacity. The weight attached to a vertex of G represents the number of channels which correspond to wires in wired communications. The problem is to find the exact location of the base station such that the total length of the radio propagation path is minimum or the total capacity is maximum.

6.2.4.1. *Median problem by distance*

In a connected graph $G = (V, E)$ with positive real number weights attached to each of its vertices and edges. Let $w(v_j)$ and $l(e_i)$ be the weight of a vertex v_j and the length of an edge e_i in G, respectively. Let the length of a path define as the sum of the weights of all edges in the path. If the length of a path in paths between v_i and v_j is shortest in G, the length is called the distance $d(v_i, v_j)$ between v_i and v_j. Let us consider points on edges and for example, see points x_i and x_j, etc. in Fig. 6.2.4.1.1. Let $Z = \{v_1, v_2, \ldots x_1, x_2, \ldots\}$ be the set of vertices and points. The distance between two elements of Z, $d(v_i, v_j)$ or $d(v_i, x_j)$ is defined as the length of a shortest path between v_i and v_j or v_i and x_j, etc. in G.

$$t(x_i) = \sum_{v_j \in V} d(x_i, v_j) w(v_j) \tag{6.2.4.1.1}$$

where $t(x_i)$ is called the *transmission number* of x_i.

Let

$$s_1 \triangleq \min_{x_i \in Z} t(x_i), \tag{6.2.4.1.2}$$

then a point with s_1 is called an *absolute median* of G.

Let

$$t(X) = \sum_{v_j \in V} d(X, v_j) w(v_j), \tag{6.2.4.1.3}$$

where $X \subset Z$ and $d(X, v_j) = \min_{v_i \in X} d(v_i, v_j)$.

Let

$$s_p \triangleq \min_{X \subset Z} [t(X) | \, |X| = p], \tag{6.2.4.1.4}$$

then the points with s_p is called an *absolute p-median* of G.

If the elements of Z are only vertices, that is, Z does not include points on edges,

$$t(v_i) = \sum_{\substack{v_j \in V \\ v_j \neq v_i}} d(v_i, v_j) w(v_j), \tag{6.2.4.1.5}$$

$$s_1 \triangleq \min_{v_i \in V} t(v_i), \tag{6.2.4.1.6}$$

$$t(X) = \sum_{\substack{v_j \in V \\ v_j \notin X}} d(X, v_j) w(v_j), \tag{6.2.4.1.7}$$

where $X \subset V$ and $d(X, v_j) = \min_{\substack{v_i \in X \\ v_j \notin X}} d(v_i, v_j)$, and

$$s_p \triangleq \min_{X \subset V} [t(X) | \, |X| = p]. \tag{6.2.4.1.8}$$

Then, a vertex with s_1 and the vertices with s_p are called a *median* and a *p-median* of G, respectively, and $t(v_i)$ is called the *transmission number* of a vertex v_i.

Example. In Fig. 6.2.4.1.1, $G = (V, E)$, where $V = \{v_1, v_2, v_3, v_4.v_5, v_6, v_7, v_8, v_9, v_{10}, v_{11}\}$, $E = \{e_1, e_2, e_3, e_4, e_5, e_6, e_7, e_8, e_9, e_{10}, e_{11}, e_{12}\}$ and $e_1 = (v_1, v_2)$, $e_2 = (v_2, v_3)$, $e_3 = (v_3, v_4)$, $e_4 = (v_3, v_8)$, $e_5 = (v_4, v_5)$, $e_6 = (v_4, v_6)$, $e_7 = (v_5, v_6)$, $e_8 = (v_6, v_7)$, $e_9 = (v_6, v_8)$, $e_{10} = (v_8, v_9)$, $e_{11} = (v_9, v_{10})$, $e_{12} = (v_9, v_{11})$.

The ijth entry of the following matrix D shows the distance $d(v_i, v_j)$.

$$D = \begin{array}{c} \\ v_1 \\ v_2 \\ v_3 \\ v_4 \\ v_5 \\ v_6 \\ v_7 \\ v_8 \\ v_9 \\ v_{10} \\ v_{11} \end{array} \begin{array}{c} \begin{array}{ccccccccccc} v_1 & v_2 & v_3 & v_4 & v_5 & v_6 & v_7 & v_8 & v_9 & v_{10} & v_{11} \end{array} \\ \left[\begin{array}{ccccccccccc} \cdot & 1 & 2 & 3 & 4 & 4 & 5 & 3 & 4 & 5 & 5 \\ 1 & \cdot & 1 & 2 & 3 & 3 & 4 & 2 & 3 & 4 & 4 \\ 2 & 1 & \cdot & 1 & 2 & 2 & 3 & 1 & 2 & 3 & 3 \\ 3 & 2 & 1 & \cdot & 1 & 1 & 2 & 2 & 3 & 4 & 4 \\ 4 & 3 & 2 & 1 & \cdot & 1 & 2 & 2 & 3 & 4 & 4 \\ 4 & 3 & 2 & 1 & 1 & \cdot & 1 & 1 & 2 & 3 & 3 \\ 5 & 4 & 3 & 2 & 2 & 1 & \cdot & 2 & 3 & 4 & 4 \\ 3 & 2 & 1 & 2 & 2 & 1 & 2 & \cdot & 1 & 2 & 2 \\ 4 & 3 & 2 & 3 & 3 & 2 & 3 & 1 & \cdot & 1 & 1 \\ 5 & 4 & 3 & 4 & 4 & 3 & 4 & 2 & 1 & \cdot & 2 \\ 5 & 4 & 3 & 4 & 4 & 3 & 4 & 2 & 1 & 2 & \cdot \end{array}\right] \end{array}.$$

For simplicity, let the elements of Z be only vertices and all of weights of edges and vertices be 1, that is, $l(e_i) = 1$ and $w(v_j) = 1$ for $e_i \in E$ and $v_j \in V$. Then,

$$t(v_i) = \sum_{\substack{v_j \in V \\ v_j \neq v_i}} d(v_i, v_j).$$

$$t(v_1) = 36,\ t(v_2) = 27,\ t(v_3) = 20,\ t(v_4) = 23,\ t(v_5) = 26,$$

$$t(v_6) = 21,\ t(v_7) = 30,\ t(v_8) = 18,\ t(v_9) = 23,\ t(v_{10}) = 32,$$

$$t(v_{11}) = 32.$$

Since $s_1 \triangleq \min_{v_i \in V} t(v_i) = 18$, the vertex v_8 is a median of G.

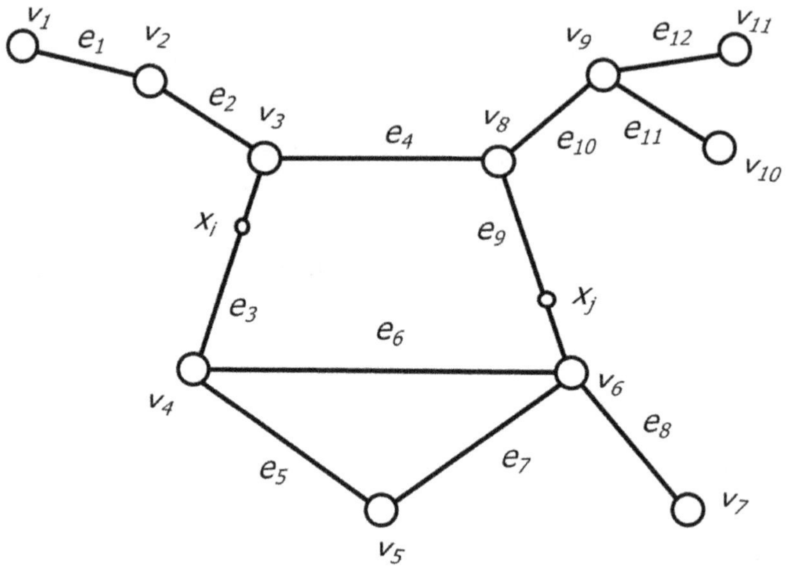

Fig. 6.2.4.1.1. A graph $G = (V, E)$ for explanation of the median problem.

In $s_p \triangleq \min_{X \subset V}[t(X) | |X| = p]$, let $X = \{v_4, v_9\} \subset V$, then $s_2 = 13$. Thus, v_4 and v_9 are 2-median. **(End of Example)**

6.2.4.2. *Median problem by gauge*

In a connected graph $G = (V, E)$ with positive real number, weights are attached to each of its vertices and edges. Let $w(v_j)$ and $c(e_i)$ be the weight of a vertex v_j and the capacity of an edge e_i in G, respectively. Let the capacity of a cutset define as the sum of the weights of all edges in the cutset. If the capacity of a cutset in cutsets between v_i and v_j is minimum in G, the capacity is called the max-flow, the capacity or gauge $g(v_i, v_j)$ between v_i and v_j. Let

$$t_c(v_i) = \sum_{\substack{v_j \in V \\ v_j \neq v_i}} g(v_i, v_j) w(v_j), \tag{6.2.4.2.1}$$

where $t_c(v_i)$ is the *transmission number for max-flow* of a vertex v_i. Let

$$s_{c1} \triangleq \min_{v_i \in V} t_c(v_i). \tag{6.2.4.2.2}$$

Then, a vertex with s_{c1} are called a *median for vulnerability* of G.

Let $X \subset V$ and $g(X, v_j) = \min_{\substack{v_j \in X \\ v_j \notin X}} g(v_i, v_j)$,

$$t_c(X) = \sum_{\substack{v_j \in V \\ v_j \notin X}} g(X, v_j) w(v_j), \tag{6.2.4.2.3}$$

and

$$s_{cp} \triangleq \min_{X \subset V} [t_c(X) \mid |X| = p]. \tag{6.2.4.2.4}$$

Then, a vertex with s_{cp} is called a *p-median for vulnerability* of G.

We have focused on the minimum of the transmission number in (6.2.4.10). Next, let us focus on its maximum value which does not represent the vulnerability, but rather the sum of the maximum flows that can be sent to other vertices. That is,

$$s_{cm1} \triangleq \max_{v_i \in V} t_c(v_i), \tag{6.2.4.2.5}$$

where a vertex with s_{cm1} are called a *median for max-flow* of G.

Let $X \subset V$ and $g(X, v_j) = \max_{\substack{v_i \in X \\ v_j \notin X}} g(v_i, v_j)$,

$$t_{cm}(X) = \sum_{\substack{v_j \in V \\ v_j \notin X}} g(X, v_j) w(v_j), \tag{6.2.4.2.6}$$

and

$$s_{cp} \triangleq \max_{X \subset V} [t_{cm}(X) \mid |X| = p]. \tag{6.2.4.2.7}$$

Then, a vertex with s_{cp} is called a *p-median for max-flow* of G.

Example. Let us consider $G = (V, E)$ of Fig. 6.2.4.1.1 again. where $V = \{v_1, v_2, v_3, v_4, v_5, v_6, v_7, v_8, v_9, v_{10}, v_{11}\}$, $E = \{e_1, e_2, e_3, e_4, e_5, e_6, e_7, e_8, e_9, e_{10}, e_{11}, e_{12}\}$ and $e_1 = (v_1, v_2), e_2 = (v_2, v_3), e_3 = (v_3, v_4), e_4 = (v_3, v_8), e_5 = (v_4, v_5), e_6 = (v_4, v_6), e_7 = (v_5, v_6), e_8 = (v_6, v_7), e_9 = (v_6, v_8), e_{10} = (v_8, v_9), e_{11} = (v_9, v_{10}), e_{12} = (v_9, v_{11})$.

The ijth entry of the following matrix G shows the gauge $g(v_i, v_j)$.

$$G = \begin{array}{c} \\ v_1 \\ v_2 \\ v_3 \\ v_4 \\ v_5 \\ v_6 \\ v_7 \\ v_8 \\ v_9 \\ v_{10} \\ v_{11} \end{array} \begin{array}{c} \begin{array}{ccccccccccc} v_1 & v_2 & v_3 & v_4 & v_5 & v_6 & v_7 & v_8 & v_9 & v_{10} & v_{11} \end{array} \\ \left[\begin{array}{ccccccccccc} \cdot & 1 & 1 & 1 & 1 & 1 & 1 & 1 & 1 & 1 & 1 \\ 1 & \cdot & 1 & 1 & 1 & 1 & 1 & 1 & 1 & 1 & 1 \\ 1 & 1 & \cdot & 2 & 2 & 2 & 1 & 2 & 1 & 1 & 1 \\ 1 & 1 & 2 & \cdot & 2 & 3 & 1 & 2 & 1 & 1 & 1 \\ 1 & 1 & 2 & 2 & \cdot & 2 & 1 & 2 & 1 & 1 & 1 \\ 1 & 1 & 2 & 3 & 2 & \cdot & 1 & 2 & 1 & 1 & 1 \\ 1 & 1 & 1 & 1 & 1 & 1 & \cdot & 1 & 1 & 1 & 1 \\ 1 & 1 & 2 & 2 & 2 & 2 & 1 & \cdot & 1 & 1 & 1 \\ 1 & 1 & 1 & 1 & 1 & 1 & 1 & 1 & 1 & \cdot & 1 \\ 1 & 1 & 1 & 1 & 1 & 1 & 1 & 1 & 1 & \cdot & 1 \\ 1 & 1 & 1 & 1 & 1 & 1 & 1 & 1 & 1 & 1 & \cdot \end{array}\right] \end{array}.$$

For simplicity, let all of weights of edges and vertices be 1, that is, $c(e_i) = 1$ and $w(v_j) = 1$ for $e_i \in E$ and $v_j \in V$. Then, from (6.2.4.2.1),

$$t_c(v_i) = \sum_{\substack{v_j \in V \\ v_j \neq v_i}} g(v_i, v_j).$$

$$t_c(v_1) = 10,\ t_c(v_2) = 10,\ t_c(v_3) = 14,\ t_c(v_4) = 15,\ t_c(v_5) = 14,$$
$$t_c(v_6) = 15,\ t_c(v_7) = 10,\ t_c(v_8) = 14,\ t_c(v_9) = 10,\ t_c(v_{10}) = 10,$$
$$t_c(v_{11}) = 10.$$

Since $s_{c1} \triangleq \min_{v_i \in V} t(v_i) = 10$, the vertex v_9 is one of medians for vulnerability of G.

In $s_p \triangleq \min_{X \subset V}[t(X) | |X| = p]$, let $X = \{v_4, v_9\} \subset V$, then $s_2 = 13$. Thus, v_4 and v_9 are 2-median.

From (6.2.4.2.5), $s_{cm1} \triangleq \max_{v_i \in V} t_c(v_i)$.

$s_{cm1} \triangleq \max_{v_i \in V} t_c(v_i) = 15$. Hence, v_4 and v_6 are medians for max-flow of G. **(End of Example)**

6.2.5. *Generalizations of centrality in graphs and networks*

In the previous section, we explained some functions, e.g., a function for a center or a median that measure the *centrality* of vertices in a graph. Among them, there was a function in which the function value became smaller as the centrality was larger, and conversely, the function value became larger as the centrality was larger, and they were mixed. For example, the transmission number of (6.2.4.1.1) is a function in which the function value become smaller as the centrality is larger, and the transmission number for max-flow of (6.2.4.2.1) is a function in which the function value become larger as the centrality is larger.

In this section, we premise that the function value becomes smaller as it seems to be central and characterize the function that expresses centrality using the reduction of metric space, which corresponds to the reduction of distance between two specific vertices. In a connected graph $G = (V, E)$ with positive real number, weights are attached to each of its vertices and edges. Let the edge weight be length. Suppose that the length of the edge between two specific vertices v_i and v_j decreases, or the change is equivalent to the addition of a new edge between v_i and v_j. Note that if the distance between the vertices other than v_i and v_j decreases due to the decrease of length between v_i and v_j, the shortest path between the two vertices always goes through v_i and v_j.

Let us start the discussion from the following two assumptions as a characteristic of the vertex that is intuitively considered to have a high degree of centrality in G.

(A) The degree of centrality of the vertices v_i and v_j in which the length between v_i and v_j has decreased, increases.
(B) For an increase in the degree of centrality of a vertex, the increment of the vertices v_i and v_j is greater than the increment of the vertex in their vicinity.

6.2.5.1. *The reduction of metric space*

Let $(S; \rho)$ be a metric space where $\rho: S \times S \rightarrow \bar{\boldsymbol{R}}^+$ is a distance function (S is a nonempty set and $\bar{\boldsymbol{R}}^+$ is a set of nonnegative real number). For $s_1, s_2, s_3 \in S$, $\rho(s_1, s_2) \geq 0$ and the following three conditions are satisfied.

$$\text{(i)}\quad \rho(s_1, s_2) = 0 \Leftrightarrow s_1 = s_2, \tag{6.2.5.1.1}$$

$$\text{(ii)}\quad \rho(s_1, s_2) = \rho(s_2, s_1), \tag{6.2.5.1.2}$$

$$\text{(iii)}\quad \rho(s_1, s_3) \leq \rho(s_1, s_2) + \rho(s_2, s_3). \tag{6.2.5.1.3}$$

$\rho(s_1, s_2)$ is called the *distance* between points s_1 and s_2.

If the following condition of (i)′ is satisfied instead of the condition of (i),

$$\text{(i)}' \quad \rho(s_1, s_2) = 0. \tag{6.2.5.1.4}$$

$(S; \rho)$ is called a *pseudo-metric space* and (i)((i)′)–(iii) is called the axiom of metric for distance.

Let us consider reducing the distance $\rho(p, q)$ between points p and q to $\lambda \leq \rho(p, q)$. That is,

$$\rho(p, q) \to \lambda \tag{6.2.5.1.5}$$

If the distance between points other than p and q doses not change after reducing the distance $\rho(p, q)$, the axiom of metric for distance may not hold. For reducing the distance $\rho(p, q)$ by (6.2.5.1.5), let us consider the new distance $\rho'(s_1, s_2)$ between points s_1 and s_2 ($s_1, s_2 \in S$) as follows:

$$\begin{aligned}\rho'(s_1, s_2) = \min\{&\rho(s_1, s_2), \rho(s_1, p) + \rho(s_2, q) \\ &+ \lambda, \rho(s_1, q) + \rho(s_2, p) + \lambda\},\end{aligned} \tag{6.2.5.1.6}$$

where $0 \leq \lambda \leq \rho(p, q)$.

Then we get the following theorem.

Theorem 6.2.5.1. *The space $(S; \rho')$ obtained by* (6.2.5.1.6) *is a metric space or a pseudo-metric space (for $\lambda = 0$).*

Proof. Let us prove that $(S; \rho')$ holds (i) or (i)′, (ii) and (iii).

[a] (i) or $(i)'$:

Let $s_1 = s_2$ in (6.2.5.1.6). Since $\rho(s_1, s_2) = 0$, it is obvious that $\rho'(s_1, s_2) = 0$.

Conversely, let $\rho'(s_1, s_2) = 0$ and $\lambda > 0$. From (6.2.5.1.6),

$$\rho(s_1, s_2) = 0, \tag{6.2.5.1.7}$$

$$\rho(s_1, p) + \rho(s_2, q) + \lambda = 0 \tag{6.2.5.1.8}$$

or

$$\rho(s_1, q) + \rho(s_2, p) + \lambda = 0. \tag{6.2.5.1.9}$$

For (6.2.5.1.7), $s_1 = s_2$ from (i). For (6.2.5.1.8), since each term in this formula is a nonnegative real number, each term is zero. Since $\lambda > 0$, this formula does not hold. Similarly, the case of (6.2.5.1.9) does not hold. If $\lambda = 0$, it is a pseudo-metric space.

[b] (ii) :
Since $\rho(s_1, s_2) = \rho(s_2, s_1)$ from (ii) in (6.2.5.1.6),

$$\begin{aligned}\rho'(s_1, s_2) &= \min\{\rho(s_1, s_2), \rho(s_1, p) + \rho(s_2, q) + \lambda, \rho(s_1, q)\\ &\quad +\rho(s_2, p) + \lambda\}\\ &= \min\{\rho(s_2, s_1), \rho(s_2, p) + \rho(s_1, q) + \lambda, \rho(s_2, q)\\ &\quad +\rho(s_1, p) + \lambda\}\\ &= \rho'(s_2, s_1). \end{aligned} \tag{6.2.5.1.10}$$

Therefore ρ' holds (ii).

[b] (iii) :
From (6.2.5.1.6),

$$\begin{aligned}\rho'(s_1, s_3) &= \min\{\rho(s_1, s_3), \rho(s_1, p) + \rho(s_3, q) + \lambda, \rho(s_1, q)\\ &\quad +\rho(s_3, p) + \lambda\},\end{aligned} \tag{6.2.5.1.11}$$

$$\begin{aligned}\rho'(s_1, s_2) &= \min\{\rho(s_1, s_2), \rho(s_1, p)\\ &\quad +\rho(s_2, q) + \lambda, \rho(s_1, q) + \rho(s_2, p) + \lambda\}\end{aligned} \tag{6.2.5.1.12}$$

and

$$\begin{aligned}\rho'(s_2, s_3) &= \min\{\rho(s_2, s_3), \rho(s_2, p) + \rho(s_3, q) + \lambda, \rho(s_2, q)\\ &\quad +\rho(s_3, p) + \lambda\}.\end{aligned} \tag{6.2.5.1.13}$$

Using (6.2.5.1.11)–(6.2.5.1.13), let us prove that

$$\rho'(s_1, s_3) \leq \rho'(s_1, s_2) + \rho'(s_2, s_3). \tag{6.2.5.1.14}$$

There are various cases to be proved depending on which values $\rho'(s_1, s_3)$, $\rho'(s_1, s_2)$ and $\rho'(s_2, s_3)$ take. For convenience of description, let $s_{i,j,k}(i, j, k = 1\text{--}3)$ be the case such that the ith term of $\rho'(s_1, s_3)$ in (6.2.5.1.11), the jth term of $\rho'(s_1, s_2)$ in (6.2.5.1.12) and the kth term of $\rho'(s_2, s_3)$ in (6.2.5.1.13) are selected. There are 27 cases for $s_{i,j,k}(i, j, k = 1{\sim}3)$, that is, $s_{1,1,1}, s_{1,1,2}, \ldots, s_{3,3,3}$. The symmetry of p and q, and s_1 and s_3 means that it is sufficient to prove the following 10 cases instead of 27 cases.

⟨1⟩ $s_{1,1,1}$:

Since $\rho'(s_1, s_3) = \rho(s_1, s_3), \rho'(s_1, s_2) = \rho(s_1, s_2)$ and $\rho'(s_2, s_3) = \rho(s_2, s_3)$,

$$\rho'(s_1, s_3) \leq \rho'(s_1, s_2) + \rho'(s_2, s_3).$$

⟨2⟩ $s_{1,1,2}, s_{1,1,3}, s_{1,2,1}, s_{1,3,1}$:

Let us prove only $s_{1,1,2}$ because of the symmetry of p and q, etc.

$$\rho'(s_1, s_3) = \rho(s_1, s_3), \tag{6.2.5.1.15}$$

$$\rho'(s_1, s_2) = \rho(s_1, s_2), \tag{6.2.5.1.16}$$

$$\rho'(s_2, s_3) = \rho(s_2, p) + \rho(s_3, q) + \lambda. \tag{6.2.5.1.17}$$

From (6.2.5.1.11) and (6.2.5.1.15), $\rho(s_1, s_3) \leq \rho(s_1, p) + \rho(s_3, q) + \lambda$ and substituting (6.2.5.1.17) into this inequality gives the following.

$$\rho(s_1, s_3) \leq \rho(s_1, p) + \rho'(s_2, s_3) - \rho(s_2, p). \tag{6.2.5.1.18}$$

From (iii), $\rho(s_1, p) - \rho(p, s_2) \leq \rho(s_1, s_2)$ and substitute this into (6.2.5.1.18). Then, $\rho(s_1, s_3) \leq \rho(s_1, s_2) + \rho'(s_2, s_3)$. From (6.2.5.1.15)–(6.2.5.1.16), $\rho'(s_1, s_3) \leq \rho'(s_1, s_2) + \rho'(s_2, s_3)$.

⟨3⟩ $s_{1,2,2}, s_{1,3,3}$:

Let us prove only $s_{1,2,2}$ because of the symmetry of p and q.

$$\rho'(s_1, s_3) = \rho(s_1, s_3), \tag{6.2.5.1.19}$$

$$\rho'(s_1, s_2) = \rho(s_1, p) + \rho(s_2, q) + \lambda, \tag{6.2.5.1.20}$$

$$\rho'(s_2, s_3) = \rho(s_2, p) + \rho(s_3, q) + \lambda. \tag{6.2.5.1.21}$$

From (6.2.5.1.11) and (6.2.5.1.19), $\rho(s_1, s_3) \leq \rho(s_1, p) + \rho(s_3, q) + \lambda$ and substituting (6.2.5.1.16) into this inequality gives the following.

$$\rho(s_1, s_3) \leq \rho'(s_1, s_2) - \rho(s_2, q) + \rho(s_3, q). \tag{6.2.5.1.22}$$

Since each distance and λ are nonnegative real number, $0 \leq \rho(s_2, p) + \rho(s_2, q) + \lambda$. Adding $\rho(s_3, q)$ to both sides of this inequality, $\rho(s_3, q) - \rho(s_2, q) \leq \rho(s_2, p) + \rho(s_3, q) + \lambda$. Substituting (6.2.5.1.21) into this inequality gives the following:

$$\rho(s_3, q) - \rho(s_2, q) \leq \rho'(s_2, s_3). \tag{6.2.5.1.23}$$

From (6.2.5.1.22) and (6.2.5.1.23), $\rho(s_1, s_3) \leq \rho'(s_1, s_2) + \rho'(s_2, s_3)$. From (6.2.5.1.19), $\rho'(s_1, s_3) \leq \rho'(s_1, s_2) + \rho'(s_2, s_3)$.

$\langle 4 \rangle$ $s_{1,2,3}$, $s_{1,3,2}$:

Let us prove only $s_{1,2,3}$ because of the symmetry of p and q.

$$\rho'(s_1, s_3) = \rho(s_1, s_3), \tag{6.2.5.1.24}$$

$$\rho'(s_1, s_2) = \rho(s_1, p) + \rho(s_2, q) + \lambda, \tag{6.2.5.1.25}$$

$$\rho'(s_2, s_3) = \rho(s_2, p) + \rho(s_3, q) + \lambda. \tag{6.2.5.1.26}$$

From (6.2.5.1.3), $\rho(s_1, s_3) \leq \rho(s_1, p) + \rho(s_3, p)$. Since $\rho(s_2, q) + \lambda \geq 0$,

$$\rho(s_1, s_3) \leq \{\rho(s_1, p) + \rho(s_2, q) + \lambda\} + \{\rho(s_3, p) + \rho(s_2, q) + \lambda\}. \tag{6.2.5.1.27}$$

Substituting (6.2.5.1.25) and (6.2.5.1.26) into (6.2.5.1.27) gives $\rho(s_1, s_3) \leq \rho'(s_1, s_2) + \rho'(s_2, s_3)$. Hence, from (6.2.5.1.24), $\rho'(s_1, s_3) \leq \rho'(s_1, s_2) + \rho'(s_2, s_3)$.

$\langle 5 \rangle$ $s_{2,1,1}$, $s_{3,1,1}$:

Let us prove only $s_{2,1,1}$ because of the symmetry of p and q.

$$\rho'(s_1, s_3) = \rho(s_1, p) + \rho(s_3, q) + \lambda, \tag{6.2.5.1.28}$$

$$\rho'(s_1, s_2) = \rho(s_1, s_2), \tag{6.2.5.1.29}$$

$$\rho'(s_2, s_3) = \rho(s_2, s_3). \tag{6.2.5.1.30}$$

From (6.2.5.1.11) and (6.2.5.1.28),

$$\rho'(s_1, s_3) \leq \rho(s_1, s_3). \tag{6.2.5.1.31}$$

From (6.2.5.1.3) and (6.2.5.1.29)–(6.2.5.1.30),

$$\begin{aligned} \rho(s_1, s_3) &\leq \rho(s_1, s_2) + \rho(s_2, s_3) \\ &\leq \rho'(s_1, s_2) + \rho'(s_2, s_3). \end{aligned} \tag{6.2.5.1.32}$$

Hence, from (6.2.5.1.31)–(6.2.5.1.32), $\rho'(s_1, s_3) \leq \rho'(s_1, s_2) + \rho'(s_2, s_3)$.

$\langle 6 \rangle$ $s_{2,2,1}, s_{3,3,1}, s_{2,1,2}, s_{3,1,3}$:

Let us prove only $s_{2,2,1}$ because of the symmetry of p and q etc.

$$\rho'(s_1, s_3) = \rho(s_1, p) + \rho(s_3, q) + \lambda, \tag{6.2.5.1.33}$$

$$\rho'(s_1, s_2) = \rho(s_1, p) + \rho(s_2, q) + \lambda, \tag{6.2.5.1.34}$$

$$\rho'(s_2, s_3) = \rho(s_2, s_3). \tag{6.2.5.1.35}$$

From (6.2.5.1.3), $\rho(s_3, q) \leq \rho(q, s_2) + \rho(s_2, s_3)$. Adding $\rho(s_1, q) + \lambda \geq 0$ to both sides of this inequality,

$\rho(s_3, q) + \rho(s_1, q) + \lambda \leq \rho(q, s_2) + \rho(s_1, q) + \lambda + \rho(s_2, s_3)$. Hence, from (6.2.5.1.33)–(6.2.5.1.35), $\rho'(s_1, s_3) \leq \rho'(s_1, s_2) + \rho'(s_2, s_3)$.

$\langle 7 \rangle$ $s_{2,3,1}, s_{3,2,1}, s_{2,1,3}, s_{3,1,2}$:

Let us prove only $s_{2,3,1}$ because of the symmetry of p and q, etc.

$$\rho'(s_1, s_3) = \rho(s_1, p) + \rho(s_3, q) + \lambda, \tag{6.2.5.1.36}$$

$$\rho'(s_1, s_2) = \rho(s_1, q) + \rho(s_2, p) + \lambda, \tag{6.2.5.1.37}$$

$$\rho'(s_2, s_3) = \rho(s_2, s_3). \tag{6.2.5.1.38}$$

From (6.2.5.1.11) and (6.2.5.1.36), $\rho'(s_1, s_3) \leq \rho(s_1, q) + \rho(s_3, p) + \lambda$. Substituting (6.2.5.1.37) into this inequality gives

$$\rho'(s_1, s_3) \leq \rho'(s_1, s_2) + \rho(s_3, p) - \rho(s_2, p). \tag{6.2.5.1.39}$$

From (6.2.5.1.3), $\rho(s_3, p) - \rho(s_2, p) \leq \rho(s_2, s_3)$. Hence, from (6.2.5.1.38) and (6.2.5.1.39), $\rho'(s_1, s_3) \leq \rho'(s_1, s_2) + \rho'(s_2, s_3)$.

$\langle 8 \rangle$ $s_{2,2,2}, s_{3,3,3}$:

Let us prove only $s_{2,2,2}$ because of the symmetry of p and q.

$$\rho'(s_1, s_3) = \rho(s_1, p) + \rho(s_3, q) + \lambda, \tag{6.2.5.1.40}$$

$$\rho'(s_1, s_2) = \rho(s_1, p) + \rho(s_2, q) + \lambda, \tag{6.2.5.1.41}$$

$$\rho'(s_2, s_3) = \rho(s_2, p) + \rho(s_3, q) + \lambda. \tag{6.2.5.1.42}$$

Since each distance and λ are nonnegative real number, $0 \le \rho(s_2, p) + \rho(s_2, q) + \lambda$. Adding $\rho(s_1, p) + \rho(s_3, q) + \lambda$ to both sides of this inequality,

$$\rho(s_1, p) + \rho(s_3, q) + \lambda \le \{\rho(s_1, p) + \rho(s_2, q) + \lambda\} \\ + \{\rho(s_2, p) + \rho(s_3, q) + \lambda\}. \tag{6.2.5.1.43}$$

Hence, from (6.2.5.1.40)–(6.2.5.1.43), $\rho'(s_1, s_3) \le \rho'(s_1, s_2) + \rho'(s_2, s_3)$.

⟨9⟩ $s_{2,3,2}$, $s_{3,2,3}$, $s_{2,2,3}$, $s_{3,3,2}$:

Let us prove only $s_{2,3,2}$ because of the symmetry of p and q, etc.

$$\rho'(s_1, s_3) = \rho(s_1, p) + \rho(s_3, q) + \lambda, \tag{6.2.5.1.44}$$

$$\rho'(s_1, s_2) = \rho(s_1, q) + \rho(s_2, p) + \lambda, \tag{6.2.5.1.45}$$

$$\rho'(s_2, s_3) = \rho(s_2, p) + \rho(s_3, q) + \lambda. \tag{6.2.5.1.46}$$

From (6.2.5.1.11) and (6.2.5.1.44), $\rho(s_1, p) + \rho(s_3, q) + \lambda \le \rho(s_1, s_3)$. From (6.2.5.1.3), $\rho(s_1, s_3) \le \rho(s_1, q) + \rho(s_3, q)$. Hence,

$$\rho(s_1, p) + \lambda \le \rho(s_1, q). \tag{6.2.5.1.47}$$

Similarly from (6.2.5.1.11), (6.2.5.1.45) and (6.2.5.1.3),

$$\rho(s_1, q) + \lambda \le \rho(s_1, p). \tag{6.2.5.1.48}$$

From (6.2.5.1.47) and (6.2.5.1.48), $\lambda \le 0$. However, since λ is a nonnegative real number,

$$\lambda = 0. \tag{6.2.5.1.49}$$

Hence, from (6.2.5.1.47) and (6.2.5.1.48),

$$\rho(s_1, p) = \rho(s_1, q). \tag{6.2.5.1.50}$$

From (6.2.5.1.44)–(6.2.5.1.46), (6.2.5.1.49) and (6.2.5.1.1.50), $\rho'(s_1, s_2) + \rho'(s_2, s_3) = \rho(s_1, q) + \rho(s_2, p) + \rho(s_2, p) + \rho(s_3, q) = \rho'(s_1, s_3) + 2\rho(s_2, p)$. Since $2\rho(s_2, p) \ge 0$, $\rho'(s_1, s_3) \le \rho'(s_1, s_2) + \rho'(s_2, s_3)$.

$\langle 10\rangle$ $s_{2,3,3}$, $s_{3,2,2}$:

Let us prove only $s_{2,3,3}$ because of the symmetry of p and q.

$$\rho'(s_1, s_3) = \rho(s_1, p) + \rho(s_3, q) + \lambda, \quad (6.2.5.1.51)$$

$$\rho'(s_1, s_2) = \rho(s_1, q) + \rho(s_2, p) + \lambda, \quad (6.2.5.1.52)$$

$$\rho'(s_2, s_3) = \rho(s_2, q) + \rho(s_3, p) + \lambda. \quad (6.2.5.1.53)$$

From (6.2.5.1.51) and (6.2.5.1.3), $\rho(s_1, p) + \rho(s_3, q) + \lambda \leq \rho(s_1, s_3) \leq \rho(s_1, p) + \rho(s_3, q)$. Hence,

$$\rho(s_1, p) + \lambda \leq \rho(s_1, q). \quad (6.2.5.1.54)$$

Similarly,

$$\rho(s_3, q) + \lambda \leq \rho(s_3, p). \quad (6.2.5.1.55)$$

Similarly, from (6.2.5.1.52) and (6.2.5.1.3),

$$\rho(s_1, q) + \lambda \leq \rho(s_1, p). \quad (6.2.5.1.56)$$

Similarly, from (6.2.5.1.53) and (6.2.5.1.3),

$$\rho(s_3, p) + \lambda \leq \rho(s_3, q). \quad (6.2.5.1.57)$$

From (6.2.5.1.54), (6.2.5.1.56) and $\lambda \geq 0$,

$$\lambda = 0. \quad (6.2.5.1.58)$$

From (6.2.5.1.54), (6.2.5.1.56) and (6.2.5.1.58),

$$\rho(s_1, p) = \rho(s_1, q). \quad (6.2.5.1.59)$$

From (6.2.5.1.55), (6.2.5.1.57) and (6.2.5.1.58),

$$\rho(s_3, p) = \rho(s_3, q). \quad (6.2.5.1.60)$$

From (6.2.5.1.51)–(6.2.5.1.53) and (6.2.5.1.58)–(6.2.5.1.60),

$\rho'(s_1, s_2) + \rho'(s_2, s_3) = \rho(s_1, q) + \rho(s_2, p) + \rho(s_2, q) + \rho(s_3, p) = \rho(s_1, q) + \rho(s_3, p) + \rho(s_2, p) + \rho(s_2, q) = \rho(s_1, p) + \rho(s_3, q) + \rho(s_2, p) + \rho(s_2, q) = \rho'(s_1, s_3) + \rho(s_2, p) + \rho(s_2, q)$.

Since $\rho(s_2, p) + \rho(s_2, q) \geq 0$, $\rho'(s_1, s_3) \leq \rho'(s_1, s_2) + \rho'(s_2, s_3)$. □

The space $(S; \rho')$ obtained by (6.2.5.1.6) from $(S; \rho)$ is called the *contraction* of two points p and q of the space $(S; \rho)$. A new space obtained by repeating this contraction is called a contracted space.

6.2.5.2. *The centrality function in metric space*

Let S of $(S; \rho)$ be a countable set of points. Let us consider a mapping σ that associates each element of S with a nonnegative real number unrelated to ρ.

$$\sigma : S \to \overline{\boldsymbol{R}}^{+}. \tag{6.2.5.2.1}$$

This can be thought of as the weight of the point, which is independent of the relative distance relationship of the points. The metric space associated with σ is expressed as $(S; \rho, \sigma)$.

There are various methods to express the degree of the centrality of the points of $(S; \rho, \sigma)$. Here, let us consider a real-valued function f_s defined in $(S; \rho, \sigma)$ which expresses the degree of the centrality of a point. Since f_s is a function of S, ρ, σ and r where r is a point $r \in S$, it is expressed as $f_s(r, \rho, \sigma)$. Let f_s in the space $(S; \rho')$ obtained by (6.2.5.1.6) from $(S; \rho)$ be $f_s(r, \rho', \sigma)$. Let

$$\Delta(r) = f_s(r, \rho, \sigma) - f_s(r, \rho', \sigma). \tag{6.2.5.2.2}$$

Let us consider the following conditions.

(i) $\Delta(p) \geq 0,\ \Delta(q) \geq 0.$ (6.2.5.2.3)

(ii) For a point r where $\rho(p, r) \leq \rho(q, r)$,

$$\Delta(p) \geq \Delta(r) \tag{6.2.5.2.4}$$

and for a point r where $\rho(p, r) \geq \rho(q, r)$,

$$\Delta(q) \geq \Delta(r). \tag{6.2.5.2.5}$$

If f_s satisfies conditions (i) and (ii), f_s is called a *centrality function* of $(S; \rho, \sigma)$. If f_s satisfies condition (i), f_s is called a *semi-centrality function* of $(S; \rho, \sigma)$. The conditions (i) and (ii) are called an axiomatic system of a centrality function. The axiom system of this centrality function means that under the premise that the smaller the value of f_s, the larger the degree of the centrality, when the distance between the points p and q is reduced, the degree of the centrality of the points p and q increases (condition (i)), and the degree of the increase is larger than the vicinity (condition (i)). These conditions (i) and (ii) correspond to (A) and (B) in Section 6.2.5, respectively.

Let us show simple examples.

$$f_s^a(r, \rho, \sigma) = \sigma(r) \tag{6.2.5.2.6}$$

is a centrality function.

$$f_s^b(r, \rho, \sigma) = \max_{t \in S} \rho(r, t) \tag{6.2.5.2.7}$$

is a semi-centrality function. This is an eccentricity of a vertex in a network.

6.2.5.3. *Characterizing a form of centrality function*

Various functions that satisfy the axiom system (i) and (ii) ((6.2.5.2.3)–(6.2.5.2.5)) of the centrality function are considered. In this section, we will limit our discussion to the following form of the function.

$$f_s^L(r, \rho, \sigma) = \sum_{\xi \in D} \varphi(\xi) \sum_{t \in S_\xi} (r) \sigma(t) \tag{6.2.5.3.1}$$

where $D = \{\rho(s_1, s_2) | s_1, s_2 \in S\}$,

$$\varphi: \overline{\boldsymbol{R}}^+ \to \overline{\boldsymbol{R}}^+ \tag{6.2.5.3.2}$$

and

$$S_\xi(r) = \{t | t \in S, \rho(r, t) = \xi\}. \tag{6.2.5.3.3}$$

f_s^L (r, ρ, σ) of (6.2.5.3.1) is determined if φ is specified, but whether $f_s^L(r, \rho, \sigma)$ is a centrality function depends on S, ρ and σ. It is important to find out what kind of function $f_s^L(r, \rho, \sigma)$ is, such as a centrality function for any S, ρ or any S, σ.

First, let us find the condition for $f_s^L(r, \rho, \sigma)$ to be a centrality function for any S and ρ, even if σ takes a value limited to a certain range.

In the following, for $r \in S$, $\sigma(r)$ is bounded not to take a sufficiently large value, and for subset $S' \subset S$ which has large cardinality,

$$\sum_{t \in S' \subset S} \sigma(t) \to \infty. \tag{6.2.5.3.4}$$

Then, we obtain the following theorem.

Theorem 6.2.5.3.1. *Under condition* (6.2.5.3.4), *the necessary and sufficient condition for* f_s^L *to be a semi-centrality function for any* $(S; \rho)$ *is as follows*:

For $\mu' \geq \mu \geq \lambda$,

$$\varphi(\mu) \leq \varphi(\mu'). \tag{6.2.5.3.5}$$

Proof. *Sufficiency*: Sufficiency is evident from the fact that the space $(S; \rho')$ obtained by (6.2.5.1.6) from $(S; \rho)$, that is, by the contraction of two points p and q of the space $(S; \rho)$.

Necessity: Let us consider the following $(S; \rho, \sigma)$ and $(S; \rho', \sigma)$.

$$S = \{r_p, p, q, s_1, s_2, \ldots, s_m\} \tag{6.2.5.3.6}$$

where

$$\begin{aligned}
&\rho(r_p, p) = x, \\
&\rho(r_p, q) = x + y, \quad (y \geq 0), \\
&\rho(r_p, s_i) = x + y + z, \quad (1 \leq i \leq m), \\
&\rho(p, q) = y, \\
&\rho(p, s_i) = y + z, \quad (1 \leq i \leq m), \\
&\rho(q, s_i) = z, \quad (1 \leq i \leq m), \\
&\rho(s_i, s_j) = 2z, \quad (i \neq j, 1 \leq i\, j \leq m),
\end{aligned}$$

$$\rho'(r_p, p) = x,$$

$$\rho'(r_p, q) = x + \lambda, \quad (0 \le \lambda \le y),$$

$$\rho'(r_p, s_i) = x + \lambda + z, \quad (1 \le i \le m),$$

$$\rho'(p, q) = \lambda,$$

$$\rho'(p, s_i) = \lambda + z, \quad (1 \le i \le m),$$

$$\rho'(q, s_i) = z, \quad (1 \le i \le m),$$

and

$$\rho'(s_i, s_j) = 2z, \quad (i \ne j, 1 \le i, j \le m).$$

Equation (6.2.5.3.6) satisfies the axiom of metric for distance, and (6.2.5.1.6). For $(S; \rho, \sigma)$ and $(S; \rho', \sigma)$,

$$f_s^L(p, \rho, \sigma) = \varphi(0)\sigma(p) + \varphi(x)\sigma(r_p) + \varphi(y)\sigma(q) + \varphi(y+z)\sum_{i=1}^{m}\sigma(s_i), \tag{6.2.5.3.7}$$

$$f_s^L(p, \rho', \sigma) = \varphi(0)\sigma(p) + \varphi(x)\sigma(r_p) + \varphi(\lambda)\sigma(q) + \varphi(\lambda+z)\sum_{i=1}^{m}\sigma(s_i). \tag{6.2.5.3.8}$$

From (6.2.5.3.7) and (6.2.5.3.8),

$$\Delta(p) = \{\varphi(y) - \varphi(\lambda)\}\sigma(q) + \{\varphi(y+z) - \varphi(\lambda+z)\}\sum_{i=1}^{m}\sigma(s_i). \tag{6.2.5.3.9}$$

Let us consider the condition that $\Delta(p)$ of (6.2.5.3.9) is nonnegative for any $(S; \rho)$. $\Delta(p)$ of (6.2.5.3.9) has two terms, that is, the first term is the item regarding the weight of the point p and the second term is the item regarding the weight of the points $s_i (1 \le i \le m)$. In the case of $\sigma(q) \ne 0$ and $\sigma(s_i) = 0$ for s_i $(1 \le i \le m)$, and in the case of $\sigma(s_i) \ne 0$ (using (6.2.5.71)) for a sufficiently large m, the necessary conditions for $\Delta(p)$ to be nonnegative are as follows:

$$\varphi(\lambda) \leq \varphi(y), \tag{6.2.5.3.10}$$

$$\varphi(\lambda + z) \leq \varphi(y + z). \tag{6.2.5.3.11}$$

Therefore, we obtain (6.2.5.3.5). □

Theorem 6.2.5.3.2. *Under condition* (6.2.5.3.4), *the necessary and sufficient conditions for f_s^L to be a centrality function for any $(S; \rho)$ are as follows*:

For $\mu' \geq \mu \geq \lambda$,

$$\varphi(\mu) \leq \varphi(\mu'). \tag{6.2.5.3.12}$$

For $\mu \geq \lambda + \varepsilon, \varepsilon \geq 0$,

$$2\varphi(\mu) \geq \varphi(\mu - \varepsilon) + \varphi(\mu + \varepsilon). \tag{6.2.5.3.13}$$

Proof. *Sufficiency*: Let φ satisfies (6.2.5.3.12) and (6.2.5.3.13). We classify the set S into the following three subsets.

$$S_p = \{r | \rho(p, r) < \rho(q, r)\}, \tag{6.2.5.3.14}$$

$$S_0 = \{r | \rho(p, r) = \rho(q, r)\}, \tag{6.2.5.3.15}$$

$$S_q = \{r | \rho(p, r) > \rho(q, r)\}. \tag{6.2.5.3.16}$$

Let r_p, r_0 and r_q be representative elements of S_p, S_0 and S_q, respectively. Suppose that the distance between r_p and r_q in $(S; \rho, \sigma)$ decreases in $(S; \rho', \sigma)$.

$$\rho(r_p, r_q) \geq \rho(r_p, p) + \rho(r_q, q) + \lambda \leq \rho(r_p, q) + \rho(r_q, p) + \lambda. \tag{6.2.5.3.17}$$

From (6.2.5.1.6)

$$\rho'(r_p, r_q) = \rho(r_p, p) + \rho(r_q, q) + \lambda. \tag{6.2.5.3.18}$$

Using triangle inequality,

$$\rho(r_p, p) + \rho(p, r_q) \geq \rho(r_p, r_q) \geq \rho(r_p, p) + \rho(r_q, q) + \lambda,$$

$$\rho(p, r_q) \geq \rho(r_q, q) + \lambda, \tag{6.2.5.3.19}$$

From (6.2.5.3.19), (6.2.5.3.14) and (6.2.5.3.16),

$$\begin{aligned}\rho'(p, r_q) &= \min\{\rho(p, r_q), \rho(p, p) + \rho(q, r_q) + \lambda, \rho(q, q) \\ &\quad + \rho(r_q, p) + \lambda\} \\ &= \rho(q, r_q) + \lambda. \end{aligned} \tag{6.2.5.3.20}$$

Let $\rho(r_p, p) = w, \rho(p, r_q) = x, \rho(r_q, q) = y$ and $\rho(r_p, r_q) = z$, and let $\partial_{r_q} \Delta(r_p)$ and $\partial_{r_q} \Delta(p)$ be the contributions of r_q to $\Delta(r_p)$ and $\Delta(p)$, respectively. From (6.2.5.3.18) and (6.2.5.3.20),

$$\partial_{r_q} \Delta(r_p) = \{\varphi(z) - \varphi(w + y + \lambda)\}\sigma(r_q), \tag{6.2.5.3.21}$$

$$\partial_{r_q} \Delta(p) = \{\varphi(x) - \varphi(y + \lambda)\}\sigma(r_q), \tag{6.2.5.3.22}$$

$$\begin{aligned}\partial_{r_q} \Delta(p) - \partial_{r_q} \Delta(r_p) &= \{\varphi(x) - \varphi(y + \lambda) - \varphi(z) \\ &\quad + \varphi(w + y + \lambda)\sigma(r_q). \end{aligned} \tag{6.2.5.3.23}$$

Using triangle inequality $\rho(r_p, p) + \rho(p, r_q) \geq \rho(r_p, r_q)$ and from (6.2.5.3.17), we obtain that $z \geq \lambda$ and $w + x \geq z \geq \lambda$. Hence, from (6.2.5.3.12)

$$\varphi(w + x) \geq \varphi(z). \tag{6.2.5.3.24}$$

From (6.2.5.3.23) and (6.2.5.3.24),

$$\begin{aligned}\partial_{r_q} \Delta(p) - \partial_{r_q} \Delta(r_p) &\geq \{\varphi(x) - \varphi(y + \lambda) - \varphi(w + x) \\ &\quad + \varphi(w + y + \lambda)\}\sigma(r_q). \end{aligned} \tag{6.2.5.3.25}$$

In (6.2.5.3.13), for $\xi > \mu \geq \lambda + \varepsilon$,

$$\varphi(\mu + \xi - \varepsilon) - \varphi(\mu + \xi) \geq \varphi(\mu - \varepsilon) - \varphi(\mu) \tag{6.2.5.3.26}$$

In (6.2.5.3.25), let $\mu = x$, $\mu - \varepsilon = y + \lambda$ and $\xi = w$. Then,

$$\varphi(w + y + \lambda) - \varphi(y + \lambda) \geq \varphi(y + \lambda) - \varphi(x) \tag{6.2.5.3.27}$$

Since $\sigma(r_q) \geq 0$, from (6.2.5.3.25),

$$\partial_{r_q} \Delta(p) - \partial_{r_q} \Delta(r_p) \geq 0. \tag{6.2.5.3.28}$$

Suppose that the distance between r_p and r_q in $(S; \rho, \sigma)$ does not change in $(S; \rho', \sigma)$. Then, $\partial_{r_q} \Delta(r_p) = 0$. From (6.2.5.3.22), (6.2.5.3.19) and (6.2.5.3.12), for any $r_q \in S_q$,

$$\partial_{r_q} \Delta\,(p) \geq 0. \tag{6.2.5.3.29}$$

In particular, the following holds even when $r_q = q(y = 0)$.

$$\partial_q \Delta\,(p) \geq 0. \tag{6.2.5.3.30}$$

Hence, for any $r_q \in S_q$,

$$\partial_{r_q} \Delta\,(p) - \partial_{r_q} \Delta(r_p) \geq 0. \tag{6.2.5.3.31}$$

Next, let us consider the distance between r_x and r_p for $r_x, r_p \in S_p \cup S_0$.

From $\rho(r_p, p) + \rho(r_x, q) \geq \rho(r_p, p) + \rho(r_x, p)$ and triangle inequality

$$\rho(r_p, p) + \rho(p, r_x) \geq \rho(r_p, r_x)$$

$$\rho(r_p, p) + \rho(r_x, q) + \lambda \geq \rho(r_p, p) + \rho(r_x, p) + \lambda \geq \rho(r_p, r_x). \tag{6.2.5.3.32}$$

Hence, from (6.2.5.1.6),

$$\rho'(r_p, r_x) = \rho(r_p, r_x), \tag{6.2.5.3.33}$$

that is, the distance the distance between r_x and r_p does not change in $(S; \rho', \sigma)$. Since (6.2.5.3.33) holds for $r_p = p$,

$$\partial_{r_x} \Delta(r_p) = 0, \tag{6.2.5.3.34}$$

$$\partial_{r_x} \Delta(p) = 0 \tag{6.2.5.3.35}$$

and

$$\partial_{r_x} \Delta(r_p) - \partial_{r_x} \Delta(r_p) = 0. \tag{6.2.5.3.36}$$

Since $\Delta(r_p)$ is the sum of $\partial_{r_q} \Delta(r_p)$ and $\partial_{r_x} \Delta\left(r_p\right)$, etc., from (6.2.5.3.28), (6.2.5.3.31) and (6.2.5.3.36),

$$\Delta(p) \geq 0, \tag{6.2.5.3.37}$$

$$\Delta(p) \geq \Delta(r_p). \tag{6.2.5.3.38}$$

Similarly,

$$\Delta(q) \geq 0, \tag{6.2.5.3.39}$$

$$\Delta(q) \geq \Delta(r_q). \tag{6.2.5.3.40}$$

Next, let us consider the distance from r_0 to other points. Choosing $r_x \in S_p \bigcup S_0$ as an other point does not lose generality.

From $\rho(r_0, p) + \rho(r_x, q) \geq \rho(r_0, q) + \rho(r_x, p) = \rho(r_0, p) + \rho(r_x, p)$ and triangle inequality

$$\rho(r_0, p) + \rho(p, r_x) \geq \rho(r_0, r_x)$$

$$\rho(r_0, p) + \rho(r_x, q) + \lambda \geq \rho(r_0, q) + \rho(r_x, p) + \lambda \geq \rho(r_0, r_x). \tag{6.2.5.3.41}$$

Hence,

$$\rho'(r_0, r_x) = \rho(r_0, r_x), \tag{6.2.5.3.42}$$

that is,

$$\Delta(r_0) = 0 \tag{6.2.5.3.43}$$

Therefore, if φ satisfies (6.2.5.3.12) and (6.2.5.3.13), f_s^L is a centrality function.

Necessity: Let us use again (6.2.5.3.6) of Theorem 6.2.5.3.1 as $(S; \rho, \sigma)$ and $(S; \rho', \sigma)$. The necessity of (6.2.5.3.12) is obvious from Theorem 6.2.5.3.1. Since r_p in (6.2.5.3.6) satisfies the condition of (6.2.5.3.14), $r_p \in S_p$ and

$$f_s^L(r_p, \rho, \sigma) = \varphi(0)\sigma(r_p) + \varphi(x)\sigma(p) + \varphi(x+y)\sigma(q) + \varphi(x+y+z)\sum_{i=1}^{m}\sigma(s_i), \tag{6.2.5.3.44}$$

$$f_s^L(r_p, \rho', \sigma) = \varphi(0)\sigma(r_p) + \varphi(x)\sigma(p) + \varphi(x+\lambda)\sigma(q) + \varphi(x+\lambda+z)\sum_{i=1}^{m}\sigma(s_i). \tag{6.2.5.3.45}$$

From (6.2.5.3.44) and (6.2.5.3.45),

$$\Delta(r_p) = \{\varphi(x+y) - \varphi(x+\lambda)\}\sigma(q) + \{\varphi(x+y+z) - \varphi(x+\lambda+z)\}\sum_{i=1}^{m}\sigma(s_i). \tag{6.2.5.3.46}$$

From (6.2.5.3.9) and (6.2.5.3.46),

$$\begin{aligned}\Delta(p) - \Delta(r_p) = {} & \{\varphi(y) - \varphi(\lambda) - \varphi(x+y) + \varphi(x+\lambda)\}\sigma(q) \\ & + \{\varphi(y+z) - \varphi(\lambda+z) - \varphi(x+y+z) \\ & + \varphi(x+\lambda+z)\}. \end{aligned} \tag{6.2.5.3.47}$$

Let us consider the condition that $\Delta(p) \geq \Delta(r_p)$ for any $(S;\rho)$. $\Delta(p) - \Delta(r_p)$ of (6.2.5.3.47) has two terms, that is, the first term is the item regarding the weight of the point q, and the second term is the item regarding the weight of the points $s_i (1 \leq i \leq m)$. In the case of $\sigma(q) \neq 0$ and $\sigma(s_i) = 0$ for $s_i (1 \leq i \leq m)$, and in the case of $\sigma(s_i) \neq 0$ (using (6.2.5.3.4)) for a sufficiently large m, the necessary conditions for $\Delta(p) \geq \Delta(r_p)$ are as follows:

$$\varphi(y) - \varphi(\lambda) - \varphi(x+y) + \varphi(x+\lambda) \geq 0, \tag{6.2.5.3.48}$$

$$\varphi(y+z) - \varphi(\lambda+z) - \varphi(x+y+z) + \varphi(x+\lambda+z) \geq 0. \tag{6.2.5.3.49}$$

Let $y = \lambda + \varepsilon$, $x = \varepsilon$, $\varepsilon > 0$, then from (6.2.5.3.48),

$$2\varphi(\lambda+\varepsilon) \geq \varphi(\lambda) + \varphi(\lambda + 2\varepsilon). \tag{6.2.5.3.50}$$

Since the (6.2.5.3.49) must hold for $z > 0$, for any $\mu \geq \lambda + \varepsilon$, $\varepsilon > 0$,

$$2\varphi(\mu) \geq \varphi(\mu - \varepsilon) + \varphi(\mu + \varepsilon). \tag{6.2.5.3.51}$$

Similarly, we obtain the necessity for $\Delta(q)\Delta(r_q)$, etc. □

Corollary 6.2.5.3.2. *Let*

$$f_s^{L_1}(r,\rho) = \sum_{\xi \in D} \varphi(\xi)\,|S_\xi(r)|. \tag{6.2.5.3.52}$$

The necessary and sufficient conditions for $f_s^{L_1}$ to be a centrality function for any $(S;\rho)$ are (6.2.5.3.12) *and* (6.2.5.3.13).

Proof. $f_s^{L_1}$ is the case where $\sigma(s_i) = 1$, $(s_i \in S)$ in f_s^L of (6.2.5.3.1). It is obvious that if f_s^L and $f_s^{L_1}$ are satisfied only (6.2.5.3.12), they are semi-centrality functions. □

In Theorems 6.2.5.3.1 and 6.2.5.3.2, we assume (6.2.5.3.4) which is the condition for large number of points. Let us consider the necessary and sufficient conditions f_s^L to be a centrality function for any σ even if the number of points is small. In this case, since it is necessary to hold for any σ, instead of taking the large enough m, σ is taken big enough in the proof of the necessity of Theorems 6.2.5.3.1 and 6.2.5.3.2. This shows that the same conditions as Theorems 6.2.5.3.1 and 6.2.5.3.2 are necessary. Since the proof of sufficiency is similar to Theorems 6.2.5.3.1 and 6.2.5.3.2, the following theorem is obtained.

Theorem 6.2.5.3.3. *The necessary and sufficient conditions for f_s^L to be a centrality function for any $(S;\rho,\sigma)$ are* (6.2.5.3.12) *and* (6.2.5.3.13).

From Theorem 6.2.5.3.3, a function $\varphi(x)$ of f_s^L is a *convex function* with a *monotonous increase*.

For example,

$$\varphi(x) = x,\ \varphi(x) = \sqrt{x},\quad \varphi(x) = 1 - e^x,\ \text{etc.} \tag{6.2.5.3.53}$$

Theorem 6.2.5.4 is applicable to measure the centrality in a graph or a network.

6.2.5.4. *Some centrality functions in a graph*

In a connected graph $G = (V, E)$ with positive real number weights attached to each of its vertices and edges. Let the length of a path define as the sum of the weights of all edges in the path. For simplicity, let all of weights of edges and vertices be 1, then the length of a path is defined as the number of edges in the path. If the length of a path in paths between

v_i and v_j is shortest in G, the length is called the distance between v_i and v_j. The graph has the following properties (a) and (b).

(a) The distance the between vertices in a graph is metric form (6.2.3.1.1)–(6.2.3.1.3).
(b) If the addition of one new path reduces the distance between two points in the graph, the shortest path between the two points goes through the new path.

Since a graph has the properties (a) and (b), the concept of generalizations of centrality in graphs and network in Section 6.2.5.2 are applicable. For examples,

(1) $\varepsilon(v_i) = \max_{v_j \in V} d(v_i, v_j)$ defined as the eccentricity of a vertex in (6.2.3.1.4) is a semi-centrality function. This is shown in (6.2.5.2.7).
(2) $t(x_i) = \sum_{v_j \in V} d\left(x_i, v_j\right) w(v_j)$ defined as the transmission number of a vertex in (6.2.4.1.1) is a centrality function. This is an example of (6.2.5.3.1) where $\varphi(\xi) = \xi$.
(3) In a graph, the degree of a vertex v_i, that is, $d(v_i)$ is a semi-centrality function.

Next, let us consider a connected graph $G = (V, E)$ with positive real number weights attached to each of its vertices and edges. Let $w(v_j)$ and $c(e_i)$ be the weight of a vertex v_j and the capacity of an edge e_i in G, respectively. Let the capacity of a cutset define as the sum of the weights of all edges in the cutset. If the capacity of a cutset in cutsets between v_i and v_j is minimum in G, the capacity is called the max-flow, the capacity or gauge $g(v_i, v_j)$ between v_i and v_j. For simplicity, let all of weights of edges and vertices be 1, then the capacity of a minimum cutsets between v_i and v_j is defined as the number of edges in the cutset. The equation set (6.2.3.2.1)–(6.2.3.2.3) called axiom of metric for gauge is shown below again.

$$g(v_i, v_j) = \infty, \text{ if and only if } v_i = v_j, \tag{6.2.3.2.1}$$

$$g(v_i, v_j) = g(v_j, v_i), \tag{6.2.3.2.2}$$

$$g(v_i, v_j) \geq \min\{g(v_i, v_k), g(v_k, v_j)\}. \tag{6.2.3.2.3}$$

If G is a nondirected graph (briefly a graph), G satisfies equations (6.2.3.2.1)–(6.2.3.2.3). If G is a directed graph, G may not satisfy equation (6.2.3.2.2). In a directed graph, gauge $g(v_i, v_j)$ represents the max-flow from v_i to v_j, and (6.2.3.2.1) and (6.2.3.2.3) are satisfied in the graph.

Lemma 6.2.5.4.1. *In a graph G, let us consider three vertices v_i, v_j and v_k such that*

$$g(v_i, v_j) \geq g(v_i, v_k) \geq g(v_k, v_j). \tag{6.2.5.4.1}$$

Then,

$$g(v_i, v_k) = g(v_k, v_j). \tag{6.2.5.4.2}$$

Proof. Since the gauge in G satisfies (6.2.3.2.1)–(6.2.3.2.3), $g(v_k, v_j) \geq \min\{g(v_i, v_k), g(v_i, v_j)\}$. From (6.2.3.2.2) and (6.2.5.4.1),

$$g(v_k, v_j) \geq g(v_i, v_k). \tag{6.2.5.4.3}$$

From (6.2.3.2.3), $g(v_i, v_k) \geq \min\{g(v_k, v_j), g(v_i, v_j)\}$ and from (6.2.3.2.2) and (6.2.5.4.1),

$$g(v_i, v_k) \geq g(v_k, v_j). \tag{6.2.5.4.4}$$

Therefore, from (6.2.5.4.3) and (6.2.5.4.4), $g(v_i, v_k) = g(v_k, v_j)$. □

Let M_{sc} be the number greater than the sum of all edge capacitors in $G = (V, E)$ and let us define $\hat{g}(v_i, v_j)$ by $g(v_i, v_j)$ as follows:

$$\hat{g}(v_i, v_j) = M_{sc} - g(v_i, v_j), \quad \text{for } v_i, v_j \in V. \tag{6.2.5.4.5}$$

Since M_{sc} is the number greater than the sum of all edge capacitors in $G = (V, E)$ in (6.2.5.4.5),

$$\hat{g}(v_i, v_j) = M_{sc} - g(v_i, v_j) \geq 0, \tag{6.2.5.4.6}$$

except for $v_i = v_j$. Let us define

$$\hat{g}(v_i, v_i) = 0 \quad \text{for } v_i \in V. \tag{6.2.5.4.7}$$

If all of weights of edges are 1, $M_{sc} > |E|$.

Theorem 6.2.5.4.1. *In a connected graph $G = (V, E)$ called flow graph with positive real number, weights are attached to each of its vertices and edges, $\hat{g}(v_i, v_j)$ is metric, that is,*

$$\hat{g}(v_i, v_j) = 0 \Leftrightarrow v_i = v_j, \tag{6.2.5.4.8}$$

$$\hat{g}(v_i, v_j) = \hat{g}(v_j, v_i), \tag{6.2.5.4.9}$$

$$\hat{g}(v_i, v_j) \leq \hat{g}(v_i, v_k) + \hat{g}(v_k, v_j). \tag{6.2.5.4.10}$$

Proof. From (6.2.5.4.5)–(6.2.5.4.7), it is obvious that (6.2.5.4.8) and (6.2.5.4.9) hold.

From Lemma 6.2.5.4.1, for v_i, v_j and v_k

$$g(v_i, v_j) \geq g(v_i, v_k) = g(v_k, v_j). \tag{6.2.5.4.11}$$

From (6.2.5.4.5),

$$\begin{aligned}\hat{g}(v_i, v_k) + \hat{g}(v_k, v_j) - \hat{g}(v_i, v_j) &= \{M_{sc} - g(v_i, v_k)\} + \{M_{sc} - g(v_k, v_j)\} \\ &\quad -\{M_{sc} - g(v_i, v_j)\} \\ &= \{M_{sc} - g(v_i, v_k)\} + \{g(v_i, v_j) \\ &\quad - g(v_k, v_j)\}. \end{aligned} \tag{6.2.5.4.12}$$

In (6.2.5.4.12), $\{M_{sc} - g(v_i, v_k)\}$ and $\{g(v_i, v_j) - g(v_k, v_j)\}$ are nonnegative from (6.2.5.4.6) and (6.2.5.4.11), that is,

$$\hat{g}(v_i, v_k) + \hat{g}(v_k, v_j) - \hat{g}(v_i, v_j) \geq 0.$$

Therefore, $\hat{g}(v_i, v_j) \leq \hat{g}(v_i, v_k) + \hat{g}(v_k, v_j)$. □

From Theorem 6.2.5.4.1, the concept of generalizations of centrality in graphs and network in Section 6.2.5.2 are applicable to nondirected flow graph. For example,

(4) Let us define the eccentricity of a vertex as (1).

$$\hat{\varepsilon}(v_i) = \max_{v_j \in V} \hat{g}(v_i, v_j) \tag{6.2.5.4.13}$$

This corresponds to (6.2.3.2.4) and is a semi-centrality function shown in (6.2.5.2.7).

(5) Let us define the transmission number for max-flow as (2).

$$\hat{t}_c(v_i) \sum_{\substack{v_j \in V \\ v_j \neq v_i}} \hat{g}(v_i, v_j) w(v_j), \tag{6.2.5.4.14}$$

This corresponds to (6.2.4.2.1) and this is a centrality function shown in (6.2.5.3.1). This is an example of (6.2.5.3.1) where $\varphi\,(\xi) = \xi$.

In a connected directed flow graph, (6.2.5.4.9) and (6.2.5.4.10) may not be satisfied and $\hat{g}(v_i, v_j)$ is not metric. Thus, the concept of generalizations of centrality in graphs and network in Section 6.2.5.2 is not applicable to a connected directed flow graph. However, it is known that the similar generalization as Section 6.2.5.2 is possible using the concept of space modification, which is also applicable to directed flow graphs. (The Concept of Space Modification and a Theory of Function Measuring the Centrality (or Mediality) of a Point in a Network is in [25]).

Example. Figure 6.2.5.4.1 is a graph reducing the distance $d(v_3, v_9)$ by adding an edge e_{13} in a graph of Fig. 6.2.4.1.1.

In Fig. 6.2.5.4.1, $G' = (V, E)$, where $V = \{v_1, v_2, v_3, v_4, v_5, v_6, v_7, v_8, v_9, v_{10}, v_{11}\}$ $E = \{e_1, e_2, e_3, e_4, e_5, e_6, e_7, e_8, e_9, e_{10}, e_{11}, e_{12}\}$ and $e_1 = (v_1, v_2), e_2 = (v_2, v_3), e_3 = (v_3, v_4), e_4 = (v_3, v_8),\ e_5 = (v_4, v_5), e_6 = (v_4, v_6), e_7 = (v_5, v_6), e_8 = (v_6, v_7), e_9 = (v_6, v_8),\ e_{10} = (v_8, v_9), e_{11} = (v_9, v_{10}), e_{12} = (v_9, v_{11}), e_{13} = (v_3, v_9)$.

Let $\boldsymbol{D}'_{\mathbf{s}}$ be a matrix of distance of G' in Fig. 6.2.5.4.1. For simplicity, all of weights of edges and vertices be 1, that is, $l(e_i) = 1$ and $w(v_j) = 1$ for $e_i \in E$ and $v_j \in V$. Then,

$$t'(v_i) = \sum_{v_j \neq v_i} v_j \in V d'\left(v_i, v_j\right),$$

$$t'(v_1) = 33, t'(v_2) = 24, t'(v_3) = 17, t'(v_4) = 20, t'(v_5) = 26, t'(v_6) = 21,$$

$$t'(v_7) = 30, t'(v_8) = 18, t'(v_9) = 19, t'(v_{10}) = 28, t'(v_{11}) = 28.$$

Since $s'_1 \triangleq \min_{v_i \in V} t'(v_i) = 17$, the vertex v_3 is a median of G' in Fig. 6.2.5.4.1. Since a median of G in Fig. 6.2.4.1.1 is the vertex v_8, the

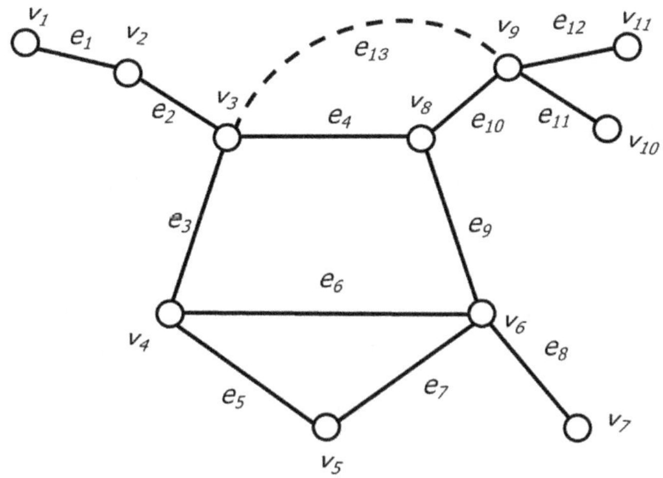

Fig. 6.2.5.4.1. A graph G' reducing the distance $d(v_3, v_9)$ by adding an edge e_{13} in a graph of Fig. 6.2.4.1.1.

position of median is changed from v_8 to v_3.

$$
\boldsymbol{D}'_s = \begin{array}{c} v_1 \\ v_2 \\ v_3 \\ v_4 \\ v_5 \\ v_6 \\ v_7 \\ v_8 \\ v_9 \\ v_{10} \\ v_{11} \end{array}
\overset{\begin{array}{ccccccccccc} v_1 & v_2 & v_3 & v_4 & v_5 & v_6 & v_7 & v_8 & v_9 & v_{10} & v_{11} \end{array}}{\begin{bmatrix}
\cdot & 1 & 2 & 3 & 4 & 4 & 5 & 3 & 3 & 4 & 4 \\
1 & \cdot & 1 & 2 & 3 & 3 & 4 & 2 & 2 & 3 & 3 \\
2 & 1 & \cdot & 1 & 2 & 2 & 3 & 1 & 1 & 2 & 2 \\
3 & 2 & 1 & \cdot & 1 & 1 & 2 & 2 & 3 & 4 & 4 \\
4 & 3 & 2 & 1 & \cdot & 1 & 2 & 2 & 3 & 4 & 4 \\
4 & 3 & 2 & 1 & 1 & \cdot & 1 & 1 & 2 & 3 & 3 \\
5 & 4 & 3 & 2 & 2 & 1 & \cdot & 2 & 3 & 4 & 4 \\
3 & 2 & 1 & 2 & 2 & 1 & 2 & \cdot & 1 & 2 & 2 \\
3 & 2 & 1 & 2 & 3 & 2 & 3 & 1 & \cdot & 1 & 1 \\
4 & 3 & 2 & 3 & 4 & 3 & 4 & 2 & 1 & \cdot & 2 \\
5 & 3 & 2 & 3 & 4 & 3 & 4 & 2 & 1 & 2 & \cdot
\end{bmatrix}}.
$$

Let $\boldsymbol{F}'_m$ be a matrix of max-flow of G' in Fig. 6.2.5.4.1, where for simplicity, all of weights of edges and vertices be 1, that is, $c(e_i) = 1$ and $w(v_j) = 1$

for $e_i \in E$ and $v_j \in V$, and the ijth entry of matrix $\boldsymbol{F}'_m$ shows the gauge $g'(v_i, v_j)$.

$$t'_c(v_1) = 10,\ t'_c(v_2) = 10,\ t'_c(v_3) = 16,\ t'_c(v_4) = 16,\ t'_c(v_5) = 15,\ t'_c(v_6) = 16,$$
$$t'_c(v_7) = 10,\ t'_c(v_8) = 16,\ t'_c(v_9) = 15,\ t'_c(v_{10}) = 10,\ t'_c(v_{11}) = 10.$$

From (6.2.5.4.5), let $M_{sc} = 15$.

$$\hat{g}'(v_i, v_j) = M_{sc} - g'(v_i, v_j) = 15 - g'(v_i, v_j).$$

Let $\hat{\boldsymbol{F}}'_m$ be a matrix of G' in Fig. 6.2.5.4.1, where the ijth entry of the matrix shows the gauge $\hat{g}'(v_i, v_j)$. From (6.2.5.4.14),

$$\hat{t}'_c(v_1) = 140,\ \hat{t}'_c(v_2) = 140,\ \hat{t}'_c(v_3) = 134,\ \hat{t}'_c(v_4) = 134,\ \hat{t}'_c(v_5) = 135,$$
$$\hat{t}'_c(v_6) = 134,\ \hat{t}'_c(v_7) = 140,\ \hat{t}'_c(v_8) = 134,\ \hat{t}'_c(v_9) = 135,$$
$$\hat{t}'_c(v_{10}) = 140,\ \hat{t}'_c(v_{11}) = 140.$$

$$F'_m = \begin{array}{c} \\ v_1 \\ v_2 \\ v_3 \\ v_4 \\ v_5 \\ v_6 \\ v_7 \\ v_8 \\ v_9 \\ v_{10} \\ v_{11} \end{array} \begin{array}{c} \begin{array}{ccccccccccc} v_1 & v_2 & v_3 & v_4 & v_5 & v_6 & v_7 & v_8 & v_9 & v_{10} & v_{11} \end{array} \\ \left[\begin{array}{ccccccccccc} \cdot & 1 & 1 & 1 & 1 & 1 & 1 & 1 & 1 & 1 & 1 \\ 1 & \cdot & 1 & 1 & 1 & 1 & 1 & 1 & 1 & 1 & 1 \\ 1 & 1 & \cdot & 2 & 2 & 2 & 1 & 3 & 2 & 1 & 1 \\ 1 & 1 & 2 & \cdot & 2 & 3 & 1 & 2 & 2 & 1 & 1 \\ 1 & 1 & 2 & 2 & \cdot & 2 & 1 & 2 & 2 & 1 & 1 \\ 1 & 1 & 2 & 3 & 2 & \cdot & 1 & 2 & 2 & 1 & 1 \\ 1 & 1 & 1 & 1 & 1 & 1 & \cdot & 1 & 1 & 1 & 1 \\ 1 & 1 & 3 & 2 & 2 & 2 & 1 & \cdot & 2 & 1 & 1 \\ 1 & 1 & 2 & 2 & 2 & 2 & 1 & 2 & \cdot & 1 & 1 \\ 1 & 1 & 1 & 1 & 1 & 1 & 1 & 1 & 1 & \cdot & 1 \\ 1 & 1 & 1 & 1 & 1 & 1 & 1 & 1 & 1 & 1 & \cdot \end{array}\right] \end{array}.$$

$$\hat{F}'_m = \begin{array}{c} \\ v_1 \\ v_2 \\ v_3 \\ v_4 \\ v_5 \\ v_6 \\ v_7 \\ v_8 \\ v_9 \\ v_{10} \\ v_{11} \end{array} \begin{array}{c} \begin{array}{ccccccccccc} v_1 & v_2 & v_3 & v_4 & v_5 & v_6 & v_7 & v_8 & v_9 & v_{10} & v_{11} \end{array} \\ \left[\begin{array}{ccccccccccc} \cdot & 14 & 14 & 14 & 14 & 14 & 14 & 14 & 14 & 14 & 14 \\ 14 & \cdot & 14 & 14 & 14 & 14 & 14 & 14 & 14 & 14 & 14 \\ 14 & 14 & \cdot & 13 & 13 & 13 & 14 & 12 & 13 & 14 & 14 \\ 14 & 14 & 13 & \cdot & 13 & 12 & 14 & 13 & 13 & 14 & 14 \\ 14 & 14 & 13 & 13 & \cdot & 13 & 14 & 13 & 13 & 14 & 14 \\ 14 & 14 & 13 & 12 & 13 & \cdot & 14 & 13 & 13 & 14 & 14 \\ 14 & 14 & 14 & 14 & 14 & 14 & \cdot & 14 & 14 & 14 & 14 \\ 14 & 14 & 12 & 13 & 13 & 13 & 14 & \cdot & 13 & 14 & 14 \\ 14 & 14 & 13 & 13 & 13 & 13 & 14 & 13 & \cdot & 14 & 14 \\ 14 & 14 & 14 & 14 & 14 & 14 & 14 & 14 & 14 & \cdot & 14 \\ 14 & 14 & 14 & 14 & 14 & 14 & 14 & 14 & 14 & 14 & \cdot \end{array}\right] \end{array}.$$

Since $s'_{cm1} \triangleq \max_{v_i \in V} t'_c(v_i) = 16$ and $\hat{s}'_{cm1} \triangleq \min_{v_i \in V} \hat{t}'_c(v_i) = 134$, the vertices v_3, v_4, v_6 and v_8 are medians for max-flow of G'.

The medians for max-flow of G in Fig. 6.2.4.1.1 are the vertices v_4 and v_6, and the medians for max-flow of G' by adding an edge e_{13} in a graph of Fig. 6.2.4.1.1 are the vertices v_3, v_4, v_6 and v_8. **(End of Example)**

6.2.5.5. *Some other centrality functions in a graph*

On the *Internet*, a directed graph that represents a web page as a vertex and a link as an edge is called a *web graph*. In the web graph, the centrality of vertices may be considered with the following intuitive properties. In a web graph (directed graph) $G = (V, E)$,

(a) Vertices that are linked from many vertices are highly central.
(b) Vertices that are linked from vertices with high centrality are highly central.
(c) The smaller the number of edges connected from a vertex with higher centrality, the higher the centrality of the linked vertex.

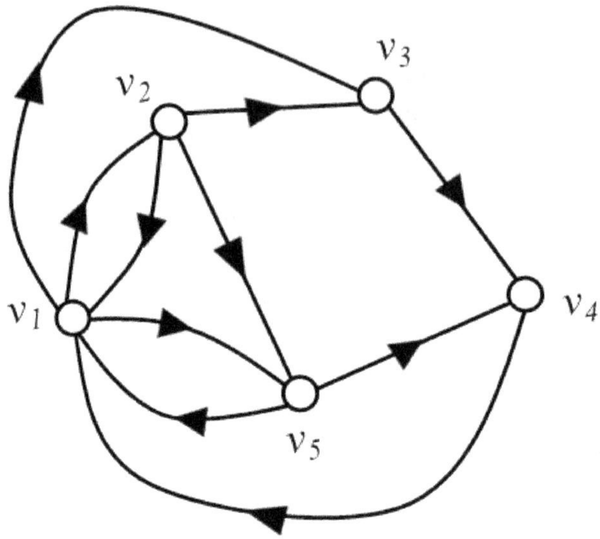

Fig. 6.2.5.5.1. A web graph $G = (V, E)$.

Let $f_w(v_i)$ be the centrality of the vertex v_i according to these (a)–(c).

$$f_w(v_i) = \sum_{(v_j, v_i) \in E} \frac{f_w(v_j)}{d^+(v_j)} \tag{6.2.5.5.1}$$

Let us show a simple example for (6.2.5.5.1). In Fig. 6.2.5.5.1, $G = (V, E)$ is a web graph and the adjacency matrix X_d is as follows:

$$X_d = \begin{array}{c} \\ v_1 \\ v_2 \\ v_3 \\ v_4 \\ v_5 \end{array} \begin{array}{c} \begin{array}{ccccc} v_1 & v_2 & v_3 & v_4 & v_5 \end{array} \\ \begin{bmatrix} 0 & 1 & 1 & 0 & 1 \\ 1 & 0 & 1 & 0 & 1 \\ 0 & 0 & 0 & 1 & 0 \\ 1 & 0 & 0 & 0 & 0 \\ 1 & 0 & 0 & 1 & 0 \end{bmatrix} \end{array}. \tag{6.2.5.5.2}$$

Let X_d^T be the transposed matrix X_d.

$$X_d^T = \begin{array}{c} \\ v_1 \\ v_2 \\ v_3 \\ v_4 \\ v_5 \end{array} \begin{array}{c} \begin{array}{ccccc} v_1 & v_2 & v_3 & v_4 & v_5 \end{array} \\ \begin{bmatrix} 0 & 1 & 0 & 1 & 1 \\ 1 & 0 & 0 & 0 & 0 \\ 1 & 1 & 0 & 0 & 0 \\ 0 & 0 & 1 & 0 & 1 \\ 1 & 1 & 0 & 0 & 0 \end{bmatrix} \end{array}. \tag{6.2.5.5.3}$$

Then, for example, at v_1,

$$\begin{aligned} f_w(v_1) &= f_w(v_2)/d^+(v_2) + f_w(v_4)/d^+(v_4) + f_w(v_5)/d^+(v_5) \\ &= f_w(v_2)/3 + f_w(v_4)/1 + f_w(v_5)/2. \end{aligned}$$

Let A be the matrix obtained by dividing the components of each column j of the matrix X_d^T by $d^+(v_j)$.

$$A = \begin{array}{c} \\ v_1 \\ v_2 \\ v_3 \\ v_4 \\ v_5 \end{array} \begin{array}{c} \begin{array}{ccccc} v_1 & v_2 & v_3 & v_4 & v_5 \end{array} \\ \begin{bmatrix} 0 & 1/3 & 0 & 1 & 1/2 \\ 1/3 & 0 & 0 & 0 & 0 \\ 1/3 & 1/3 & 0 & 0 & 0 \\ 0 & 0 & 1 & 0 & 1/2 \\ 1/3 & 1/3 & 0 & 0 & 0 \end{bmatrix} \end{array}. \tag{6.2.5.5.4}$$

Then,

$$F = AF, \tag{6.2.5.5.5}$$

where

$$F = \begin{bmatrix} f_w(v_1) \\ f_w(v_2) \\ f_w(v_3) \\ f_w(v_4) \\ f_w(v_5) \end{bmatrix}. \tag{6.2.5.5.6}$$

Equation (6.2.5.5.5) is an equation representing the *eigenvector* F with the matrix A. The eigenvector F obtained from (6.2.5.5.5) is as follows:

$$F = \begin{bmatrix} 0.72 \\ 0.24 \\ 0.32 \\ 0.48 \\ 0.32 \end{bmatrix}. \tag{6.2.5.5.7}$$

From (6.2.5.5.7), v_1 has the highest centrality, followed by v_4.

Web graphs are known to be growing. A graph obtained by growing while adding edges between two vertices with a certain probability for a given set of vertices is called a *random graph*. For a random graph, the line chart with the horizontal axis as the degree of vertices and the vertical axis as the number of vertices has a peak somewhere and becomes smaller before and after that. On the other hand, in a huge web graph, there is no peak and it is called a *complex network*. One of the characteristics of complex networks is that there are a small number of vertices with very large degrees called *hubs*. This network is also called a *scale-free network* and does not have a scale that represents the network. Specifically, it is a network in which the probability that the degree is d is proportional to $d^{-\gamma}$. Since a scale-free network has hubs, the distance between two vertices does not increase so much even if the number of vertices and edges increases or the network grows. For example, in a graph in which vertices are randomly arranged on a plane and the vertices that are close to each other are preferentially connected by edges, the average distance between the two vertices increases as the plane increases. In the case of a scale-free network, the increase in the average distance is gradual. This property is said to be *small world phenomenon*, as when two people, who were thought to have no contact at all, are connected through a few friends, they say, "*It's a small world.*"

6.2.6. *Fault diagnosis by measuring transmission numbers*

In location problems, the out-transmission and in-transmission numbers are important indices to evaluate a directed graph in which each edge is associated with a positive real number called the length of the edge and each vertex is associated with a positive real number called the weight of the vertex. The generalized indices of the out-transmission and in-transmission numbers are used in this section. The indices are nonlinear functions of the distances between the vertices which represent the total cost or loss of communication service among the vertices. By a fault, we mean a change in the length of an edge. A fault diagnosis problem is discussed in a directed graph in which the above indices are measurable at all times. This is useful for ensuring stable communication services among base stations in the mobile communication systems.

6.2.6.1. *Out-transmission number and in-transmission number and their generalizations*

Let $G = (V, E)$ be a directed and connected graph with a set V of vertices and a set E of edges, where with each edge and each vertex a positive real number called the length of the edge and the weight of the vertex, respectively, is associated, and $l(e_k)$ and $w(v_j)$ denote the length of $e_k \in E$ and the weight of $v_j \in V$, respectively. We sometimes refer to this weighted graph as a network $N(G, l, w)$.

Let $t_o(v_j)$ and $t_i(v_j)$ for $v_j \in V$ be out-transmission number and in-transmission number, respectively. Then

$$t_o(v_j) = \sum_{v_k \in V} d\left(v_j, v_k\right) w\left(v_k\right), \tag{6.2.6.1.1}$$

$$t_i(v_j) = \sum_{v_k \in V} d\left(v_k, v_j\right) w(v_k), \tag{6.2.6.1.2}$$

where $d(v_i, v_j)$ is the length of a shortest path from vertex v_i to vertex v_j called the distance from v_i to v_j. Let $|V| = n$ and let $\boldsymbol{D} = [d_{ij}]$ be an $n \times n$ matrix, where $d_{ij} = d(v_i, v_j)$. $\boldsymbol{D}$ is called the *distance matrix*. Let

$$\boldsymbol{T_o} = \begin{bmatrix} t_o(v_1) \\ t_o(v_2) \\ \vdots \\ t_o(v_n) \end{bmatrix}, \quad \boldsymbol{T_i} = \begin{bmatrix} t_i(v_1) \\ t_i(v_2) \\ \vdots \\ t_i(v_n) \end{bmatrix}, \quad \boldsymbol{W} = \begin{bmatrix} w(v_1) \\ w(v_2) \\ \vdots \\ w(v_n) \end{bmatrix}. \tag{6.2.6.1.3}$$

Then, (6.2.6.1) and (6.2.6.2) can be rewritten as follows:

$$\boldsymbol{T_o} = \boldsymbol{DW}, \tag{6.2.6.1.4}$$

$$\boldsymbol{T_i} = \boldsymbol{D}^t\boldsymbol{W}, \tag{6.2.6.1.5}$$

where $\boldsymbol{D}^t$ is the transpose of $\boldsymbol{D}$.

We generalize the out-transmission and in-transmission numbers, as follows:

$$h_o(v_j) = \sum_{v_k \in V} F_k(d(v_j, v_k), w(v_k)), \tag{6.2.6.1.6}$$

$$h_i(v_j) = \sum_{v_k \in V} F_k(d(v_k, v_j), w(v_k)), \tag{6.2.6.1.7}$$

where $F_k(x, y)$ is a real-valued function of x and y. If $F_k(x, y) = xy$, then $h_o = t_o$, and $h_i = t_i$. Let

$$\boldsymbol{H_o} = \begin{bmatrix} h_o(v_1) \\ h_o(v_2) \\ \vdots \\ h_o(v_n) \end{bmatrix}, \quad \boldsymbol{H_i} = \begin{bmatrix} h_i(v_1) \\ h_i(v_2) \\ \vdots \\ h_i(v_n) \end{bmatrix}. \tag{6.2.6.1.8}$$

Now $h_o(v_j)$ and $h_i(v_j)$ are called the *generalized out-transmission and in-transmission numbers* of a vertex v_j of $N(G, l, w)$, respectively.

Example. Figure 6.2.6.1.1 is a directed and connected graph $G = (V, E)$ with edge weights and vertex weights called $N(G, l, w)$, where $l(e_k)$ and $w(e_i)$ denote the length of $e_k \in E$ and the weight of $v_j \in V$, respectively.

$V = \{v_1, v_2, v_3, v_4, v_5\}$, $E = \{e_1, e_2, e_3, e_4, e_5, e_6, e_7\}$ and $e_1 = (v_3, v_4)$, $e_2 = (v_4, v_2)$, $e_3 = (v_4, v_5)$, $e_4 = (v_2, v_3)$, $e_5 = (v_5, v_3)$, $e_6 = (v_2, v_1)$, $e_7 = (v_1, v_2)$. $l(e_1) = l(e_2) = l(e_3) = l(e_4) = l(e_5) = l(e_6) = l(e_7)$. $w(v_1) = 1$, $w(v_2) = 2$, $w(v_3) = 3$, $w(v_4) = 4$, $w(v_5) = 5$.

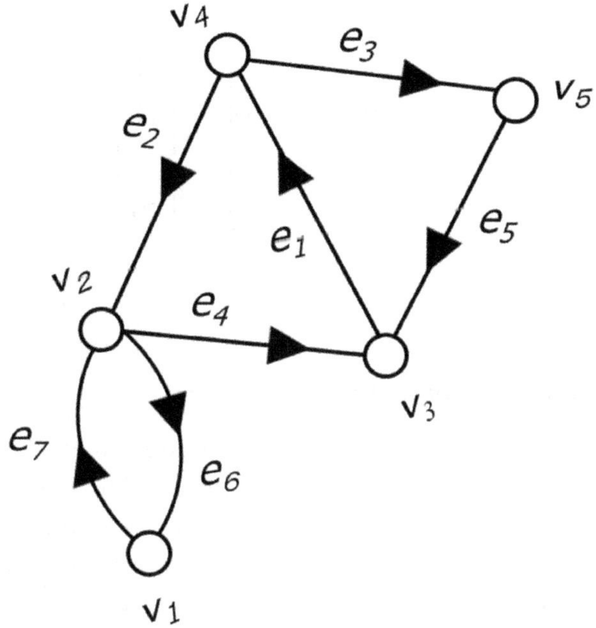

Fig. 6.2.6.1.1. A directed and connected graph $G(\ N(G, l, w))$ with weights.

Then, matrices $\boldsymbol{D}$, $\boldsymbol{T}_o$ and $\boldsymbol{T}_i$ of N are given below:

$$\boldsymbol{D} = \begin{bmatrix} \cdot & 1 & 2 & 3 & 4 \\ 1 & \cdot & 1 & 2 & 3 \\ 3 & 2 & \cdot & 1 & 2 \\ 2 & 1 & 2 & \cdot & 1 \\ 4 & 3 & 1 & 2 & \cdot \end{bmatrix}, \tag{6.2.6.1.9}$$

$$\boldsymbol{T_o} = \boldsymbol{DW} = \boldsymbol{D} \begin{bmatrix} 1 \\ 2 \\ 3 \\ 4 \\ 5 \end{bmatrix} = \begin{bmatrix} 40 \\ 27 \\ 21 \\ 15 \\ 21 \end{bmatrix}, \tag{6.2.6.1.10}$$

$$T_i = D^t W = D^t \begin{bmatrix} 1 \\ 2 \\ 3 \\ 4 \\ 5 \end{bmatrix} = \begin{bmatrix} 39 \\ 26 \\ 17 \\ 20 \\ 20 \end{bmatrix} \qquad (6.2.6.1.11)$$

(End of Example)

6.2.6.2. *Assumptions and a Fault of an Edge*

For example, an edge in a network corresponds to a communication link in a communication system or to a road in a road distribution network. The length of an edge in the network represents the transmission or transportation time of the link or the length of the road, and the distance between two vertices represents the transportation time between the two vertices. The generalized out-transmission and in-transmission numbers represent the total cost or loss of service among the vertices. A fault of a link or a road in these systems generally causes the increase of the transmission time (cost or loss) or the transportation time (cost or loss). In this section, by a fault in a directed network, we mean any increase in length of an edge with respect to its nominal length which can cause an increase of the length of at least one of the shortest paths between two vertices. We make the following assumptions for a network $N(G, l, w)$:

(1) The generalized out-transmission and/or in-transmission numbers of every vertex in N are measurable at all times.
(2) A fault occurs at an edge and the fault of an edge means the increase of the edge length. Fault does not occur at two or more edges at a time. The weight of any vertex does not change.
(3) A network is defined to be *fault-free* with respect to the generalized out-transmission and/or in-transmission numbers if the generalized out-transmission and/or in-transmission numbers of every vertex of the network remain unchanged.

If a network N is evaluated only by the out-transmission number or the in-transmission number of vertices, the assumption is reasonable.

A network is said to be *k-edge fault detectable* if one is able to determine a set of edges which consists of k edges containing a fault edge, from the measurements of generalized out-transmission and/or in-transmission numbers of the network. Vertices u and v of a fault edge $e_1 = (u, v)$ are called *fault vertices*. A network is said to be *k-vertex fault detectable* if one can determine a set of vertices which consists of k vertices containing a fault vertex, from the measurements of generalized out-transmission and/or in-transmission numbers of the network.

6.2.6.3. *Increment of distance by increase of the length of an edge*

Let $N'(G, l', w)$ be a network obtained from $N(G, l, w)$ by increasing an edge length $l(e_m)$, where $e_m = (v_p, v_q)$. That is,

$$l'(e_m) = l(e_m) + \delta, \quad (\delta > 0,\ l(e_m) > 0), \tag{6.2.6.3.1}$$

$$l'(e_j) = l(e_j), \quad (j \neq m, e_j \in E). \tag{6.2.6.3.2}$$

Let $d'(v_i,v_j)$ be the distance from v_i to v_j in N'. Then, we obtain the following lemmas.

Lemma 6.2.6.3.1. *For any vertices* $v_i, v_j.v_k \in V$,

$$d(v_i, v_k) \leq d(v_i, v_j) + d(v_j, v_k), \tag{6.2.6.3.3}$$

$$d'(v_i,\ v_k) \leq d'(v_i,\ v_j) + d'(v_j, v_k). \tag{6.2.6.3.4}$$

Lemma 6.2.6.3.2. *For any vertices* $v_i, v_j \in V$,

$$\begin{aligned} d(v_j, v_i) &= \min\{d'(v_j, v_i), d'(v_j, v_p) + d'(v_q, v_i) + l'(e_m) - \delta\} \\ &= \min\{d'(v_j, v_i), d'(v_j, v_p) + d'(v_q, v_i) + d'(v_p, v_q) - \delta'\}, \end{aligned} \tag{6.2.6.3.5}$$

where $\delta \geq \delta' \geq 0$.

From the above lemmas, we get the following theorems.

Theorem 6.2.6.3.1. *If for any two distinct vertices v_i and v_j of V*

$$d'(v_j, v_i) > d(v_j, v_i), \tag{6.2.6.3.6}$$

then

$$d(v_j, v_i) > d(v_p, v_i), \quad (p \neq j) \tag{6.2.6.3.7}$$

and

$$d'(v_p, v_i) - d(v_p, v_i) \geq d'(v_j, v_i) - d(v_j, v_i). \tag{6.2.6.3.8}$$

Proof. Since the distance between any two vertices does not decrease by the modification (6.2.6.3.1)–(6.2.6.3.2) of N, for any $v_i \in V$

$$d'(v_p, v_i) \geq d(v_p, v_i). \tag{6.2.6.3.9}$$

Assume that

$$d'(v_p, v_i) = d(v_p, v_i). \tag{6.2.6.3.10}$$

Then, from equation (6.2.6.3.5),

$$d(v_p, v_i) = d'(v_p, v_i) \leq d'(v_q, v_i) + l'(e_m) - \delta. \tag{6.2.6.3.11}$$

Using Equations (6.2.6.3.5) and (6.2.6.3.6) for any vertex v_j $(j \neq i,\ j \neq p)$,

$$d'(v_j, v_i) > d(v_j, v_i) = d'(v_j, v_p) + d'(v_q, v_i) + l'(e_m) - \delta. \tag{6.2.6.3.12}$$

From Equations (6.2.6.3.11) and (6.2.6.3.12),

$$d'(v_j, v_i) > d(v_j, v_i) \geq d'(v_j, v_p) + d'(v_p, v_i). \tag{6.2.6.3.13}$$

This is contradictory to (6.2.6.3.4). Hence, $d'(v_p, v_i) \neq d(v_p, v_i)$ and in (6.2.6.3.9)

$$d'(v_p, v_i) > d(v_p, v_i). \tag{6.2.6.3.14}$$

Then, from Equations (6.2.6.3.5) and (6.2.6.3.14),

$$d(v_p, v_i) = d'(v_q, v_i) + l'(e_m) - \delta. \tag{6.2.6.3.15}$$

Using (6.2.6.3.5), (6.2.6.3.6) and (6.2.6.3.15),

$$d(v_j, v_i) = d'(v_j, v_p) + d(v_p, v_i). \tag{6.2.6.3.16}$$

Since $d'(v_j,\ v_p) > 0,\ (p \neq j),$ it follows that (6.2.6.3.17)

$$d(v_j, v_i) > d(v_p, v_i). \tag{6.2.6.3.18}$$

Hence we get (6.2.6.3.7).

Using Equation (6.2.6.3.4),

$$d'(v_j, v_i) - d(v_j, v_i) \leq \{d'(v_j, v_p) + d'(v_p, v_i)\} - d(v_j, v_i). \tag{6.2.6.3.19}$$

From (6.2.6.3.16) and (6.2.6.3.19),

$$d'(v_j, v_i) - d(v_j, v_i) \leq d'(v_p, v_i) - d(v_p, v_i). \tag{6.2.6.3.20}$$

Hence we get (6.2.6.3.8). □

Similarly, we get the following theorem.

Theorem 6.2.6.3.2. *If for any two distinct vertices v_i and v_j of V*

$$d'(v_j, v_i) > d(v_j, v_i), \tag{6.2.6.3.21}$$

then

$$d(vj, v_i) > d(v_j, v_q), \quad (q \neq i) \tag{6.2.6.3.22}$$

and

$$d'(v_j, v_q) - d(v_j, v_q) \geq d'(v_j, v_i) - d(v_j, v_i). \tag{6.2.6.3.23}$$

We also get the following theorem using Lemmas 6.2.6.3.1 and 6.2.6.3.2.

Theorem 6.2.6.3.3. *Let v_i be a vertex such that $d'(v_i, v_p) > d'(v_i, v_q)$. Then, for any vertex $v_j \in V$,*

$$d'(v_i, v_j) = d(v_i, v_j). \tag{6.2.6.3.24}$$

Let v_i be a vertex such that $d'(v_p, v_i) > d'(v_q, v_i)$, Then, for any vertex $v_j \in V$,

$$d'(v_j, v_i) = d(v_j, v_i). \tag{6.2.6.3.25}$$

Proof. We show (6.2.6.3.24). In general, since the distance between any two vertices does not decrease by the modification (6.2.6.3.1)–(6.2.6.3.2), $d'(v_j, v_i) \geq d(v_j, v_i)$. Assume that $d'(v_j, v_i) > d(v_j, v_i)$. From (6.2.6.3.1)–(6.2.6.3.2) and (6.2.6.3.5),

$$d'(v_i, v_j) > d(v_i, v_j) = d'(v_i, v_p) + d'(v_q, v_j) + l'(e_m) - \delta > d'(v_i, v_p) + d'(v_q, v_j). \tag{6.2.6.3.26}$$

Since $d'(v_i, v_p) > d'(v_i, v_q)$ from the hypothesis,

$$d'(v_i, v_j) > d'(v_i, v_q) + d'(v_q, v_j) \tag{6.2.6.3.27}$$

This is contradictory to (6.2.6.3.4). Hence we get Equation (6.2.6.3.24). Similarly we get Equation (6.2.6.3.25). □

6.2.6.4. *Detection of a fault edge*

In Section 6.2.6.2, we explain the physical meaning of generalized out-transmission and in-transmission numbers of a network of a communication system or a road distribution network. A function $F_k(x, y)$ represents its cost or loss. In general, when the distance between two vertices increases, the transmission cost will increase. Therefore, it is reasonable to assume that $F_k(x, y)$ for any vertex $v_k \in V$ is an *increasing function* of x. There are many kinds of increasing functions as shown in Fig. 6.2.6.4.1. The function $F(x)$ of Fig. 6.2.6.4.1(a) is a *linear function.* The function $F(x)$ of Fig. 6.2.6.4.1(b) is a *concave function* and the function $F(x)$ of Fig. 6.2.6.4.1(c) is a *convex function.* The function of Fig. 6.2.6.4.1(d) is neither concave nor convex. Note that a linear function is both convex and concave.

If $F_k(x, y)$ represents transmission cost, it is frequently a concave function as Fig. 6.2.6.4.1(b) with respect to x. Thus, we assume that $F_k(x, y)$ for any vertex $v_k \in V$ is an increasing concave function with respect to x.

Let $h_o'(v_j)$ and $h_i'(v_j)$ be the generalized out-transmission and in-transmission numbers of N', respectively. Let us define

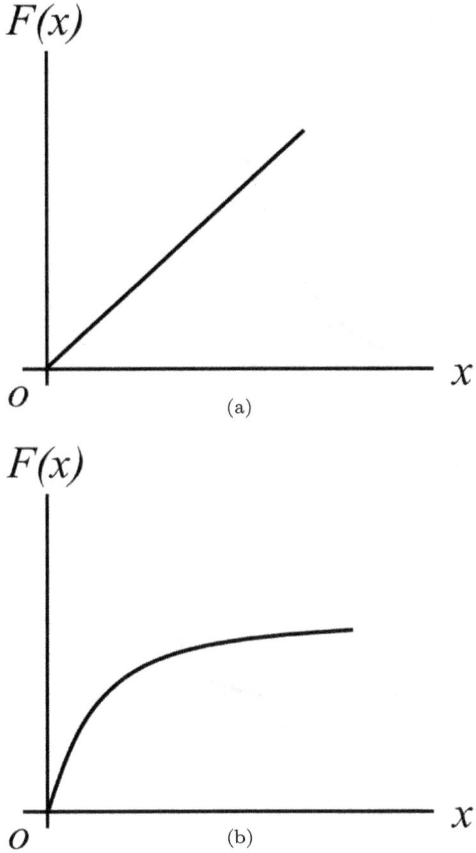

Fig. 6.2.6.4.1. Some types of increasing functions. (a) An increasing and linear function. (b) An increasing and concave function.

$$\Delta h_o(v_j) = h'_o(v_j) - h_o(v_j), \tag{6.2.6.4.1}$$

$$\Delta h_i(v_j) = h'_i(v_j) - h_i(v_j). \tag{6.2.6.4.2}$$

Then, we obtain the following theorem.

Theorem 6.2.6.4.1. *If $F_k(x, y)$, ($v_k \in V$), are increasing and concave functions with respect to x, then, for any vertex $v_j \in V$,*

$$\Delta h_o(v_p) \geq \Delta h_o(v_j), \tag{6.2.6.4.3}$$

$$\Delta h_o(v_q) = 0. \tag{6.2.6.4.4}$$

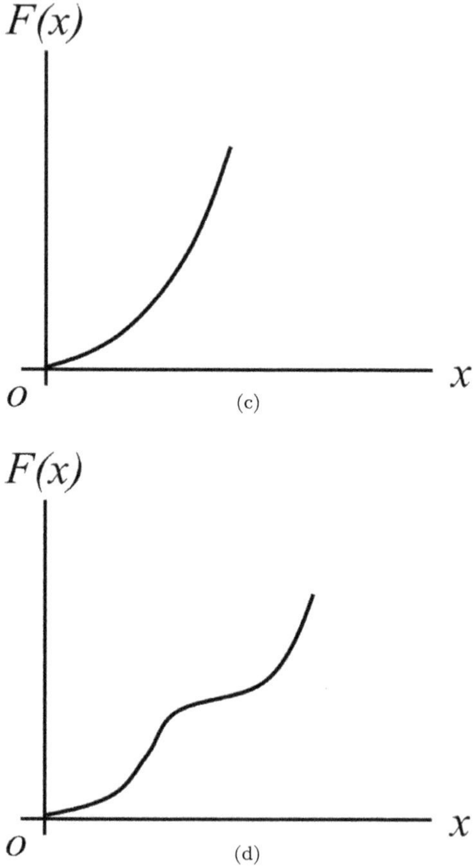

Fig. 6.2.6.4.1. (*Continued*). (c) An increasing and convex function. (d) An increasing function.

Proof. From the definition of generalized out-transmission number,

$$\begin{aligned}\Delta h_o(v_p) - \Delta h_o(v_j) = \sum_{v_k \in V} [\{F_k(d'(v_p, v_k)\, w(v_k)) \\ - F_k(d\left(v_p, v_k\right), w(v_k))\} \\ - \{F_k(d'(v_j, v_k), w(v_k)) \\ - F_k(d(v_j, v_k)\, w(v_k))\}]. \quad (6.2.6.4.5)\end{aligned}$$

Let

$$Y_k = \{F_k(d'(v_p, v_k), w(v_k)) - F_k(d(v_p, v_k)\, w(v_k))\}$$
$$-\{F_k(d'(v_j, v_k), w(v_k)) - F_k(d(v_j, v_k)\, w(v_k))\}. \quad (6.2.6.4.6)$$

In general, $d'(v_j, v_k) \geq d(v_j, v_k)$. In (6.2.6.4.6), if $d'(v_j, v_k) = d(v_j, v_k)$, Y_k is nonnegative since $d'(v_p, v_k) \geq d(v_p, v_k)$ and $F_k(x, y)$ is an increasing function with respect to x.

Next, in (6.2.6.4.6), suppose $d'(v_j, v_k) > d(v_j, v_k)$. Let

$$d(v_p, v_k) = x_1, \quad d'(v_p, v_k) = x_1', \quad (6.2.6.4.7)$$

$$d(v_j, v_k) = x_2, \quad d'(v_j, v_k) = x_2'. \quad (6.2.6.4.8)$$

From Theorem 6.2.5.6,

$$x_2 > x_1, \quad (6.2.6.4.9)$$

$$x_1' - x_1 \geq x_2' - x_2. \quad (6.2.6.4.10)$$

Let

$$x_1' - x_1 = a > 0, \quad (6.2.6.4.11)$$

$$x_2' - x_2 = b > 0. \quad (6.2.6.4.12)$$

Then,

$$Y_k = \{F_k(x_1 + a, w(v_k)) - \{F_k(x_1, w(v_k))\}$$
$$- \{F_k(x_2 + b, w(v_k)) - F_k(x_2, w(v_k))\}. \quad (6.2.6.4.13)$$

Since $a \geq b$ and $F_k(x, y)$ is an increasing function with respect to x in (6.2.6.4.13),

$$Y_k \geq \{F_k(x_1 + a, w(v_k)) - \{F_k(x_1\, w(v_k))\}$$
$$- \{F_k(x_2 + a, w(v_k)) - F_k(x_2, w(v_k))\}. \quad (6.2.6.4.14)$$

Since $x_2 > x_1$ and $F_k(x, y)$ is a concave function in (6.2.6.4.14), $Y_k \geq 0$. Since $\Delta h_o(v_p) - \Delta h_o(v_j) = \sum_{v_k \in V} Y_k$ from (6.2.6.4.5), we obtain (6.2.6.4.3). In general $d'(v_q, v_p) > d'(v_q, v_q) = 0$.

From (6.2.6.3.24) of Theorem 6.2.6.3.3, $d'(v_q, v_k) = d(v_q, v_k)$ for any vertex $v_k \in V$. Hence $\Delta h_o(v_q) = 0$.

Similarly, we obtain the following theorem.

Theorem 6.2.6.4.2. *If $F_k(x, y)$, $(v_k \in V)$, are increasing and concave functions with respect to x, then, for any vertex $v_j \in V$*

$$\Delta h_i(v_q) \geq \Delta h_i(v_j), \tag{6.2.6.4.15}$$

$$\Delta h_i(v_p) = 0. \tag{6.2.6.4.16}$$

Let us define the following subsets of vertices:

$$V_{mho} = \{v_k | \Delta h_o(v_k) \geq \Delta h_o(v_j), v_k, v_j \in V\}, \tag{6.2.6.4.17}$$

$$V_{mhi} = \{v_k | \Delta h_i(v_k) \geq \Delta h_i(v_j), v_k, v_j \in V\}, \tag{6.2.6.4.18}$$

$$V_{zho} = \{v_k | \Delta h_o(v_k) = 0, \ v_k \in V\}, \tag{6.2.6.4.19}$$

$$V_{zhi} = \{v_k | \Delta h_i(v_k) = 0, v_k \in V\}. \tag{6.2.6.4.20}$$

First, let us consider a network in which only generalized out-transmission numbers are measurable.

Theorem 6.2.6.4.3. *Let N be a network in which only generalized out-transmission numbers with $F_k(x, y)$, $(v_k \in V)$, which are increasing and concave functions with respect to x are measurable. Then, N is k-vertex fault detectable if min $\{|V_{mho}|, |V_{zho}|\} = k$.*

Theorem 6.2.6.4.4. *Let N be a network in which only the vertices with the maximum increment and the minimum increment of the generalized out-transmission number are distinguishable. Then, we can detect a fault edge for any N and for any increasing concave function $F_k(x, y)$, $(v_k \in V)$, with respect to x, that is N is 1-edge fault detectable if and only if $|V_{zho}| = 1$.*

Proof. *Sufficiency*: From (6.2.6.4.4) of Theorem 6.2.6.4.1, $V_{zho} = \{v_q\}$, since $|V_{zho}| = 1$. Let

$$V_s = \{v_i | d(v_i, v_p) < d(v_j, v_q), v_i, v_j \in V_{mho}\}. \tag{6.2.6.4.21}$$

Then, $V_s = \{v_p\}$, since $|V_s| = 1$ from Theorem 6.2.3.12. Therefore, we obtain a fault edge $e = (v_p, v_q)$. If N has more than one edge associated with a pair of vertices v_p and v_q, the edge with minimum length in the edges is the fault edge.

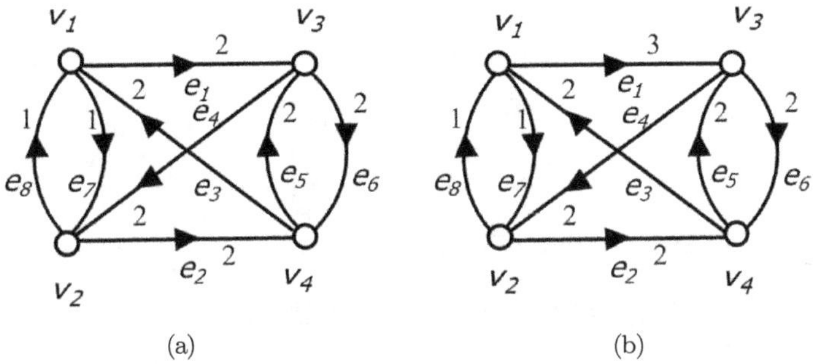

Fig. 6.2.6.4.2. Networks (a) N and (b) N'.

Necessity: We show that a network N including two vertices with zero increment is not *1*-edge fault detectable. Let $F_k(x, y) = xy$ for $v_k \in V$. These functions are increasing concave functions with respect to x. Matrices $\boldsymbol{D}, \boldsymbol{D}', \boldsymbol{H}_0, \boldsymbol{H}'_0$ and $\Delta\boldsymbol{H}_0$ of networks N and N' of Fig. 6.2.6.4.2(a) and 6.2.6.4.2(b), in which the weight given to each edge indicates the edge length of the edge and the weight of every vertex is 1, are given below:

$$\boldsymbol{D} = \begin{bmatrix} \cdot & 1 & 2 & 3 \\ 1 & \cdot & 3 & 2 \\ 3 & 2 & \cdot & 2 \\ 2 & 3 & 2 & \cdot \end{bmatrix}, \tag{6.2.6.4.22}$$

$$\boldsymbol{D}' = \begin{bmatrix} \cdot & 1 & 3 & 3 \\ 1 & \cdot & 4 & 2 \\ 3 & 2 & \cdot & 2 \\ 2 & 3 & 2 & \cdot \end{bmatrix}, \tag{6.2.6.4.23}$$

$$\boldsymbol{H_o} = \begin{bmatrix} 6 \\ 6 \\ 7 \\ 7 \end{bmatrix}, \quad \boldsymbol{H'_o} = \begin{bmatrix} 7 \\ 7 \\ 7 \\ 7 \end{bmatrix}, \tag{6.2.6.4.24}$$

$$\Delta \boldsymbol{H}_{\mathbf{o}} \triangleq \boldsymbol{H}'_{\mathbf{o}} - \boldsymbol{H}_{\mathbf{o}} = \begin{bmatrix} 1 \\ 1 \\ 0 \\ 0 \end{bmatrix}. \tag{6.2.6.4.25}$$

From (6.2.6.4.25), we obtain V_{mho} and V_{zho}.

$$V_{\mathrm{mho}} = \{v_1, v_2\}, \quad V_{\mathrm{zho}} = \{v_3, v_4\} \tag{6.2.6.4.26}$$

The number of vertices with zero increment is 2. N' of Fig. 6.2.6.4.2(b) is obtained from N by increasing the length of e_1 by 1. If N'' is obtained from N by increasing the length of e_2 by 1, then $\Delta \boldsymbol{H}_o$ obtained from N and N'' is clearly the same as (6.2.6.4.26). This means that the fault of e_1 and e_2 is not distinguishable by $\Delta \boldsymbol{H}_o$. It is simple to realize a network with $|V_{\mathrm{mho}}| = 1$ and $|V_{\mathrm{zho}}| = 2$ that is not 1-edge fault detectable. Thus, there are networks with $|V_{\mathrm{zho}}| > 1$ that are not 1-edge fault detectable. □

We obtain similar theorems of a network in which only the generalized in-transmission number is measurable.

Theorem 6.2.6.4.5. *Let N be a network in which only generalized in-transmission numbers with $F_k(x, y)$, $(v_k \in V)$, which are increasing and concave functions with respect to x are measurable. Then, N is k-vertex fault detectable if min $\{|V_{mhi}|, |V_{zhi}|\} = k$.*

Theorem 6.2.6.4.6. *Let N be a network in which only the vertices with the maximum increment and the minimum increment of the generalized in-transmission number are distinguishable. Then, we can detect a fault edge for any N and for any increasing concave function $F_k(x, y)$, $(v_k \in V)$, with respect to x, that is N is 1-edge fault detectable if and only if $|V_{zhi}| = 1$.*

Next, let us consider a network in which both generalized out-transmission numbers and generalized in-transmission numbers are measurable.

Theorem 6.2.6.4.7. *A network N in which both generalized out-transmission numbers and generalized in-transmission numbers with $F_k(x, y)$, $(v_k \in V)$, which are increasing and concave functions with respect to x, are measurable is 1-edge fault detectable.*

Proof. Let v_i be an element of V_{mho}. Let v_j be an element of V_{mhi}. Since $d'(v_p, v_q) > d(v_p, v_q)$, $\Delta h_o(v_p) = \Delta h_o(v_i)$ and $\Delta h_i(v_q) = \Delta h_i(v_j)$, increment of the distance from v_i to v_j by the fault is not zero, that is $d'(v_i, v_j) > d(v_i, v_j)$.

$$V_{sj} = \{v_k | d(v_k, v_j) < d(v_i, v_j), v_k, v_i \in V_{\text{mho}}\}. \quad (6.2.6.4.27)$$

Then, $V_{sj} = \{v_p\}$, since $|V_{sj}| = 1$ from Theorem 6.2.6.3.1. Similarly,

$$V_{si} = \{v_k | d(v_i, v_k) < d(v_i, v_j), v_k, v_j \in V_{\text{mhi}}\}. \quad (6.2.6.4.28)$$

Then we get $V_{si} = \{v_q\}$. Therefore, the fault edge is $e = (v_p, v_q)$. If N has more than one edge associated with a pair of vertices v_p and v_q, the edge with minimum length in the edges is the fault edge. □

In Theorem 6.2.6.4.7, both V_{mho} and V_{mhi} are used for detecting a fault edge. However, we can detect a fault edge without using all vertices of V_{mho} or V_{mhi}.

Corollary 6.2.6.4.7. *A fault edge is detectable if V_{mho} (or V_{mhi}) and a vertex v_j with $\Delta h_i(v_j) > 0$ (or $\Delta h_o(v_j) > 0$) are given.*

Proof. Suppose that V_{mho} and a vertex v_j with $\Delta h_i(v_j) > 0$ are given, i.e.,

$$V_s = \{v_k | d(v_k, v_j) < d(v_i, v_j), v_k, v_i \in V_{\text{mho}}\}. \quad (6.2.6.4.29)$$

Since $\Delta h_i(v_j) > 0$,

$$d'(v_p, v_j) > d(v_p, v_j). \quad (6.2.6.4.30)$$

Therefore, $V_s = \{v_p\}$ from Theorem 6.2.5.6.

Let $E_p = \{e_1, e_2, \ldots, e_m\}$ such that $e_1 = (v_p, v_{k1})$, $e_2 = (v_p, v_{k2}), \ldots, e_m = (v_p, v_{km})$ and an edge $e_i \in E_p$ is obtained in a shortest path from v_p to v_j. Assume that $m > 1$. This means that there are more than one shortest path from v_p to v_j. Then the distance $d(v_p, v_j)$ does not change by increasing the length of an edge $e_i \in E_p$. This is contradictory to (6.2.6.4.30). Hence, $m = 1$ and $E_p = \{e_1\}$, that is e_1 is the fault edge. □

Then, we obtain a following algorithm detecting a fault edge.

An algorithm of detection of a fault edge:

N, $\Delta h_o(v_j)$ and $\Delta h_i(v_j)$, $(v_j \in V)$, are given.

Step 0: Set $V_{mho} = \{v_i | \Delta h_o(v_i) \geq \Delta h_o(v_j), v_i, v_j \in V\}$.
Step 1: Select a vertex v_j such that $\Delta h_i(v_j) > 0$.
Step 2: Set $V_s = \{v_k | d(v_k, v_j) < d(v_i, v_j), v_k, v_i \in V_{\text{mho}}\}$. Then, $V_s = \{v_p\}$.
Step 3: Find a shortest path from v_p to v_j. Then, an edge $e_m = (v_p, v_q)$ on the shortest path is a fault edge.

We show two examples for this algorithm.

Example. (1) Consider the example in the proof of Theorem 6.2.6.4.4, in which the network was not ledge fault detectable by the measurements of only out-transmission numbers of all vertices.

In the network of Fig. 6.2.6.4.2(a), N and

$$\Delta \boldsymbol{H}_{\mathbf{o}} = \begin{bmatrix} 1 \\ 1 \\ 0 \\ 0 \end{bmatrix}, \tag{6.2.6.4.31}$$

$$\Delta \boldsymbol{H}_{\mathbf{i}} = \begin{bmatrix} 0 \\ 0 \\ 2 \\ 0 \end{bmatrix} \tag{6.2.6.4.32}$$

are given.

Step 0: $V_{\text{mho}} = \{v_1, v_2\}$.
Step 1: $v_j = v_3$, since $\Delta h_i(v_3) = 2 > 0$.

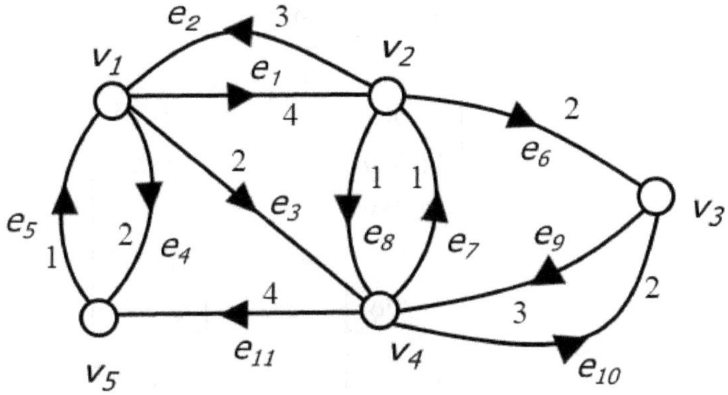

Fig. 6.2.6.4.3. A network N.

Step 2: $V_s = \{v_k | d(v_k, v_3) < d(v_i, v_3), v_k, v_i \in V_{\text{mho}}\} = \{v_1\}$. Hence, $v_p = v_1$.

Step 3: Since the shortest path from v_1 to v_3 consists of e_1, $v_q = v_3$. The fault edge is e_1.

Note that in this example, we do not need Step 3, because we can see $v_q = v_3$ in Step 1 since $V_{mho} = \{v_3\} = \{v_j | \Delta h_i(v_j) > 0, v_j \in V\}$.

(2) Consider the network N of Fig. 6.2.6.4.3 in which the real number given to each edge indicates the edge length. Let $w(v_1) = 1.0$, $w(v_2) = 2.0$, $w(v_3) = 3.0$, $w(v_4) = 4.0$ and $w(v_5) = 5.0$. Matrix $\boldsymbol{D}$ of the network is given as follows:

$$\boldsymbol{D} = \begin{bmatrix} \cdot & 3 & 5 & 2 & 2 \\ 3 & \cdot & 2 & 1 & 5 \\ 7 & 4 & \cdot & 3 & 7 \\ 4 & 1 & 2 & \cdot & 4 \\ 1 & 4 & 5 & 3 & \cdot \end{bmatrix}. \tag{6.2.6.4.33}$$

Let us consider the following functions as $F_k(x, y)$:

$$F_k(x, y) = F(x, y) = \{1 - \exp(-x)\}y \tag{6.2.6.4.34}$$

for any vertex $v_k \in V$. Then, this function is increasing and concave with respect to x.

$$\boldsymbol{H}_o = \begin{bmatrix} 12.662 \\ 11.089 \\ 11.759 \\ 9.748 \\ 9.376 \end{bmatrix}, \quad \boldsymbol{H}'_o = \begin{bmatrix} 12.738 \\ 11.039 \\ 11.782 \\ 10.225 \\ 9.399 \end{bmatrix}, \quad (6.2.6.4.35)$$

$$\boldsymbol{H}_i = \begin{bmatrix} 11.985 \\ 11.332 \\ 11.148 \\ 9.731 \\ 9.775 \end{bmatrix}, \quad \boldsymbol{H}'_i = \begin{bmatrix} 12.033 \\ 12.386 \\ 11.152 \\ 9.731 \\ 9.775 \end{bmatrix}, \quad (6.2.6.4.36)$$

$$\Delta \boldsymbol{H}_o \triangleq \boldsymbol{H}'_o - \boldsymbol{H}_o = \begin{bmatrix} 0.076 \\ 0.000 \\ 0.023 \\ 0.477 \\ 0.023 \end{bmatrix}, \quad (6.2.6.4.37)$$

$$\Delta \boldsymbol{H}_i \triangleq \boldsymbol{H}'_i - \boldsymbol{H}_i = \begin{bmatrix} 0.048 \\ 1.054 \\ 0.004 \\ 0.000 \\ 0.000 \end{bmatrix}. \quad (6.2.6.4.38)$$

Step 0: $V_{\text{mho}} = \{v_4\}$.
Step 1: Let $v_j = v_1$, since $\Delta h_i(v_1) = 0.048 > 0$.
Step 2: $V_s = \{v_k | d(v_k, v_1) < d(v_i, v_1), v_k, v_i \in V_{\text{mho}}\}$. In this example, this operation is not necessary because $|V_{\text{mho}}| = 1$. Then $V_s = \{v_4\}$, that is $v_p = v_4$.

Step 3: The shortest path from v_4 to v_1 consists of e_7 and e_2. The initial vertex of e_7 is $v_p = v_4$. Hence, the fault edge is $e_7 = (v_p, v_q) = (v_4, v_2) = e_7$.

(End of Example)

- **Summary:** Location problems for communication networks including mobile communication systems are introduced. The center problems, median problems and their generalizations are described. The fault diagnosis by measuring transmission numbers is also described.
- **Interesting challenges:** Consider how centrality indicators should be used to extract features of graph structures such as the mobile communication network, knowledge network, frame network and web network.

Bibliography

[1] W. C. Y. Lee, *Mobile Communications Engineering*, McGraw-Hill, 1998.

[2] H. Morikawa. *5 G*, Iwanami Pub., 2020.

[3] Md. Noor-A-Rahim, Zilong Liu, Haeyoung Lee, Mohammad Omar Khyam, Jianhua He, Dirk Pesch, Klaus Moessner, Walid Saad and H. Vincent Poor, "6G for Vehicle-to-Everything (V2X) Communications: Enabling Technologies, Challenges, and Opportunities". *Proc. IEEE*, vol. 110, pp. 712–734, 2022.

[4] M. Iri, *Network Flow, Transportation and Scheduling, Theory and Algorithms*, Academic Press, 1969.

[5] Deo, *Graph Theory with Applications to Engineering and Computer Science*, Prentice-Hall, 1974.

[6] G. Chartrand, L. Lesniak and P. Zhang, *Graphs & Digraphs*, Sixth Edition, Chapman and Hall/CRC, 2015.

[7] H. Ozaki, I. Shirakawa and K. Onaga, *Graph Theory*, Corona Pub., 1975.

[8] J. B. Kruskal, Jr., "On the shortest spanning subtree of graph and the traveling salesman problem," *Proc. Amer. Math. Soc.*, vol. 7, pp. 48–50, 1956.

[9] R. C. Prim, "Shortest connection networks and some generalizations," *Bell System Tech. J.*, vol. 36, pp. 1389–1401, 1957.

[10] R. E. Gomory and T. C. Hu, "Multi-terminal network flows," *J. Soc. Indust. Appl. Math.*, vol. 9, no. 4, pp. 551–570, 1961.

[11] G. Y. Handler and P. B. Mirchandani, *Location on Networks, Theory and Algorithms*, MIT Press, 1979.

[12] S. L. Hakimi, "Optimum locations of switching centers and the absolute centers and medians of a graph," *Opns. Res.*, vol. 12, pp. 450–459, 1964.

[13] S. L. Hakimi, "Optimum distribution of switching centers in a communications network and some related graph theoretic problems," *Opns. Res.*, vol. 13, pp. 462–475, 1965.

[14] S. L. Hakimi and S. M. Maheshwari, "Optimum locations of centers in networks," *Opns. Res.*, vol. 20, pp. 967–973, 1972.

[15] E. Minieka, "The centers and medians of a graph," *Opns. Res.*, vol. 25, no. 4, pp. 641–650. 1977.

[16] Y. Kajitani and T. Maruyama, "Functional expression of centrality in a graph — an application to the assessment of communication networks," *IEICE Trans.*, vol. J59-A, no. 7, pp. 531–538, 1976.

[17] S. Shinoda and M. Sengoku, "Axiomatic foundations of the theories of functions expressing the mediality of a point in metric space," *IEICE Trans.*, vol. J66-A, no. 4, pp. 352–359, 1983.

[18] G. Kishi and M. Takeuchi, "A type of centrality functions for a directed graph," *IEICE Trans.*, vol. J66-A, no. 6, pp. 470–477, 1983.

[19] H. Matsumoto, M. Sengoku and T. Abe, "Effect of diversity on frequency hopping techniques," *IEICE Trans.*, vol. J66-B, no. 12, pp. 1478–1485, 1983.

[20] H. Tamura, M. Sengoku, S. Shinoda and T. Abe, "Location problems on undirected flow networks," *IEICE Trans.*, vol. E73, no. 12, pp. 1989–1993, 1990.

[21] M. Kano, M. Sengoku and S. Shinoda," A note on a generalized transmission number and its Sabidussi difference," *IEICE Trans.*, vol. E74-A (4), pp. 684–686, 1991.

[22] T. Takahashi, K. Tochihara, C. Yamada, M. Sengoku, T. Abe and W. K. Chen, "Minimum order of unique-center graphs with specified radius and diameter," *Proc. IEEE Asia-Pacific Conference on Circuits and Systems* (*APCCAS '92*), pp. 74–78, 1992.

[23] A. Ito, H. Tamura, M. Sengoku, S. Shinoda and T. Abe, "On the realizable synchronization distance on the mark graph," *IEICE Tech. Rep.*, *CAS97*-65, vol. 97, no. 401, pp. 49–56. 1997.

[24] M. Sengoku and S. Shinoda, "A function measuring the centrality (or mediality) of edges in an undirected network with edge-capacity," *IEICE Trans.*, vol. J68-A, no. 11, pp. 1157–1165, 1985.

[25] S. Shinoda and M. Sengoku, "The concept of space modification and a theory of function measuring the centrality (or mediality) of a point in a network," *IEICE Trans.*, vol. J69-A, no. 1, pp. 42–53, 1986.

[26] M. Sengoku, K. Kobayashi and S. Shinoda, " Radius and diameter with respect to edges in a nondirected graph, and upper and lower bounds for the diameter," *IEICE Trans.*, vol. J69-A, no. 2, pp. 194–202, 1986.

[27] M. Sengoku, S. Shinoda and W. K. Chen, "A fault diagnosis in a nonlinear location network," *J. Franklin Institute*, vol. 323, no. 3, pp. 395–405, 1987.

[28] H. Imai, "Minimax line segment facility location problems," *IEICE Tech. Rep.* CAS 87-206, 1987.

[29] M. Sengoku, "Location problems on networks," *Proc. IEICE*, vol. 71, no. 6, pp. 568–572, 1988.

[30] H. Tamura, S. Goto, M. Sengoku, S. Shinoda and T. Abe, "Realization of a directed network from a subset of the distance set between vertex pairs," *IEICE Trans.*, vol. J72-A, no.11, pp. 1866–1875, 1989.

[31] M. Sengoku, S. Shinoda and T. Abe, "On eccentric sets of edges in graphs," *IEICE Trans.*, vol. E74, no. 4, pp. 687–691, 1991.

[32] H. Tamura, M. Sengoku, S. Shinoda and T. Abe, "Some covering problems in location theory on flow networks," *IEICE Trans.*, vol. E75-A, no. 6, pp. 678–684, 1992.

[33] H. Tamura, M. Sengoku, S. Shinoda and T. Abe, "Realization of a network from the upper and lower bounds of the distances (or capacities) between vertices," *Proc. IEEE Inte. Symp. Circuits and Systems (ISCAS'93)*, vol. 4, pp. 2545–2548, May 1993.

[34] K. Watanabe, M. Sengoku, H. Tamura and Y. Yamaguchi, "Realization problems of a tree with a transmission number sequence," *IEICE Trans.*, vol. E77-A, no. 3, pp. 527–533, 1994.

[35] K. Watanabe, H. Tamura and M. Sengoku, "Problems of where to locate p-sinks in a flow network," *IEICE Trans.*, vol. J78-A, no. 8, pp. 938–946, 1995.

[36] K. Watanabe, H. Tamura and M. Sengoku, "The problem of where locate p-sinks in a flow network: complexity approach," *IEICE Trans.*, vol. E79-A, no. 9, pp. 1495–1503, 1996.

[37] H. Tamura, H. Sugawara, M. Sengoku and S. Shinoda, "On a generalization of a covering problem called single cover on undirected flow networks," *IEICE Trans.*, vol. E80-A, no.3, pp. 544–550, 1997.

[38] M. Sengoku, S. Shinoda and R. Yatsuboshi, "On a function for the vulnerability of a directed flow network," *Networks*, vol. 18, no. 1, pp. 73–83, Spring, 1988.

[39] H. Tamura, H. Sugawara, M. Sengoku and S. Shinoda, "Plural cover problem on undirected flow networks," *IEICE Trans.*, vol. J81-A, no.5, pp. 863–869, 1998.

[40] M. Takeuchi, T. Ohkubo and Y. Itoh "Centrality functions of a network with capacities on vertices and edges," *IEICE Trans.*, vol. J82–A, no. 10. pp. 1563–1576, 1999.

[41] H. Tamura, K. Nakano, M. Sengoku and S. Shinoda, "On transformation of extended edge coloring problems into vertex coloring problems," *IEICE Trans.*, vol. J93-A, no. 11, pp.782–784, 2010.

[42] K. Tani, Y. Maeda, K. Nakano and M. Sengoku, "Multi agent simulation on change of customer flow line by arrangement by store clerks in home center," *IEICE Tech. Rep.*, ICTSSL-2016-79, 2017.

[43] K. Tani, Y. Maeda, K. Nakano, H. Numata and M. Sengoku, "Power interchange modeling using multi-agent system," *IEICE Tech. Rep.*, CAS-2015-71, 2016.

[44] M. Sengoku, "Service technology and ICT," *IEICE Tech. Rep.*, ICTSSL-2020-6, 2020.

[45] H. Tamura, K. Nakano and M. Sengoku, *Network Engineering*, IEICE, Corona Pub., 2020.

Chapter 7

Mobile Communication Systems (2)

7.1. Cellular Systems and Graphs

7.1.1. *Cellular systems in mobile communication systems*

The mobile communication network, which started in the 1980s, has evolved significantly every 10 years. From $1G$ to $4G$ (kG: kth Generation Mobile Communication System), mobile communication has been developed as a tool for improving communication speed and mainly for communication between people. In IoT era, where various things are connected to the network, change the way of communication, and contribute to the development of new business. 5G, 6G, ..., kG are expected to become an important foundation for it.

In comparison between mobile communication systems and general communication systems, the characteristics of mobile communication systems are cellular systems. In a cellular system, the service area is divided into many small cells in which the base station of a cell covers the mobile terminals in the area of the cell by wireless and the system enables communication around the world via base stations. The each channel (see the definition in Section 7.1.1.2) is re-used in some cells, simultaneously, which leads to the effective use of frequency.

7.1.1.1. *The cell structure*

There exist different structures of service areas (see Figs. 7.1.1.1.1–7.1.1.1.6). Assuming radio propagation is on only roads in urban area, the service area is shown in Figs. 7.1.1.1.1–7.1.1.1.3 which are called a band-shaped service area, tree-shaped service area and ring shaped service

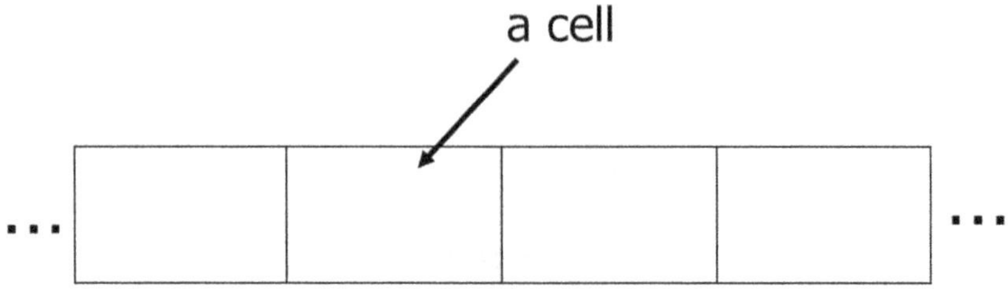

Fig. 7.1.1.1.1. Cells for a straight road (one-dimensional cell model).

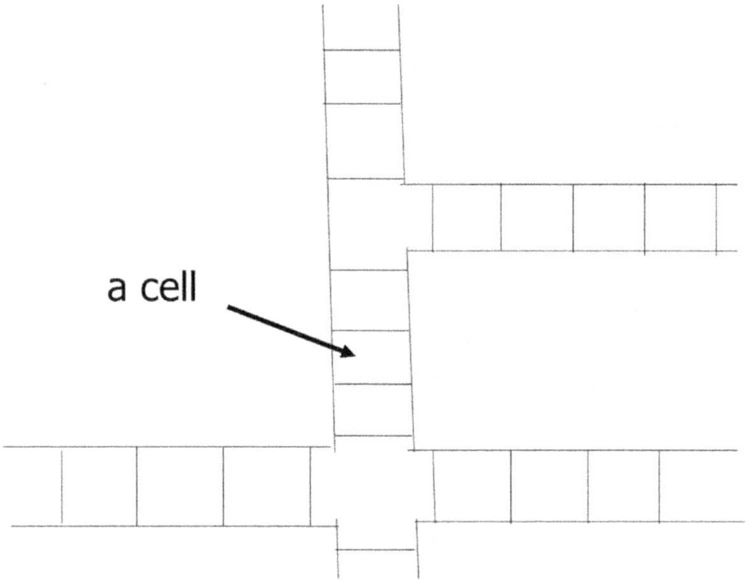

Fig. 7.1.1.1.2. Cells for straight roads.

area, respectively. Since the service area in Fig. 7.1.1.1.1 has cells for a straight road, it is sometimes called a one-dimensional cell model. In Fig. 7.1.1.1.3, a cell for a straight road is divided into three cells, called a higher cell, a middle cell and a lower cell. Large-sized cars such as buses use higher cells and small-sized cars such as private cars use middle or lower cells. Figure 7.1.1.1.4(a) shows cells for a plane in which the plane is divided into regular hexagons, and the regular hexagons are used as cells. This is called two-dimensional cell model. This model is often used because a regular hexagon is close to a circle and can cover the plane when

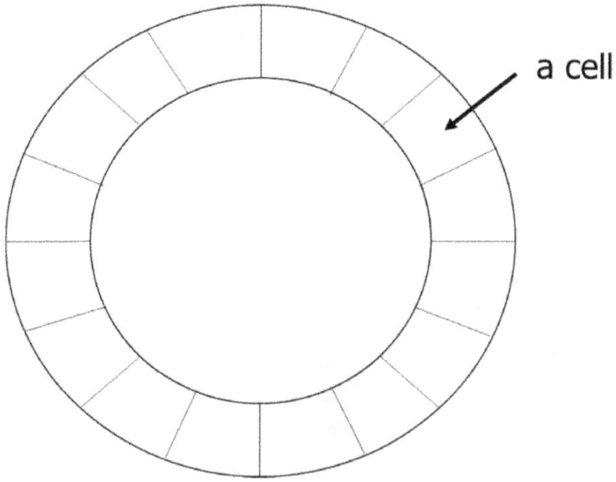

Fig. 7.1.1.1.3. Cells for a ring road.

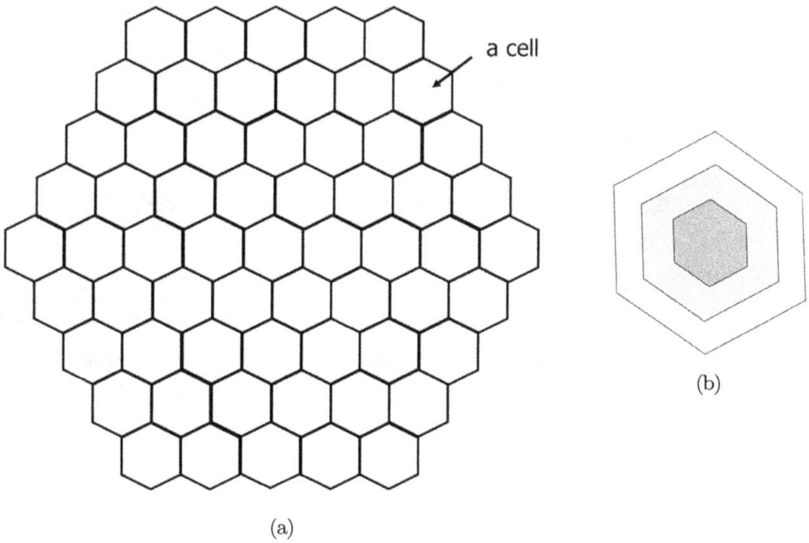

Fig. 7.1.1.1.4. Cells for a plane (two-dimensional cell model).

it is considered that the radio wave propagation on the plane is uniform and circular. Figure 7.1.1.1.4(b) shows the divided small cells of a cell in Fig. 7.1.1.1.4(a), which may be divided into three small cells as shown.

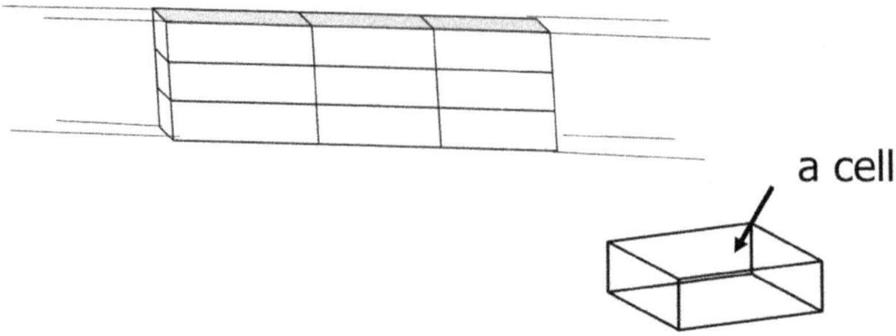

Fig. 7.1.1.1.5. Cells for a three-dimensional road.

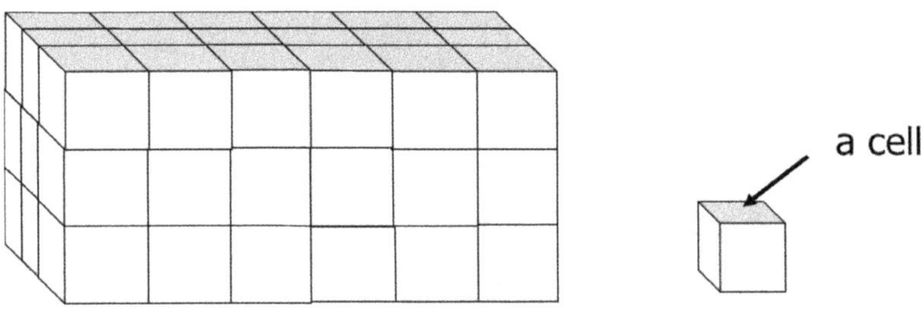

Fig. 7.1.1.1.6. A three-dimensional cell model.

Figure 7.1.1.1.6 shows a three-dimensional cell model. For example, this model is used in-building mobile communication.

The cell structure varies due to time fluctuations in the number of mobile terminals in the service area, time fluctuations in radio propagation characteristics, and time fluctuations in the vacancy of the frequency spectrum. Such a change in cell structure means a *four-dimensional cell model.* This is called *dynamic spectrum sharing* in research on ITU (International Telecommunication Union) [67,68].

Thus, as described above, there are one-dimensional to four-dimensional cell structures. These communications are sometimes called as *multidimensional mobile communications.*

7.1.1.2. *Channels and buffer cells in a cellular system*

In general, there are three types of multiplexing methods in a communication system: Frequency Division Multiple Access (FDMA), Time Division

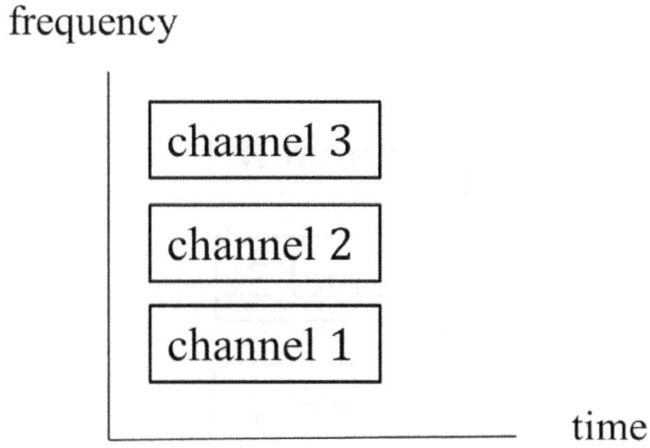

Fig. 7.1.1.1.1. Brief explanatory diagram of channels of Frequency Division Multiple Access (FDMA).

Multiple Access (TDMA) and Code Division Multiple Access (CDMA). Multiplexing is to combine multiple signals and send them on one shared transmission line or band. In addition to these three, Space Division Multiple Access (SDMA) and Polarization Division Multiple Access (PDMA) may be added. SDMA divides the space and multiplexes it, and PDMA uses the polarization antenna to multiplex it, but these use the division of space and can be included in the cellular system in mobile communication. Therefore, only the above three will be explained.

A *call* in a mobile communication system means a wireless communication request in a cell. Figure 7.1.1.2.1 shows brief explanatory diagram of channels of Frequency Division Multiple Access (FDMA) where the a channel is a frequency band for a call. Figure 7.1.1.2.2 shows brief explanatory diagram of channels of Time Division Multiple Access (TDMA) where a channel is a time slot for a call. Figure 7.1.1.2.3 shows brief explanatory diagram of channels of Code Time Division Multiple Access (CDMA) where a channel is a code for a call. Figure 7.1.1.2.3 is an example of CDMA called *frequency-hopping spread spectrum* (FHSS) which is a method of transmitting radio signals by rapidly changing the carrier frequency among many distinct frequencies occupying a large spectral band and the changes are controlled by a given code.

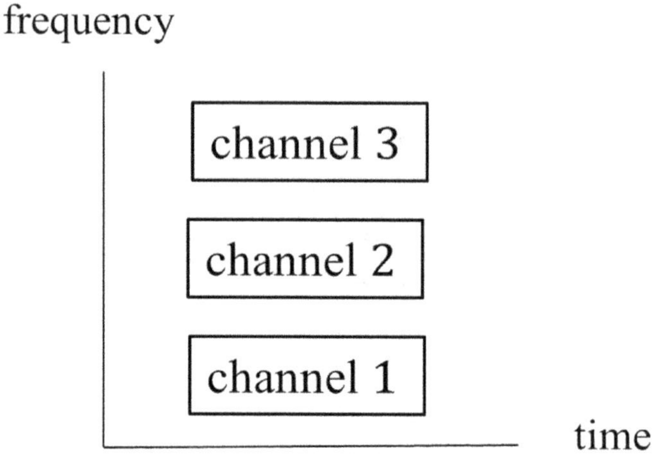

Fig. 7.1.1.2.2. Brief explanatory diagram of channels of Time Division Multiple Access (TDMA).

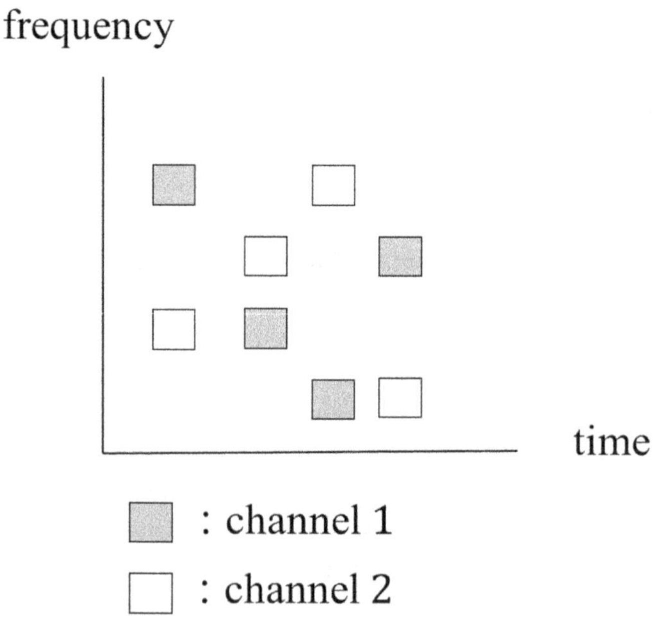

Fig. 7.1.1.2.3. Brief explanatory diagram of channels of Code Division Multiple Access (CDMA).

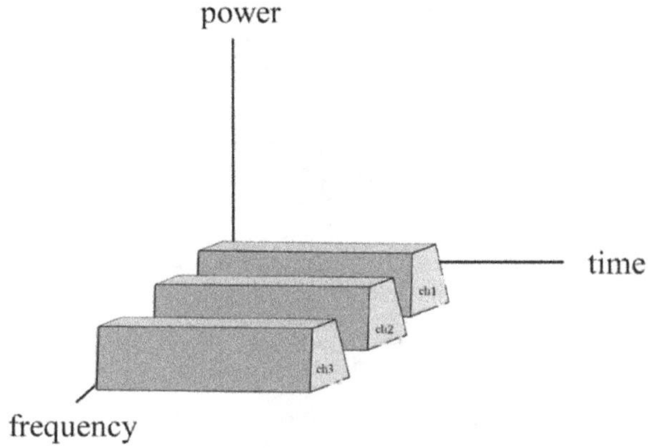

Fig. 7.1.1.2.4. Explanatory diagram with power axis for channels of Frequency Division Multiple Access (FDMA).

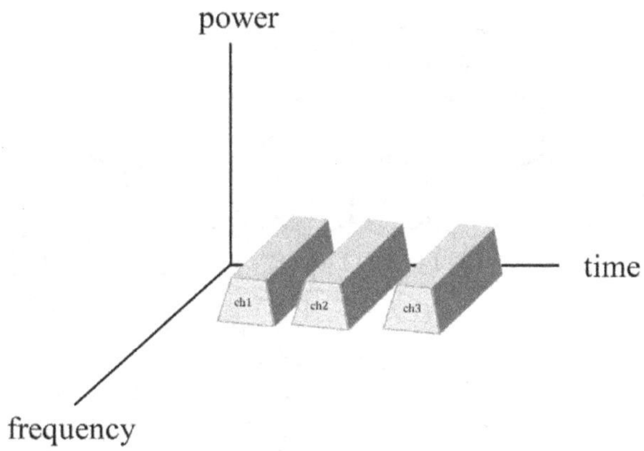

Fig. 7.1.1.2.5. Explanatory diagram with power axis for channels of Time Division Multiple Access (TDMA).

Figures 7.1.1.2.4–7.1.1.2.6 show explanatory diagrams with power axis for Channels for FDMA, TDMA and CDMA, respectively. As shown in these diagrams, all three types multiplexing have power on the frequency axis and the time axis, and each has its own characteristics. Using that feature, the type is decided in consideration of the communication system, the propagation of radio waves, etc.

If a channel is used in a cell, it cannot be used in some other cells due to interference by the channel. These other cells are called *buffer cells* of the

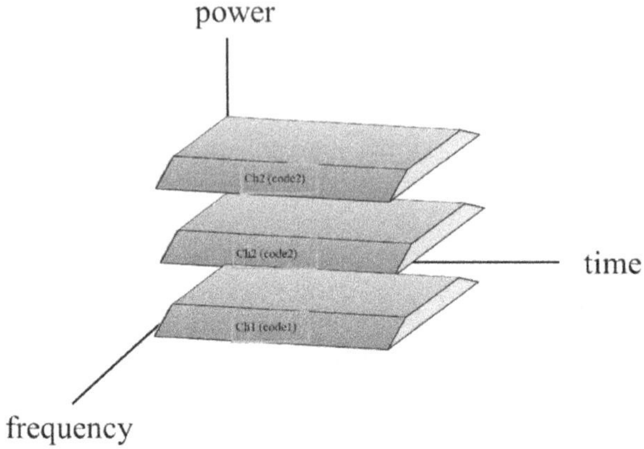

Fig. 7.1.1.2.6. Explanatory diagram with power axis for channels of Code Division Multiple Access (CDMA).

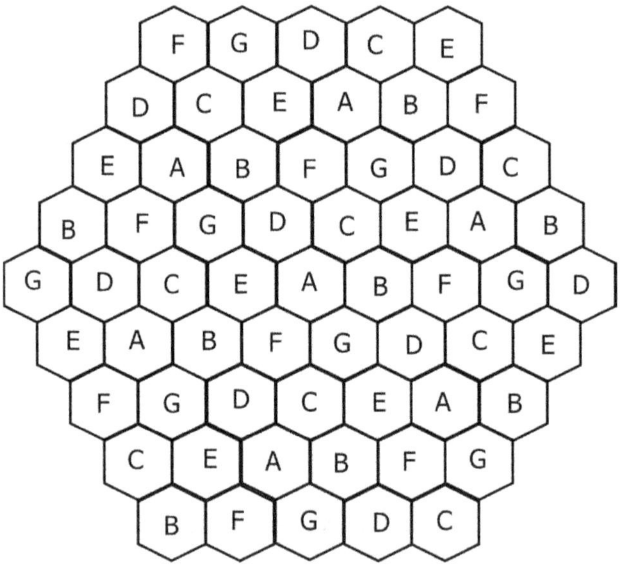

Fig. 7.1.1.2.7. A channel allocation.

cell. If the buffer cells of a given cell consist of all closer than k cells away from it, the buffering system is called the *k-belt buffering*. For example, we assume two-belt buffering system in the cells for a plane (two-dimensional cell model) of Fig. 7.1.1.1.4(a). Figure 7.1.1.2.7 shows a channel allocation for two-belt buffering system where seven channels A, B, C, D, E, F and G are used. In the cellular system of Fig. 7.1.1.2.7, i.e., channels A and B are used seven times and nine times in the service area, respectively, in order to make effective use of the frequency spectrum.

7.1.1.3. *Coloring problems of graphs and channel assignment problems*

Let us define the following graph which represents the relations between a cell and its buffer cells in a cellular system. This graph is convenient for considering the communication traffic characteristics of the system.

Let $G = (V, E)$ be, a simple graph where V and E are the set of vertices and edges in the graph G, respectively. A vertex $v_i \in V$ corresponds to a cell $z_i (i = 1, 2, \ldots, n)$, and an edge $e_k = (v_i, v_j) \in E$ represents that a cell z_i belongs to the buffer zones of z_j. As mentioned in Section 2.2, a simple graph that contains every possible edge between all the vertices is called a *complete graph* and a complete graph with n vertices is denoted as K_n. A graph $\bar{G}$, obtained by deleting the edges of a given simple graph G from a complete graph (K_n) having the same vertices, is called the *complement* of G. A complete subgraph of a graph G is called a *clique*, and it is also called a *k-clique* if it contains k vertices.

Example. Let us consider the service area of Fig. 7.1.1.3.1 as an example. Let the buffer cells of a cell consist of the contiguous cells to it, that is, one-belt buffering. Then, Fig. 7.1.1.3.2 shows the graph G and Fig. 7.1.1.3.3 shows the complement graph $\bar{G}$ of G. **(End of Example)**

From Section 4.4.6 and the above definition of graph G, the minimum number of channels required to communicate in all cells simultaneously is equal to the chromatic number of the graph G. The coloring problem of a graph is to find the minimum number of colors (chromatic number) and the coloring. In a cellular system, the channel assignment problem is to allocate a channel for a call (a communication request) in a cell, provided that a graph G showing whether the same channel can be used is given. The vertices of a clique of $\bar{G}$ represent cells that can be assigned the same channel at the same time. It is obvious that the coloring problem of a graph is used the

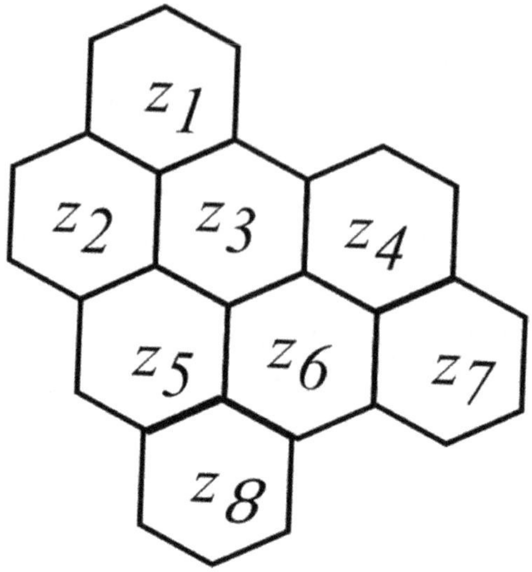

Fig. 7.1.1.3.1. A cellular system.

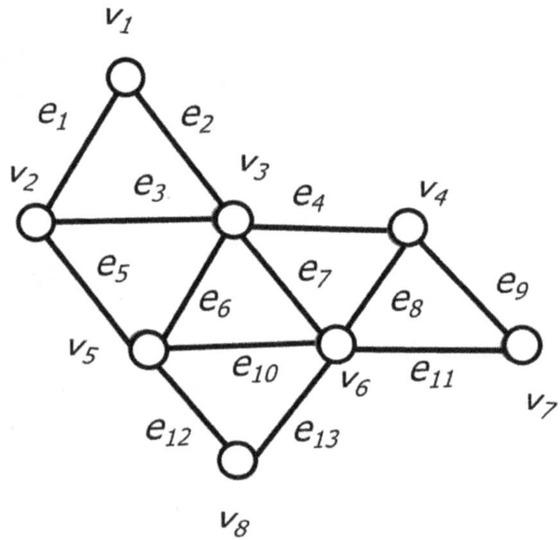

Fig. 7.1.1.3.2. The graph G of Fig. 7.1.1.3.1.

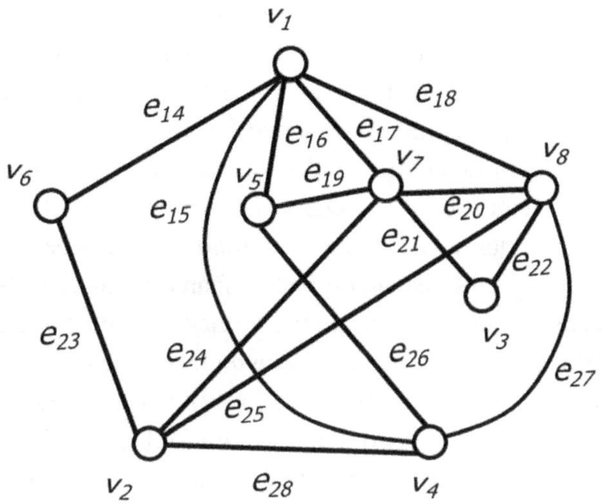

Fig. 7.1.1.3.3. The complement graph $\bar{G}$ of G (Fig. 7.1.1.3.2).

channel assignment problem. In a cellular system, the channel assignment methods are classified roughly into two types, that is, a *fixed channel assignment* (FCA) method and a *dynamic channel assignment* (DCA) method. In the fixed channel assignment method, a subset of the channels available to a cell is permanently served for use within the cell. In the most general form of dynamic channel assignment, it is assumed that any channel can be used in any cell. Let us explain a simple example using a cellular system of Fig. 7.1.1.3.1, G of Fig. 7.1.1.3.2 and $\bar{G}$ of Fig. 7.1.1.3.3. The chromatic number of G is 3 and the coloring is, e.g., as follows. (*Note*: it is known as *four color theorem* that the chromatic number of every simple planar graph is four [9]. Since the graph of Fig. 7.1.1.3.2 is planar, the chromatic number of the graph is 4 or less. Since the cellular system of Figs. 7.1.1.3.1 and 7.1.1.3.2 is small, it is easy to see that it is 3-colorable. The cellular system of Fig. 7.1.1.2.7 is two-belt buffering, the corresponding graph may not be planar and A–G of Fig. 7.1.1.2.7 show that it is 7-colorable.)

The color C_1 is assigned at v_1, v_4 and v_5. The color C_2 is assigned at v_2 and v_6, and the color C_3 is assigned at v_3, v_7 and v_8.

As a fixed channel assignment method, let a channel c_i be permanently assigned to the cells corresponding vertex v_j assigned C_i where $i = 1, 2, 3$, that is, this system has three channels. Since one channel is assigned in each cell in the system, the channel is assigned for a call in a cell. When one call is in progress (busy) in a cell z_i and a second call occurs in z_i, the second call is blocked because only one channel is assigned in cell z_i.

As a dynamic channel assignment method, let the three channels c_1, c_2, and c_3 be not assigned to each cell but be shared in all cells. Since the call randomly occurs in cells throughout the service area, the three channels will be dynamically assigned depending on which cell the call originated. This means that when one call is in progress (busy) in a cell z_i and a second call occurs in z_i, the second call may not be blocked in the dynamic channel assignment. Thus, in general a dynamic channel assignment method gives better performance than a fixed channel assignment method. In the dynamic channel allocation method, there may be multiple channel assignment candidates for the occurrence of a new call, and the performance differs depending on which channel is selected from the candidates, so various channel assignment algorithms have been proposed [6–8] etc.

7.1.1.4. *Communication traffic characteristics in a cellular system*

Figure 7.1.1.4.1 shows a model of *queueing system* where jobs or customers arrive to the queue, possibly wait some time, take some time being

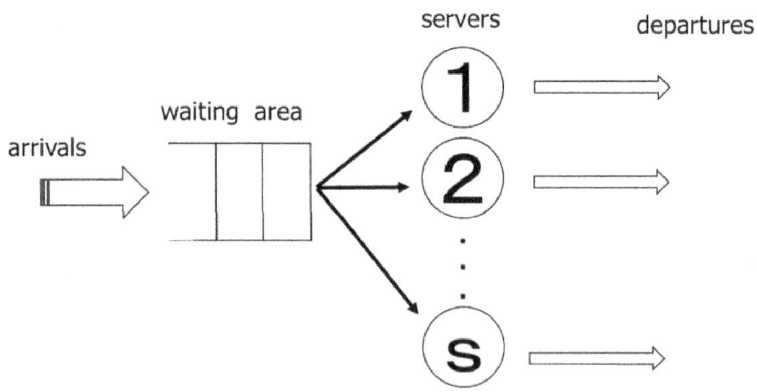

Fig. 7.1.1.4.1. A model of queueing system.

processed, and then depart from the queue. *Queueing theory* is the mathematical study of queues using the model and it has many application areas, e.g., information and communication technology, traffic engineering, computing, and particularly in industrial engineering, in the design of factories, shops, offices and hospitals, as well as in project management. Since the waiting area is often not included in the information and communication technology field including mobile communication, so only the *loss system* which does not have waiting area is considered in this section.

In the system, the number of servers is s assuming that the arrival process is Poisson with parameter λ and that each server renders a service which is exponentially distributed with parameter μ. This system is called $M/M/s$ queueing system. For the offered traffic

$$a = \lambda/\mu, \tag{7.1.1.4.1}$$

some of the traffic is blocked in which the probability is called *blocking probability*. The blocking probability $E_s(a)$ is given as follows:

$$E_s(a) = \frac{\frac{a^s}{s!}}{\sum_{n=0}^{s} \frac{a^n}{n!}}, \tag{7.1.1.4.2}$$

or if s is not an integer,

$$E_s(a) = \frac{a^s e^{-a}}{\int_a^{\infty} x^s e^{-x} dx}. \tag{7.1.1.4.3}$$

Equations (7.1.1.4.2) and (7.1.1.4.3) are known as *Erlang's Loss Formula*, or *Erlang B Formula*, published first by A. K. Erlang in 1917 [5]. The unit of offered traffic is [*erlang*] or [*erl*].

Let the system be a loss system. Let us assume that the arrivals (requests for service) form a Poisson process, and the call duration (holding time) is an exponential random variable. For simplicity, it is assumed that the mobile unit does not cross a cell boundary during a communication. (The cell size is sufficiently large and the average message length is short, so that the probability of the occurrence can be neglected. In the case of the mobile unit crosses a cell boundary during a communication, we explain it in next section).

(1) Communication Traffic Characteristics for a Fixed Channel Assignment (FCA) in a Cellular System

Let the number of channels a cell z_i in the system be s_{zi}. We make the following two mathematical assumptions for a call in a cell z_i:

(a) arrivals occur according to a Poisson process with intensity λ_i;
(b) the duration of a message (holding time) is an exponential random variable with mean $1/\mu_i$.

Let the offered traffic in cell z_i be $a_i = \lambda_i/\mu_i$. Since the number of channels in a cell is s_{zi} and it is fixed, the blocking probability of cell z_i is $Es_{zi}(a_i)$ using (7.1.1.4.2).

(2) Communication Traffic Characteristics for a Dynamic Channel Assignment (DCA) in a Cellular System

In the most general form of dynamic channel assignment, it is assumed that any channel can be used in any coverage cell. As mentioned in Section 7.1.1.3, there may be multiple channel assignment candidates for the occurrence of a new call, and the performance differs depending on which channel is selected from the candidates. This means that in the system having more than one channel, the traffic carrying capacity depends on the channel assignment algorithm, and it is difficult to obtain the strict solution of the traffic characteristics in a multichannel system. So, let us discuss the one channel operation first and next an approximation method for multichannel operations.

(1) One-channel operation

Let the number of channels n_c in the system be one. We make the following two mathematical assumptions for a call in a cell z_i:

(a) arrivals occur according to a Poisson process with intensity λ_i;
(b) the duration of a message (holding time) is an exponential random variable with mean $1/\mu_i$.

A channel can be used in some cells simultaneously, and there are many ways that a channel is used in some cells in the system. Each different way may be considered as a state of the system. Let $\Theta = \{\theta_i\}$ be the set of the states, and an element θ_i of the set consists of the vertices (of the graph G) corresponding to the cells in which a channel is busy. Then a set Θ is classified as follows:

$$\Theta = \{\Theta_0, \Theta_1, \ldots, \Theta_{k_m}\}, \tag{7.1.1.4.4}$$

$$\Theta_k = \{\theta_i | \theta_i \in \Theta, |\theta_i| = k\}, \tag{7.1.1.4.5}$$

where $|\theta_i|$ denotes the cardinal number of θ_i and k_m is its maximum number in the system. From assumptions (a) and (b), if Δt is a very short time, the probability of more than one call gained or one call lost in Δt is negligibly small. The probability of losing one call in Δt in a cell z_i is $\mu_i \Delta t$ and the probability of gaining one call in Δt in a cell z_i is $\lambda_i \Delta t$. For a state $\theta_i \in \Theta_k$, let us define the following sets $U_i^+(\theta_i)$, $U_i^-(\theta_i)$, V_i^+ and V_i^-:

$$U_i^+(\theta_i) \triangleq \{\theta_j | \theta_j \in \Theta_{k+1}, \theta_j \supset \theta_i\}, \tag{7.1.1.4.6}$$

$$U_i^-(\theta_i) \triangleq \{\theta_j | \theta_j \in \Theta_{k-1}, \theta_j \subset \theta_i\}, \tag{7.1.1.4.7}$$

$$V_i^+ \triangleq \{\theta_l - \theta_i | \theta_l \in U_i^+(\theta_i)\}, \tag{7.1.1.4.8}$$

$$V_i^- \triangleq \{\theta_i - \theta_l | \theta_l \in U_i^-(\theta_i)\}. \tag{7.1.1.4.9}$$

Let the probability that the state of the system is θ_i be $p(\theta_i)$. Then the state flow diagram of the system is shown in Fig. 7.1.1.4.2.

We assume the system to be in statistical equilibrium. Then for $k = 0$

$$p(\theta_i) \sum_{\upsilon_j \in V_i^+} \lambda_j \Delta t = \sum_{\substack{\theta_l \in U_i^+(\theta i) \\ \upsilon_j \in \theta_l - \theta_i}} p(\theta_l) \mu_j \Delta t, \tag{7.1.1.4.10}$$

for $k = k_m$

$$p(\theta_i) \sum_{\upsilon_j \in V_i^-} \mu_j \Delta t = \sum_{\substack{\theta_l \in U_i^-(\theta i) \\ \upsilon_j \in \theta_i - \theta_l}} p(\theta_l) \lambda_j \Delta t, \tag{7.1.1.4.11}$$

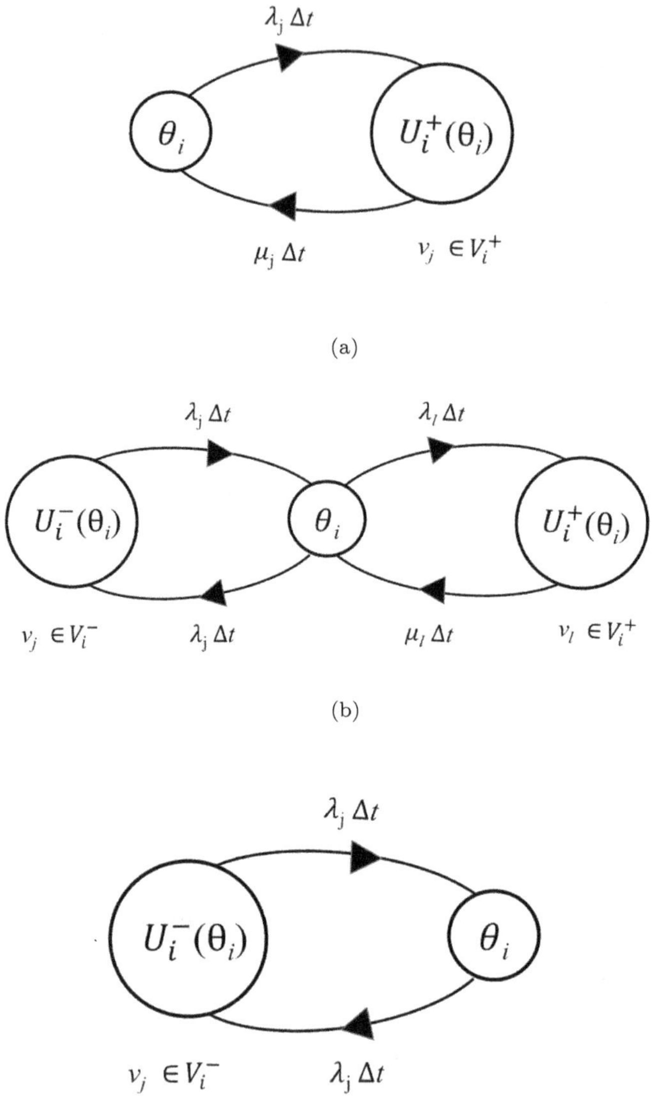

Fig. 7.1.1.4.2. State flow diagrams. (a) $k = 0$, $\theta_i = \emptyset$; (b) $k \neq 0$, $k \neq k_m$, $\theta_i = \{v_{i1}, v_{i2}, \ldots, v_{ik}\}$; (c) $k = k_m$, $\theta_i = \{v_{i1}, v_{i2}, \ldots, v_{ik}\}$.

and for $k \neq 0$ and $k \neq k_m$

$$p(\theta_i)\sum_{v_j\in V_i}\lambda_j\Delta t + p(\theta_i)\sum_{v_j\in V_i}\lambda_j\Delta t,$$

$$= \sum_{\substack{\theta_l\in U_i^+(\theta i)\\ v_j\in\theta_l-\theta_i}} p(\theta_i)\mu_j\Delta t + \sum_{\substack{\theta_l\in U_i^-(\theta i)\\ v_j\in\theta_i-\theta_l}} p(\theta_i)\lambda_j\Delta t, \qquad (7.1.1.4.12)$$

where the terms of order $(\Delta t)^2$ and higher are neglected as $\Delta t \to 0$.

To solve (7.1.1.4.10)–(7.1.1.4.12), let

$$\theta_i = \{v_{i_1}, v_{i_2}, \ldots, v_{i_k}\},$$

$$\theta_l = \{v_{i_1}, v_{i_2}, \ldots, v_{i_{k+1}}\}, \quad \theta_l \in U_i^+(\theta_i),$$

and

$$\theta_j = \{v_{i_1}, v_{i_2}, \ldots, v_{i_{k-1}}\}, \quad \theta_j \in U_i^-(\theta_i).$$

Then we can see

$$p(\theta_i)\lambda_{i_{k+1}} = p(\theta_l)\mu_{i_{k+1}}, \qquad (7.1.1.4.13)$$

$$p(\theta_i)\mu_{i_k} = p(\theta_j)\lambda_{i_k} \qquad (7.1.1.4.14)$$

are the solutions of (7.1.1.4.10)–(7.1.1.4.12). Using $a_i = \lambda_i/\mu_i$ (erlangs offered in cell z_i),

$$p(\theta_l) = a_{i_{k+1}} \cdot p(\theta_i) \qquad (7.1.1.4.15)$$

and

$$p(\theta_i) = a_{i_k} \cdot p(\theta_j). \qquad (7.1.1.4.16)$$

From (7.1.1.4.15) and (7.1.1.4.16),

$$p\,(\theta_i) = a_{ik+1} \cdot a_{ik} \cdot \ldots \cdot a_{i1} \cdot p_0, \qquad (7.1.1.4.17)$$

where p_0 is the probability that the channel is not used in any cells.

Let p_x be the probability that a channel is used in x cells simultaneously. Then,

$$p_x = \sum_{\theta_i \in \Theta x} p(\theta_i)$$
$$= \sum_{\{v_{i_1}, v_{i_2}, \ldots, v_{i_x}\} \in \Theta x} (a_{i_1} \cdot a_{i_2} \cdot \ldots \cdot a_{i_x}) \cdot p_0$$

Using the requirement $\sum_{x=0}^{k_m} p_x = 1$,

$$p_0 = \frac{1}{1 + \sum_{x=1}^{k_m} \sum_{\{v_{i_1}, v_{i_2}, \ldots, v_{i_x}\} \in \Theta x} a_{i_1} \cdot a_{i_2} \cdot \ldots \cdot a_{i_x}}. \quad (7.1.1.4.18)$$

Therefore,

$$p_x = \frac{\sum_{\theta_i \in \Theta x} a_{i_1} \cdot a_{i_2} \cdot \ldots \cdot a_{i_x}}{1 + \sum_{y=1}^{k_m} \sum_{\theta_s \in \Theta y} a_{j_1} \cdot a_{j_2} \cdot \ldots \cdot a_{j_y}}, \quad (7.1.1.4.19)$$

where $\theta_i = \{v_{i1}, v_{i2}, \ldots, v_{ix}\}$ and $\theta_s = \{v_{j1}, v_{j2}, \ldots, v_{jy}\}$. The total load carried is

$$a_c = \sum_{x=1}^{k_m} x \cdot p_x$$
$$= \frac{\sum_{x=1}^{k_m} \sum_{\theta_i \in \Theta x} a_{i_1} \cdot a_{i_2} \cdot \ldots \cdot a_{i_x}}{1 + \sum_{y=1}^{k_m} \sum_{\theta_s \in \Theta y} a_{j_1} \cdot a_{j_2} \cdot \ldots \cdot a_{j_y}}. \quad (7.1.1.4.20)$$

The probability of blocking in the system is

$$B = (a - a_c)/a, \quad (7.1.1.4.21)$$

where

$$a = \sum_{i=1}^{n} a_i.$$

Next, let us obtain the probability of blocking in each cell. Let Θ_k^j be the subset of the set $\Theta \in \Theta_k$:

$$\Theta_k^j = \{\theta_i | \upsilon_j \in \theta_i, \theta_i \in \Theta_k\}. \quad (7.1.1.4.22)$$

Let $p_{x,j}$ be the probability that the channel is used in a cell z_j and the other $x - 1$ cells simultaneously. Then,

$$p_{x,j} = \frac{\sum_{\theta_i \in \Theta_k^j} a_{i_1} \cdot a_{i_2} \cdot \ldots \cdot a_{i_x}}{1 + \sum_{y=1}^{k_m} \sum_{\theta_s \in \Theta y} a_{j_1} \cdot a_{j_2} \cdot \ldots \cdot a_{j_y}}, \quad (7.1.1.4.23)$$

where $\theta_i = \{\upsilon_{i1}, \upsilon_{i2}, \ldots, \upsilon_{ix}\}$ and $\theta_s = \{\upsilon_{j1}, \upsilon_{j2}, \ldots, \upsilon_{jy}\}$. The load carried in a cell z_j is

$$\begin{aligned} a_{c,j} &= \sum_{x=1}^{k_m} p_{x,j} \\ &= \frac{\sum_{x=1}^{k_m} \sum_{\theta_i \in \Theta_k^j} a_{i_1} \cdot a_{i_2} \cdot \ldots \cdot a_{i_x}}{1 + \sum_{y=1}^{k_m} \sum_{\theta_s \in \Theta y} a_{j_1} \cdot a_{j_2} \cdot \ldots \cdot a_{j_y}}. \end{aligned} \quad (7.1.1.4.24)$$

The probability of blocking in the cell z_j is

$$B_j = (a_j - a_{c,j})/a_j. \quad (7.1.1.4.25)$$

If the load is uniformly offered in each cell, that is, $a_i = a_j (i \neq j)$, then the load carried a_c and $a_{c,j}$ is

$$a_c = \frac{\sum_{x=1}^{k_m} (a_i)^x x \nu_x}{\sum_{y=0}^{k_m} (a_i)^y \nu_y}, \quad (7.1.1.4.26)$$

$$a_{c,j} = \frac{\sum_{x=1}^{k_m} (a_i)^x \nu_{x,j}}{\sum_{y=0}^{k_m} (a_i)^y \nu_y}, \quad (7.1.1.4.27)$$

where

$$\nu_x = |\Theta x| \quad (7.1.1.4.28)$$

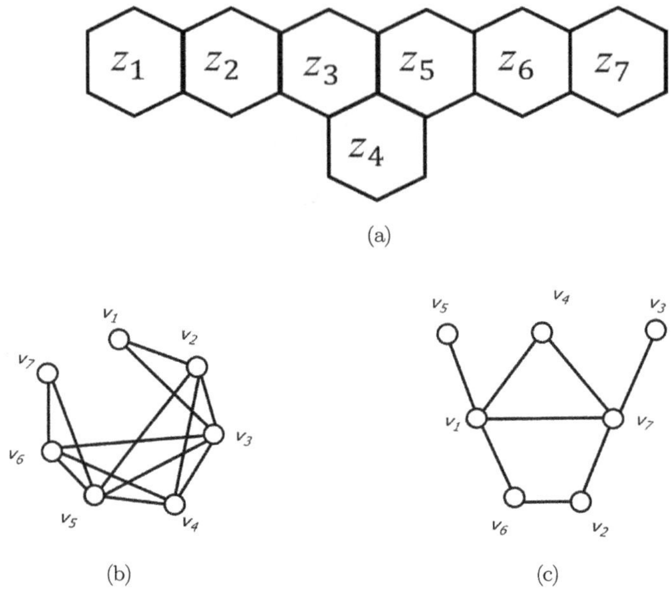

Fig. 7.1.1.4.3. An example of service areas and its graphs G and $\bar{G}$. (a) A service area. (b) The graph G. (c) The graph $\bar{G}$.

and

$$\nu_{x,j} = |\Theta_k^j|. \tag{7.1.1.4.29}$$

In the graph $\bar{G}$, an element θ_i of the set Θ corresponds to a clique. (In the graph G, it corresponds to an independent set.) The subset Θ_k of Θ is a set of x-cliques in the graph $\bar{G}$ and Θ_k^j is a set of x-cliques which contain a vertex v_j. Therefore, from (7.1.1.4.20), (7.1.1.4.21), (7.1.1.4.24), and (7.1.1.4.25), the traffic characteristics of the system are obtained by the set of cliques in the graph $\bar{G}$. If the load is uniformly offered in each cell, only the knowledge of the number of cliques and x-cliques is required.

Example. Figure 7.1.1.4.3(a) is a service area, and if the buffer zones of a given sell consist of all cells closer than two cells away from the given cell (that is, two-belt buffering), the graph G is Fig. 7.1.1.4.3(b). Figure 7.1.1.4.3(c) is the complement graph $\bar{G}$ of G.

The set of one-cliques Θ_1, the set of two-cliques Θ_2, and the set of three-cliques Θ_3 in the graph $\bar{G}$ are as follows:

$$\Theta_1 = \{\upsilon_1, \upsilon_2, \upsilon_3, \upsilon_4, \upsilon_5, \upsilon_6, \upsilon_7\},$$
$$\Theta_2 = \{\upsilon_1\upsilon_4, \upsilon_1\upsilon_5, \upsilon_1\upsilon_6, \upsilon_1\upsilon_7, \upsilon_2\upsilon_6, \upsilon_2\upsilon_7, \upsilon_3\upsilon_7, \upsilon_4\upsilon_7\},$$
$$\Theta_3 = \{\upsilon_1\upsilon_4\upsilon_7\}.$$

The graph $\bar{G}$ dose not contain four-cliques, five-cliques, etc.

Let us consider Θ_k^j for υ_1.

$$\Theta_1^1 = \{\upsilon_1\}, \nu_{1,1} = 1,$$
$$\Theta_2^1 = \{\upsilon_1\upsilon_4, \upsilon_1\upsilon_5, \upsilon_1\upsilon_6, \upsilon_1\upsilon_7\}, \quad \nu_{2,1} = 4,$$
$$\Theta_3^1 = \{\upsilon_1\upsilon_4\upsilon_7\}, \quad \nu_{3,1} = 1.$$

From (7.1.1.4.24),

$$\begin{aligned} a_{c,1} &= \frac{\sum_{x=1}^{k_m} \sum_{\theta_i \in \Theta_k^j} a_{i_1} \cdot a_{i_2} \cdot \ldots \cdot a_{i_x}}{1 + \sum_{y=1}^{k_m} \sum_{\theta_s \in \Theta y} a_{j_1} \cdot a_{j_2} \cdot \ldots \cdot a_{j_y}} \\ &= (a_1 + a_1a_4 + a_1a_5 + a_1a_6 + a_1a_7 + a_1a_4a_7)/(1 + a_1 + a_2 \\ &\quad + a_3 + a_4 + a_5 + a_6 + a_7 + a_1a_4 + a_1a_5 + a_1a_6 + a_1a_7 \\ &\quad + a_2a_6 + a_2a_7 + a_3a_7 + a_4a_7 + a_1a_4a_7). \end{aligned} \tag{7.1.1.4.30}$$

Therefore, the probability of blocking in a cell z_1,

$$\begin{aligned} B_1 &= (a_1 - a_{c,1})/a_1 = (a_1 + a_2 + a_3 + a_1a_4 + a_1a_5 + a_1a_6 + a_1a_7 \\ &\quad + a_2a_6 + a_2a_7 + a_3a_7 + a_1a_4a_7)/(1 + a_1 + a_2 + a_3 + a_4 + a_5 \\ &\quad + a_6 + a_7 + a_1a_4 + a_1a_5 + a_1a_6 + a_1a_7 + a_2a_6 + a_2a_7 \\ &\quad + a_3a_7 + a_4a_7 + a_1a_4a_7). \end{aligned} \tag{7.1.1.4.31}$$

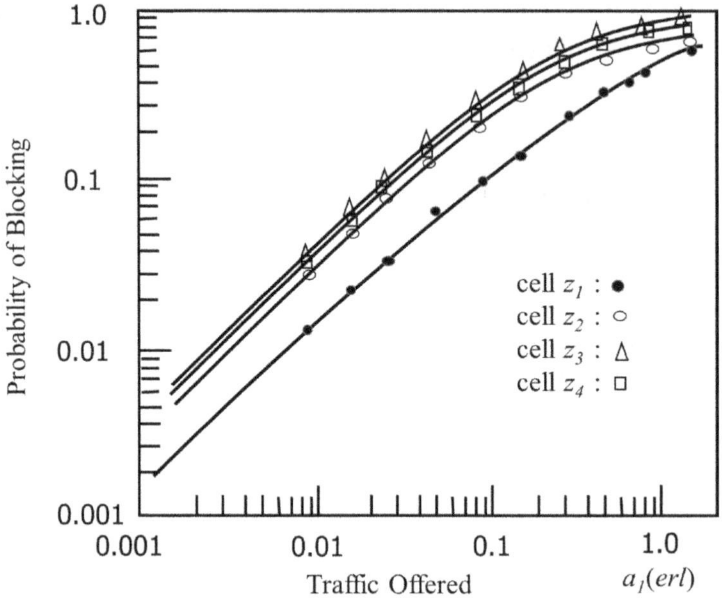

Fig. 7.1.1.4.4. Probability of blocking versus traffic offered for one channel in a service area of Fig. 7.1.1.4.3(a).

If $a_i = a_j = a_1 (i, j = 1, 2, \ldots, 7)$,

$$a_{c,1} = \frac{a_1 + 4a_1^2 + a_1^3}{1 + 7a_1 + 8a_1^2 + a_1^3} \tag{7.1.1.4.32}$$

$$\begin{aligned} B_1 &= (a_1 - a_{c,1})a_1 \\ &= \frac{3a_1 + 7a_1^2 + a_1^3}{1 + 7a_1 + 8a_1^2 + a_1^3}. \end{aligned} \tag{7.1.1.4.33}$$

B_2, B_3, ..., and B_7 are also obtained similarly. The data from a *computer simulation*, which show the relationships between the offered load (erlang/cell) and the probability of blocking in each cell, are plotted in Fig. 7.1.1.4.4. The solid lines in Fig. 7.1.1.4.4 are theoretical results using (7.1.1.4.31) where $a_2 = 0.5a_1$, $a_3 = 0.2a_1$, $a_4 = 1.5a_1$, $a_5 = 1.2a_1$, $a_6 = 1.5a_1$ and $a_7 = 0.5a_1$. **(End of Example)**

(2) Multichannel Operations

As mentioned in Section 7.1.1.4(2), it is difficult to obtain the strict solution of the traffic characteristics in a mobile radio communication system with multi-channels using dynamic channel assignments. So, it is necessary to use an approximation for obtaining the traffic characteristics of multichannel operations. In this section, two approximation methods (A) and (B) for the calculation are proposed.

(A) In the system with one channel, for the load offered a_i $(i = 1, 2, \ldots, n)$, the blocking probability B_i of cell z_i is obtained from (7.1.1.4.24) and (7.1.1.4.25). Assuming that a_i and B_i depend on the Erlang B Formula (7.1.1.4.2) or (7.1.1.4.3), we can find the equivalent number of channels s_i in a cell z_i using $B_i = E_{s_i}(a_i)$. The number s_i is not always an integer and must be one or less, because at most one channel is available. Let the system have n_c channels $(n_c > 1)$ and let B_{i,n_c} be the blocking probability of z_i in the system. If the load is uniformly offered to the dynamic channels, the blocking probability B_{i,n_c} for the load offered $n_c a_i$ in a zone z_i is $B_{i,n_c} = E_{n_c s_i}(n_c a_i)$, because the equivalent number of channels s_i for one dynamic channel is obtained from a_i. In other words, for the load offered a_i $(i = 1, 2, \ldots, n)$ to all dynamic channels (see Fig. 7.1.1.4.5(a)), the equivalent number of channels s_i' of a cell z_i for a dynamic channel of the system is obtained from a_i/n_c $(i = 1, 2, \ldots, n)$, (7.1.1.4.24), (7.1.1.4.25), and the Erlang B formula ((7.1.1.4.2) or (7.1.1.4.3)). We assume that a cell z_i is served by a complete group of $n_c s_i'$ channels (see Fig. 7.1.1.4.5(b)). Then, the blocking probability B_{i,n_c} of a cell z_i for the load offered a_i is $B_{i,n_c} = E_{n_c s_i'}(a_i)$. This approximation technique has been applied to obtaining the blocking probability in the whole system [74].

(B) In the previous method, the equivalent number of channels s_i' of a cell z_i are obtained, assuming that the traffic intensity is uniformly offered to each dynamic channel. In a multichannel system, however, the traffic offered a channel is generally overflow traffic from the other channels. So, the traffic offered is different for each channel; consequently, the equivalent number of channels for each dynamic channel is also different. The equivalent number of channels s_{ij} of a cell z_i for a dynamic channel j is obtained in a way similar

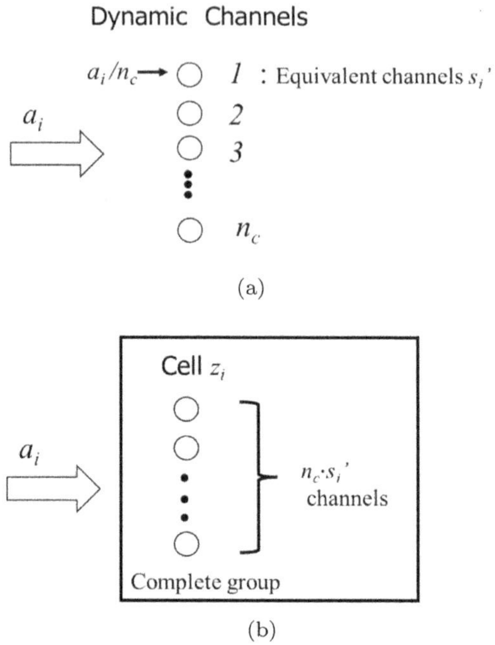

Fig. 7.1.1.4.5. Explanation of approximation (A). (a) Equivalent channels $\{s_i'\}$. (b) An equivalent switching network of a cell z_i.

to the previous method if the traffic offered to the channel is given. Although the overflow traffic does not always form a Poisson process, let us assume that the equivalent number of channels s_{ij} are obtained from (7.1.1.4.24), (7.1.1.4.25) and the traffic with the same mean (first moment) as the overflow traffic of $j-1$ channels (see Fig. 7.1.1.4.6). The approximation method for obtaining the blocking probability of the system with n_c channels is shown in Fig. 7.1.1.4.7 as a flowchart. This method may correspond to the first-available method in dynamic channel assignment methods, since the first-available method is the assignment of the first available channel encountered during the channel search from channel 1 to channel n_c and the equivalent number of channels s_{ij} are obtained from the overflow traffic of $j-1$ channels (see Fig. 7.1.1.4.6).

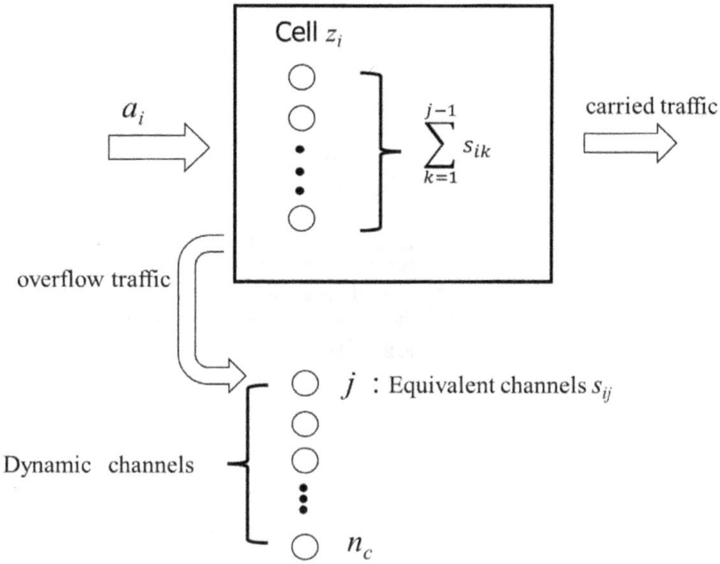

Fig. 7.1.1.4.6. Overflow traffic (approximation (B)).

Example. As an example, let us consider the system of Fig. 7.1.1.4.3 again. Let the traffic offered be uniform in each cell and let the system have ten channels. Figure 7.1.1.4.8 shows a computer simulation result using the first-available method as a dynamic channel assignment algorithm. Figures 7.1.1.4.8 and 7.1.1.4.9 are the results of the above approximations (A) and (B), respectively. Let us compare the computer simulation results with the two approximations (A) and (B). Focusing on the blocking probability in cell 4 of the three figures (Figs. 7.1.1.4.8–7.1.1.4.10) showing the change in the traffic offered versus the blocking probability, the approximation method (B) is clearly consistent with the simulation result. That is, the method (B) is better than method (A) in the degree of approximation. Although method (A) does not give as good an approximation, the calculation has the advantage of simplicity. **(End of Example)**

In the above two approximation methods, the equivalent number of channels is obtained by a one-parameter fit, based on the first moment

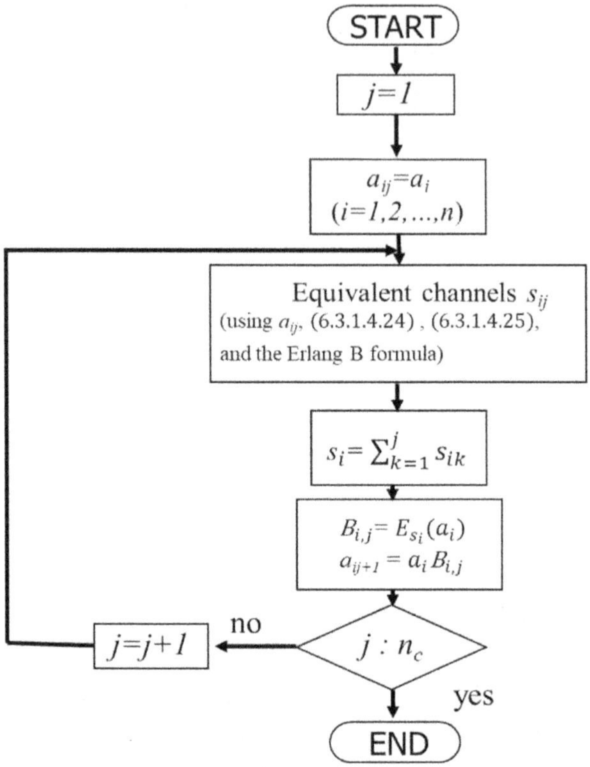

Fig. 7.1.1.4.7. A flowchart of approximation (B).

of overflow traffic. If both the first moment and the second moment (or variance) of overflow traffic are given, the equivalent number of channels may be more accurately obtained in the same way as the equivalent random method [75].

As other approximation methods, the method called *clique packing* is known. In the graph G defined in Section 7.1.1.3 for a cellular network, a clique means a set of cells which interfere with each other. In the k-belt buffering system, all cells in a clique are interference cells of each other. A clique which is not included in any other cliques is called a maximal clique. A maximal clique which has the largest number of cells is called a

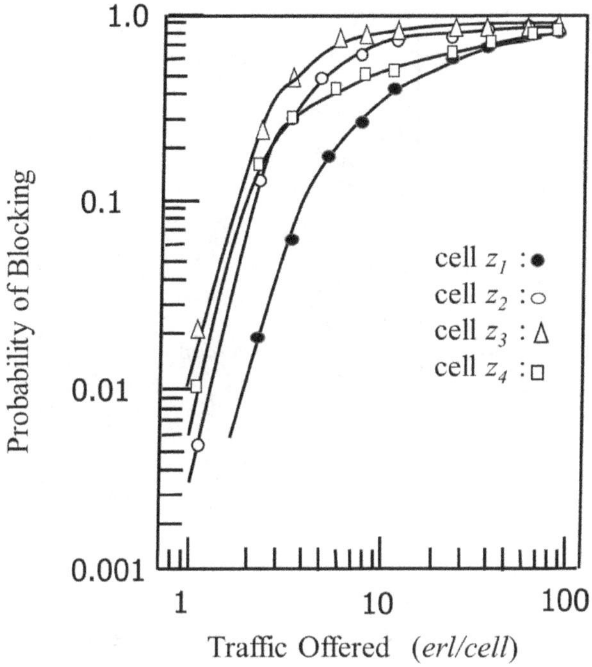

Fig. 7.1.1.4.8. Simulation results.

maximum clique. Let us consider the number of acceptable calls in a maximum clique. From the definition of a clique, channels are never reused in cells of a clique. Therefore, any DCA methods cannot accept more calls than the number of channels in a maximal than that of channels. Any DCA methods cannot accept the new call in such a situation because of the property of cliques. Thus, consider what is needed when determining the blocking probability for clique packing. It is the probability that n_c channels have already been used in the maximal clique. From this idea, the number of combinations of states to be considered can be reduced. From the approximation, we should consider only the number of channels used in the clique unit, not the number of channels in the cell unit. It is known that this approximation method gives a better result to the analysis of the DCA in a system with a large number of cells and channels [76,28,29].

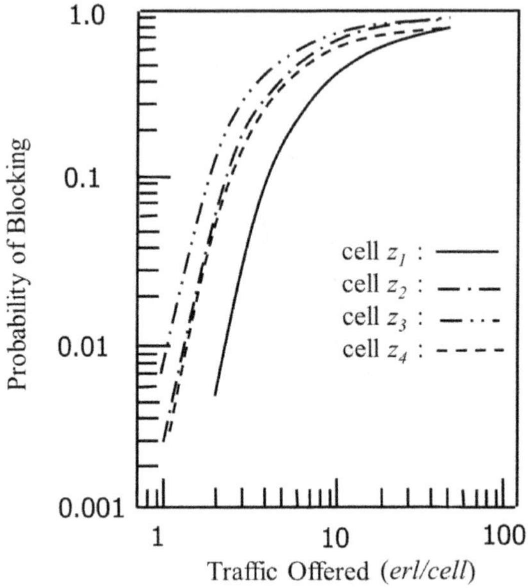

Fig. 7.1.1.4.9. Results of approximation method (A).

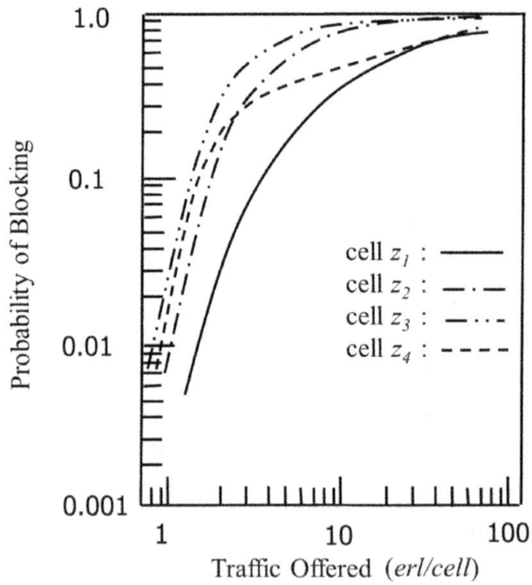

Fig. 7.1.1.4.10. Results of approximation method (B).

7.1.2. *Generalized coloring problems in mobile communication systems*

7.1.2.1. *Channel offset systems in a cellular mobile system*

(1) Channel offset systems and interference networks

In a cellular mobile system, assigning a channel for a call in a cell so as to achieve high spectral efficiency is an important problem. This assignment problem is related to the coloring problem of graph theory as shown in Section 7.1.1.3. In Section 7.1.1.2, we assumed that there would be no interference between channels, but this is not entirely free of interference. A *channel offset system* in a cellular mobile system which is premised on allowing interference rather to achieve higher spectral efficiency has been proven to have better spectral efficiency. We formulate the channel offset system design using a graph theoretical concept, which leads to the generalized coloring problem of graphs. From graph theoretical considerations, we investigate an optimal channel offset scheme in a cellular mobile system and give an upper and a lower bound for overall spectral bandwidth. These bounds are useful in designing a channel offset system. Formulating this method in graph theory leads to the generalized coloring problem of graphs. In this section, only FDMA as a channel is used for discussion. However, it should be noted that the similar discussion can be made in TDMA and CDMA as well.

A spectrum which is divided into channels is shown in Fig. 7.1.2.1.1 and let us introduce a channel offset scheme. A $1/k$ *channel offset* scheme is introduced by offsetting each channel by $1/k$ of its channel unit bandwidth. The $1/3$ channel offset scheme is shown in Fig. 7.1.2.1.2. We call the channel scheme in Fig. 7.1.2.1.1 the *offsetless scheme*. The number of channels for a

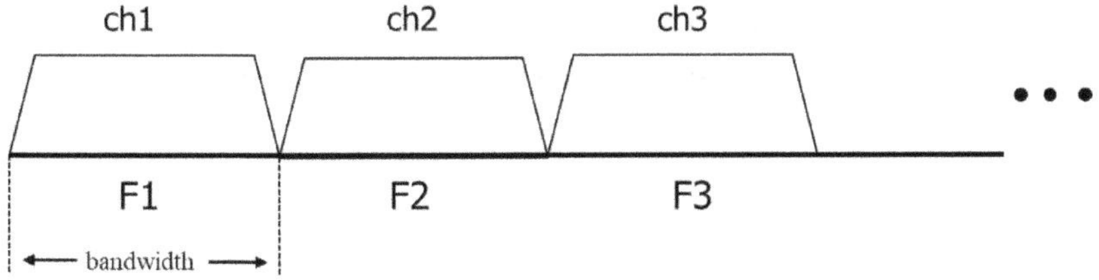

Fig. 7.1.2.1.1. A division of a frequency spectrum into channels.

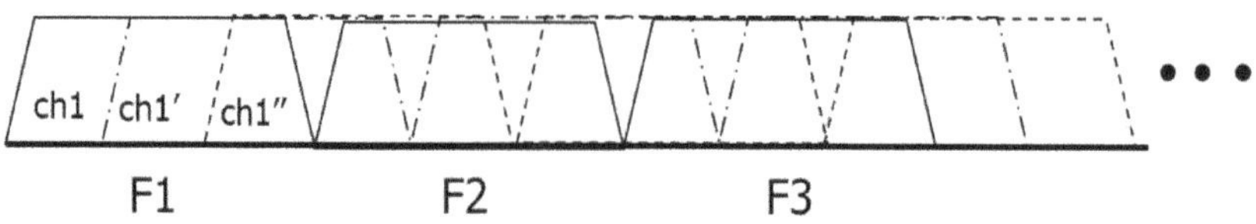

Fig. 7.1.2.1.2. The 1/3 channel offset scheme.

given bandwidth in a channel offset scheme is larger than that in the scheme (offsetless scheme) in Fig. 7.1.2.1.1.

For simplicity, let us represent the axis of frequency spectrum by the number line as shown in Fig. 7.1.2.1.3, where intervals [0,1), [1,2), [1.5, 2.5), ... represent channels. The simple expressions of the spectrum in Figs. 7.1.2.1.1 and 7.1.2.1.2 are shown in Figs. 7.1.2.1.3(a) and 7.1.2.1.3(b), respectively. The channel $[x, y)$, where $y-x = 1$, which means the channel bandwidth, is represented by x, that is, a channel x means channel $[x, y)$. The *degree* $d\ (f_i, f_j)$ of *interchannel interference* between two channels f_i and f_j is determined, depending on f_i, f_j, $|f_i - f_j|$, and so on.

The *interchannel interference* $g\ (z_i, z_j)$ is sufficient interval of channels for effective immunity to the interference between two cells z_i and z_i and we assume $0 \leq g(z_i, z_j) \leq 1$, where $g\ (z_i, z_j)$ is generally determined by the measurement of interference. That is, $g\ (z_i, z_j)$ may depend on the location of service area, the size of cells, and so on.

We assume that ① a service area, ② the interchannel interference $g\ (z_i, z_j)$ for every pair of cells in the service area, ③ the degree $d\ (f_i, f_j)$ of interchannel interference for every pair of channels and ④ the number w_i of channels assigned to cell z_i for any cell z_i are given. We define $d\ (f_i, f_j)$ as $|f_i - f_j)|$:

$$d(f_i, f_j) = |f_i - f_j|. \tag{7.1.2.1.1}$$

Let $c(\cdot)$ be a mapping from a cell into channels. Therefore, $c(z_i)$ denotes the set of channels assigned at a cell z_i. For $f_p \in c(z_i)$ and $f_q \in c(z_j)$, if

$$d(f_p, f_q) \geq g(z_i, z_j) \tag{7.1.2.1.2}$$

and

$$|c(z_i)| = w_i, \quad i = 1, 2, \ldots, \tag{7.1.2.1.3}$$

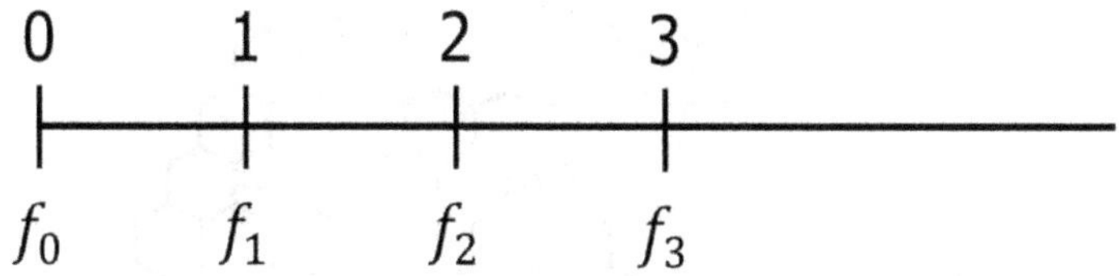

(a)

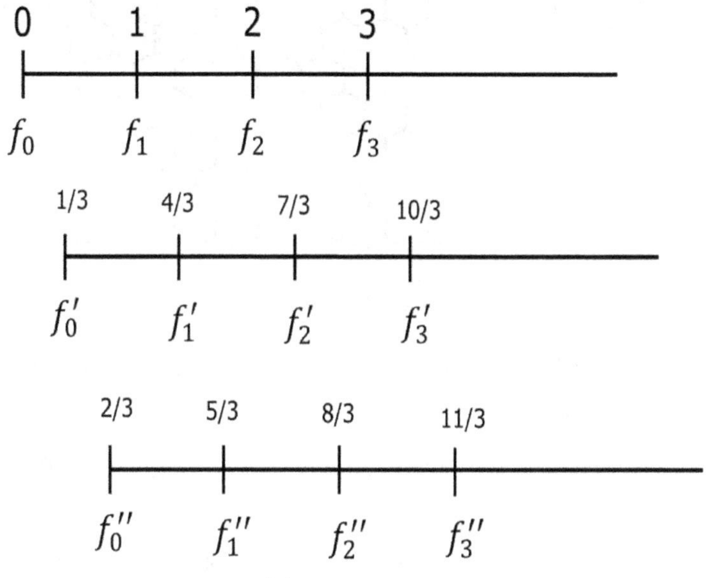

(b)

Fig. 7.1.2.1.3. The channels by the axis of frequency spectrum.

then $c(\cdot)$ is called a *feasible assignment of channels*. Let us define *total bandwidth* B as follows:

$$B = \max_{f_p, f_q \in \bigcup_i c(z_i)} |f_p - f_q + 1|. \tag{7.1.2.1.4}$$

Our purpose is to find the minimum total bandwidth B in the assignment of channels. Since the problem is defined by given condition in terms only

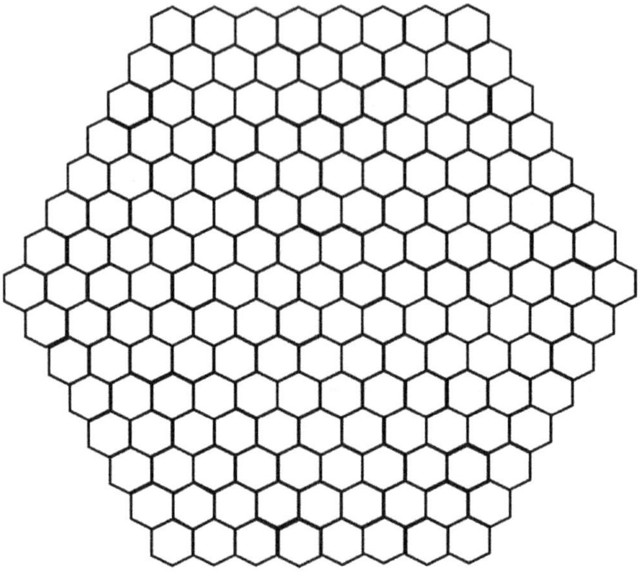

Fig. 7.1.2.1.4. A service area.

by (7.1.2.1.2) and (7.1.2.1.3), there are many different channel assignment problems. A simple example is shown.

Example 1. Consider a service area in Fig. 7.1.2.1.4. Let the cellular mobile system be a 3-belt buffering system, where

$$g(z_i, z_j) = \begin{cases} 1, & \text{if } z_i \text{ and } z_j \text{ are contiguous,} \\ 0.5, & \text{if } z_j \text{ is located at 1 or 2 cells away from } z_i, \\ 0, & \text{otherwise.} \end{cases} \tag{7.1.2.1.5}$$

The 1/2 channel offset scheme called *interleave* is shown in Fig. 7.1.2.1.5. Figure 7.1.2.1.6 illustrates the cluster of 12 cells for a feasible assignment of channels in a service area in Fig. 7.1.2.1.4 using the 1/2 channel offset scheme and a feasible assignment of channels using this cluster where the total bandwidth $B = 6.5$ is shown in Fig. 7.1.2.1.7. **(End of Example)**

Let us define a network N from a cellular mobile system. $N = (V, E; g, w)$ is a network with vertex set V and edge set E. Each vertex v_i corresponds to cell z_i and $e_k = (v_i, v_j) \in E$ if and only if $g\ (z_i, z_j) > 0$.

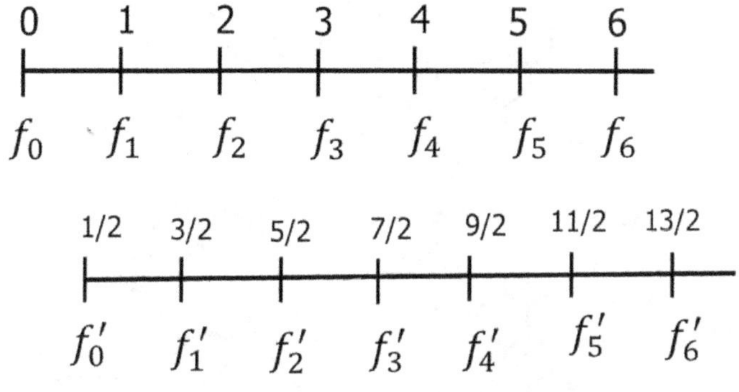

Fig. 7.1.2.1.5. The 1/2 channel offset scheme.

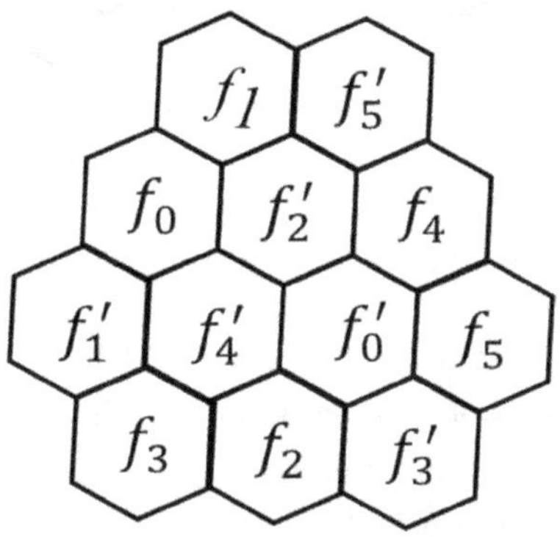

Fig. 7.1.2.1.6. The cluster of 12 cells for a feasible assignment of channels in a service area in Fig. 7.1.2.1.4.

With each edge and each vertex of N, the positive real number called the *weight of the edge* and a nonnegative integer called the *weight of the vertex* are associated, respectively. The weight of an edge $e_k = (v_i, v_j) \in E$ is $g(v_i, v_j) = g(z_i, z_j)$ and the weight of a vertex v_i is w_i. The network N is called the *interference network* of the cellular mobile system.

For given N and a channel offset scheme, let $c\ (v_i)$ be the set of real numbers called channels which are assigned to v_i, subject to the constraint (7.1.2.1.2) and (7.1.2.1.3). Then, $c(\cdot)$ is called a feasible assignment of channels to vertices of the network N. For a feasible assignment of channels of N, B of (7.1.2.1.3) is called the total bandwidth of the assignment. If $|f_p - f_q + 1| = B$, f_p and f_q are called the maximum channel $f_{\max}$ and the minimum channel $f_{\min}$, respectively.

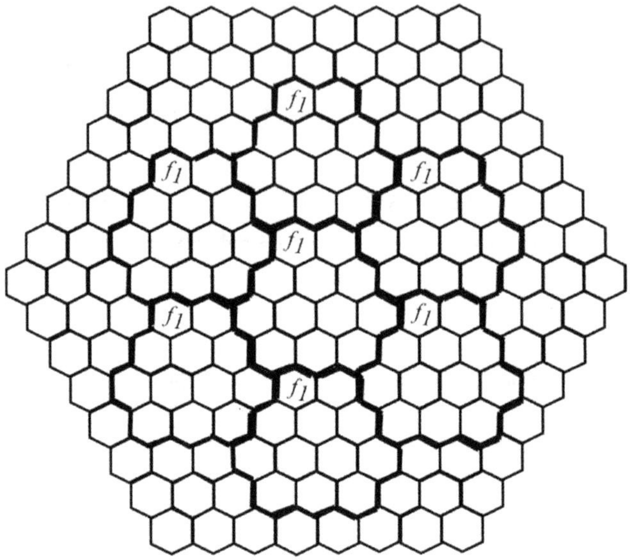

Fig. 7.1.2.1.7. A feasible assignment of channels using the cluster of Fig. 7.1.2.1.6.

The feasible assignment of channels of N becomes the assignment of channels of the cellular mobile system. One of the merits of this formulation using a network N is following. Let us consider a cellular mobile system that has different size cells and nonuniform interference from cell to cell. This system is not a uniform buffering system like k-belt buffering. However, we can formulate the channel assignment problem of the system using a network N. There are some problems considered, for example,

***Problem* 1:** For a given interference network N and a given channel offset scheme, find the minimum total bandwidth,

***Problem* 2:** For a given interference network N, find the channel offset scheme whose minimum total bandwidth is smaller than any other minimum total bandwidth of any channel offset scheme.

In N, if $w_i = 1$ for $i = 1, 2, \ldots$, let us call a feasible assignment of channels of N a *coloring of N* and we also call the total bandwidth the *color bandwidth*. The minimum color bandwidth of N is called the *chromatic bandwidth* of N.

Let N be the interference network of a cellular mobile system where $w_i = 1$ for $i = 1, 2, \ldots$ and each $g\ (z_i, z_j) = 0$ or 1. Let G be the underlying graph of N. In the case that the channel offset scheme is the offsetless scheme, a coloring of N coincides with the coloring of a graph G and clearly the chromatic bandwidth of N is equal to the chromatic number of G. So, we obtain that *Problem 1* is an *NP-hard* problem [12]. This is an intractable problem.

(2) Chromatic bandwidth of *N*

As mentioned above in (1), the number of channels for a given bandwidth in a channel offset scheme is larger than in the offsetless scheme. This means that we have more candidates of channels for a call in a channel offset scheme and the scheme may achieve higher spectral efficiency than in the offsetless scheme. In general, w_i is a nonnegative integer in the interference network N of a cellular mobile system. However, we only need to consider the case $w_i = 1$ for $i = 1, 2, \ldots$ without loss of generality. The reason follows.

If $w_i = 0$, we may delete the vertex v_i from N because we do not need to assign any channels on v_i. If $w_i > 1$, we replace a vertex v_i with a clique whose vertices consist of $v_{i_1}, v_{i_2}, \ldots, v_{i_p}$ where $p = w_i$, and all edge weights are 1. An example of the replacement in the case $w_i = 3$ is shown in Fig. 7.1.2.1.8. Let N' be a network obtained from N by the replacement for every vertex v_i such that $w_i > 1$ and by the deletion of v_i such that $w_i = 0$. Then, the channel assignment problem of N becomes

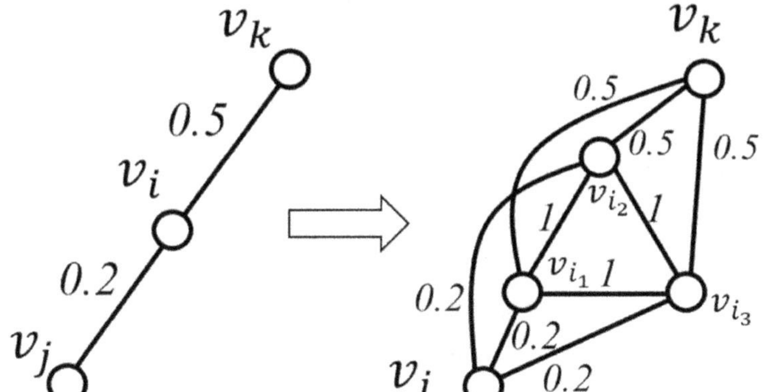

Fig. 7.1.2.1.8. Replacement of v_i with w_i-clique.

the assignment problem of one channel to each vertex in N'. Therefore, hereafter, we investigate chromatic bandwidth.

Let B_k be the chromatic bandwidth of N in the $1/k$ channel offset scheme. Note that B_1 is the chromatic bandwidth of N in the offsetless scheme. We shall examine B_k, hereafter.

Theorem 7.1.2.1.1. *Let h be a positive integer. For an interference network $N = (V, E; g, w)$, if $g\ (v_i, v_j) = t/h$ for $v_i, v_j \in V$, where t is a positive integer and $t \leq h$, then $B_h \leq B_k$ for any integer k, that is, B_k is minimum in $B_1, B_2, \ldots$.*

Proof. Let $c(\cdot)$ be a coloring realizing the chromatic bandwidth B_k. We show that there is a coloring $c'(\cdot)$ using the $1/h$ channel offset scheme, whose color bandwidth is not greater than B_k. Let $c\ (z_i) = \{f_i\}$ and $f_{\min} = 0$. For each f_i, let $f_i' = [hf_i]/h$ where for a real number x, $[x]$ represents the greatest integer not exceeding x, namely Gauss' notation. Let $c'\ (z_i) = \{f_i'\}$. Note that for any channel f_i,

$$f_i' \leq f_i. \tag{7.1.2.1.6}$$

Then, we can show that $c'(\cdot)$ is a coloring as follows. We assume that $g\ (z_i, z_j) = t/h$. Since $c(\cdot)$ is a coloring,

$$f_i - f_j \geq t/h, \tag{7.1.2.1.7}$$

$$\begin{aligned} f_i' - f_j' &= [hf_i]/h - [hf_j]/h \\ &= ([hf_i] - [hf_j])/h \\ &\geq [hf_i - hf_j]/h. \end{aligned} \tag{7.1.2.1.8}$$

From (7.1.2.1.7),

$$f_i' - f_j' \geq [t]/h = t/h = g(z_i, z_j). \tag{7.1.2.1.9}$$

Hence, the coloring $c'(\cdot)$ satisfies the condition (7.1.2.1.2). Let B' be the color bandwidth of the coloring $c'(\cdot)$. Since $f'_{\min} = f_{\min}$ and $f'_{\max} \leq f_{\max}$ from (7.1.2.1.6),

$$B' \leq B_k. \tag{7.1.2.1.10}$$

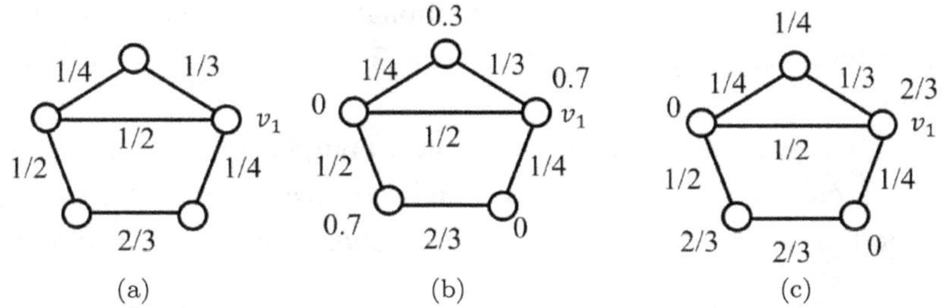

Fig. 7.1.2.1.9. An example of an interference network to explain Theorem 7.1.2.1.1.

Since $c'(\cdot)$ is one coloring using the $1/h$ channel offset scheme and B_h is the chromatic bandwidth in the $1/h$ channel offset scheme from the definition,

$$B_h \leq B'. \tag{7.1.2.1.11}$$

From (7.1.2.1.10) and (7.1.2.1.11), $B_h \leq B_k$. □

To make it easier to understand Theorem 7.1.2.1.1, here is an example.

Example 2. Figure 7.1.2.1.9(a) illustrates an interference network where the value set of $g\ (u, v)$ is $\{1/4, 1/3, 1/2, 2/3\}$. Since the value set can be represented by $\{3/12, 4/12, 6/12, 8/12\}$, the 1/12 channel offset scheme gives the chromatic bandwidth from Theorem 7.1.2.1.1. In this case, h is 12 in Theorem 7.1.2.1.1.

As examples, a coloring $c_0(\cdot)$ of the 1/10 channel offset scheme is shown in Fig. 7.1.2.1.9(b) and a coloring $c_0'(\cdot)$ of the 1/12 channel offset scheme is shown in Fig. 7.1.2.1.9(c). The relation between $c_0(\cdot)$ and $c_0'(\cdot)$ is equivalent to the relation between c(·) and c′(·) in the proof of Theorem 7.1.2.1.1. For example, $c_0(v_1) = 0.7 = 8.4/12$ and $c_0'(v_1) = 2/3 = [8.4]/12$. The color bandwidth of $c_0(\cdot)$ is 1.7 and the color bandwidth of $c_0'(v_1)$ is $2/3 - 0 + 1 = 1.66\ldots$. **(End of Example)**

Corollary 7.1.2.1.1. *For an interference network* $N = (V, E; g, w)$,

$$B_k = B_1, \quad k = 2, 3, \ldots, \tag{7.1.2.1.12}$$

if $g(v_i, v_j) = 1$ *or* 0 *for* $v_i, v_j \in V$ *of* N.

In the case that each $g(v_i, v_j)$ is a rational number, the optimal channel offset scheme can be determined by using Theorem 7.1.2.1.1, namely, we can solve *Problem 2*.

Next, we examine an upper and a lower bound for the color bandwidth in the $1/k$ channel offset scheme since it is not easy to obtain the chromatic bandwidth of N. These bounds are rough standard of channel assignments.

Lemma 7.1.2.1.2. *Let $c'(\cdot)$ be a coloring using a channel offset scheme of N. If n channels are used in this assignment (the coloring), then*

$$n \geq B_1. \tag{7.1.2.1.13}$$

Proof. Let $f_1', f_2', f_3', \ldots$ and f_n' (where $f_1' < f_2' < f_3' < \cdots < f_n'$) be the channels used in coloring $c'(\cdot)$ of N. If the coloring uses the $1/k$ channel offset scheme,

$$|f_i' - f_j'| \geq 1/k, \quad \text{for } i, j = 1, 2, \ldots, n;\ i \neq j, \tag{7.1.2.1.14}$$

where $k \geq 1$

Let $f_0 = 0$, $f_1 = 1, \ldots$ and $f_{n-1} = n - 1$. Then,

$$|f_i - f_j| \geq 1, \quad \text{for } I, j = 1, 2, \ldots, n-1;\ i \neq j. \tag{7.1.2.1.15}$$

For $i = 1, 2, \ldots, n$, assign f_{i-1} for f_i'. This coloring is denoted by c($\cdot$). Note that $f_i' = f_j'$ if and only if $f_{i-1} = f_{j-1}$. Since the coloring $c'(\cdot)$ satisfies the condition (7.1.2.1.2), $|f_i' - f_j'| \geq g(u, v)$ where $c\,(u) = f_i'$ and $c\,(v) = f_j'$. If $|f_i' - f_j'| = 0$ then $g(u, v) = 0$. So, $|f_{i-1} - f_{j-1}| \geq g\,(u, v)$. If $|f_i' - f_j'| \neq 0$ then $f_{i-1} \neq f_{j-1}$. From (7.1.2.1.15), $|f_{i-1} - f_{j-1}|1$. So, $|f_{i-1} - f_{j-1}| \geq g(u, v)$. Therefore the coloring $c(\cdot)$ satisfies the condition (7.1.2.1.2). The color bandwidth B of the coloring is $|f_{n-1} - f_o + 1|$. That is,

$$B = |f_{n-1} - f_o + 1| = n. \tag{7.1.2.1.16}$$

Note that the coloring $c(\cdot)$ is one coloring using the offsetless scheme. Since B_1 is the chromatic bandwidth in the offsetless scheme,

$$B = n \geq B_1. \tag{7.1.2.1.17}$$

□

Using Lemma 7.1.2.1.2, we obtain the following theorem.

Theorem 7.1.2.1.2. *For an interference network N and a positive integer k,*

$$B_1 \geq B_k \geq (B_1 - 1)/k + 1 \qquad (7.1.2.1.18)$$

and there exists an interference network N satisfying the equality.

Proof. Since the channels of the $1/k$ channel offset scheme include the channels of the offsetless scheme, it is obvious that $B_1, \geq B_k$.

$$\text{Suppose that } B_k < (B_1 - 1)/k + 1. \qquad (7.1.2.1.19)$$

Let f_1', f_2', ... and f_n' be the channels used in a coloring using the $1/k$ channel offset scheme of N. Since the coloring uses the $1/k$ channel offset scheme, $|f_i' - f_j'| \geq 1/k$ for $i \neq j$ and $i, j = 1, 2, \ldots, n$. This means that

$$B_k \geq (n-1)/k + 1. \qquad (7.1.2.1.20)$$

From (7.1.2.1.19) and (7.1.2.1.20), we obtain $n < B_1$, contradicting Lemma 7.1.2.1.2. Therefore, we obtain (7.1.2.1.18).

Next, let us consider a service area in Fig. 7.1.2.1.10 (a) which consists of $k^2 + 1$ cells. Its interference network where

$$g(v_i, v_j) = \begin{cases} 1 & \text{if } z_i \text{ and } z_j \text{ are contiguous}, \\ 1/k & \text{otherwise} \end{cases} \qquad (7.1.2.1.21)$$

is shown in Fig. 7.1.2.1.10(b). □

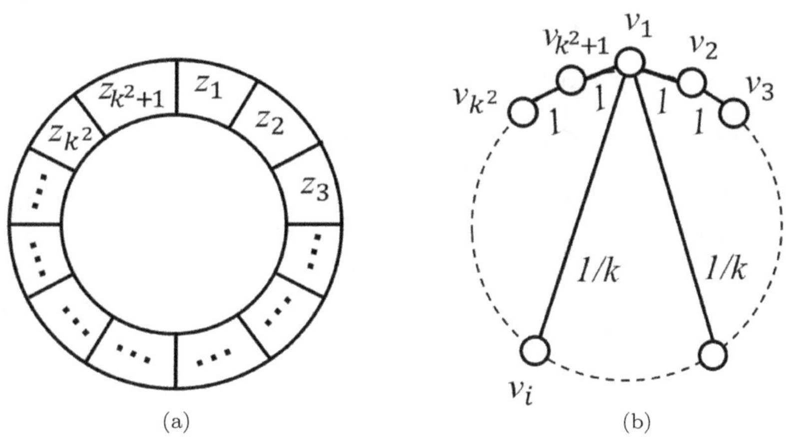

Fig. 7.1.2.1.10. A service area and the interference network.

If we assign the channels to vertices using the offsetless scheme, it is clear that

$$B_1 = k^2 + 1. \tag{7.1.2.1.22}$$

We assign the channels to vertices using the $1/k$ channel offset scheme, as

$$\begin{aligned}
c(v_1) &= \{f_1'\} = 1\\
c(v_2) &= \{f_2'\} = 2\\
&\vdots\\
c(v_i) &= \{f_i'\} = \{s_i/k\},\ (s_i = s_{i-1} + k(\mathrm{mod}\, k^2+1), 0 \le s_i \le k^2,\\
&\qquad \text{or } s_i = k \cdot i(\mathrm{mod}\, k^2+1), 0 \le s_i \le k^2)\\
&\vdots\\
c(v_{k^2+1}) &= \{f_{k^2}' + 1\} = \{0\}
\end{aligned} \tag{7.1.2.1.23}$$

In this coloring, since $0 \le s_i \le k^2$, $f_{\min} = f_{k+1}' = 0$ and $f_{\max} = f_k' = k^2/k = k$. This coloring satisfies (7.1.2.1.2) and its color bandwidth $B = (k-0)+1 = k+1 = (B_1-1)/k+1$ which is a lower bound, from (7.1.2.1.18) and (7.1.2.1.22).

In a service area in Fig. 7.1.2.1.10 (a), if

$$g(v_i, v_j) = \begin{cases} 1 & \text{if } z_i \text{ and } z_j \text{ are contiguous,}\\ 0 & \text{otherwise,}\end{cases} \tag{7.1.2.1.24}$$

then $B_1 = B_k$ (from Corollary 7.1.2.1.1) which is an upper bound. □

Let us consider some examples.

Example 3. In Example 1, the color bandwidth using the $1/2$ channel offset scheme is $B = 6.5$. The chromatic bandwidth in the offsetless scheme is 12. From (7.1.2.1.18), $B = 6.5$ is the lower bound of the $1/2$ channel offset scheme. **(End of Example)**

Example 4. A service area consisting of 10 cells is shown in Fig. 7.1.2.1.11(a). Let $g(v_i, v_j)$ be the same as (7.1.2.1.21) where $k = 3$. A coloring using (7.1.2.1.23) is shown in Fig. 7.1.2.1.11(b).

(End of Example)

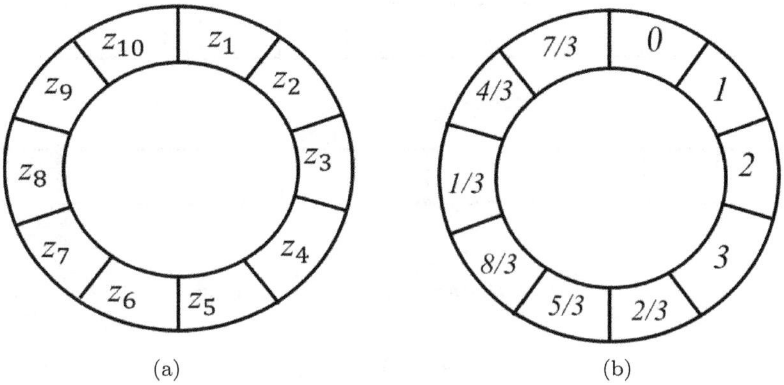

Fig. 7.1.2.1.11. A service area and a coloring.

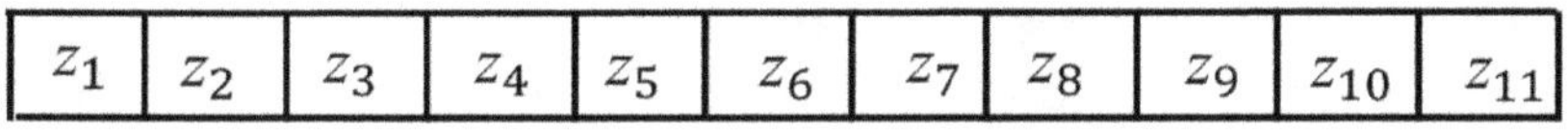

Fig. 7.1.2.1.12. A service area.

Example 5. Let us consider a path like service area in Fig. 7.1.2.1.12, where

$$g(v_i, v_j) = \begin{cases} 1, & \text{if } z_i \text{ and } z_j \text{ are contiguous,} \\ \dfrac{1}{3}, & \text{if } z_j \text{ is located at 1–6 cells away from } z_i, \\ 0, & \text{otherwise.} \end{cases} \tag{7.1.2.1.25}$$

A coloring using the offsetless scheme is shown in Fig. 7.1.2.1.13(a). Its color bandwidth is 7, which is the chromatic bandwidth in the offsetless scheme. A coloring using the 1/3 channel offset scheme is shown in Fig. 7.1.2.1.13(b). Its color bandwidth is 3, which is the chromatic bandwidth in the 1/3 channel offset scheme. Examples 4 and 5 are examples in which $B_3 = (B_1 - 1)/3 + 1$ in Theorem 7.1.2.1.2. **(End of Example)**

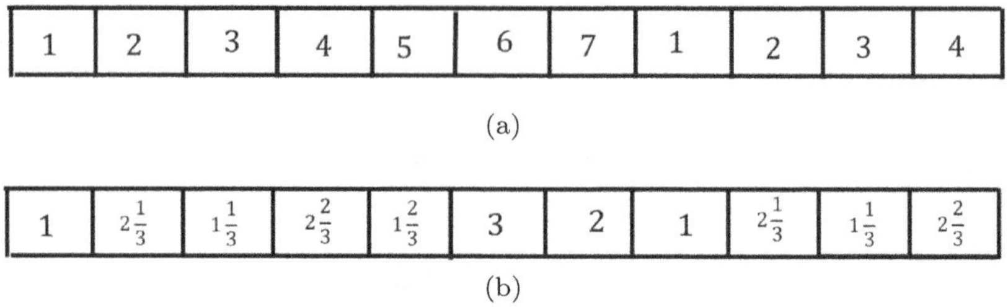

Fig. 7.1.2.1.13. Colorings of a service area of Fig. 7.1.2.1.12.

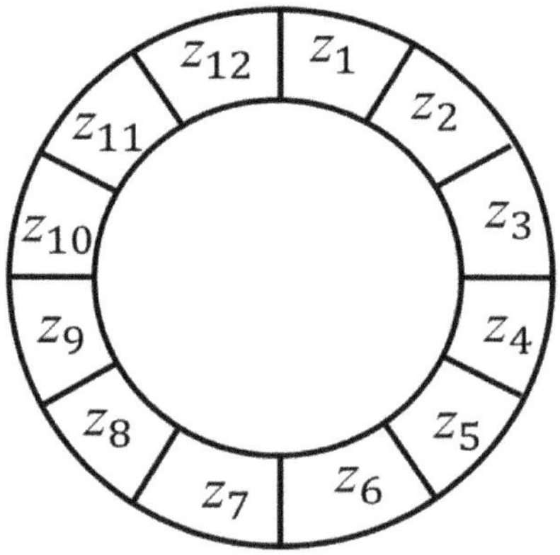

Fig. 7.1.2.1.14. A service area.

Example 6. Let us consider a ring like service area in Fig. 7.1.2.1.14, where

$$g(v_i, v_j) = \begin{cases} 1, & \text{if } z_i \text{ and } z_j \text{ are contiguous}, \\ 1/2, & \text{if } z_j \text{ is located at 1–5 cells away from } z_i, \\ 0, & \text{otherwise}. \end{cases} \tag{7.1.2.1.26}$$

A coloring using the offsetless scheme is shown in Fig. 7.1.2.1.15(a). Its color bandwidth is 6, which is the chromatic bandwidth in the offsetless scheme. A coloring using the 1/2 channel offset scheme is shown in Fig. 7.1.2.1.15(b). Its color bandwidth is $7/2 = 3.5$, which is the chromatic bandwidth in the 1/2 channel offset scheme. This is an example in which $B_2 = (B_1 - 1)/2 + 1$ in Theorem 7.1.2.1.2. **(End of Example)**

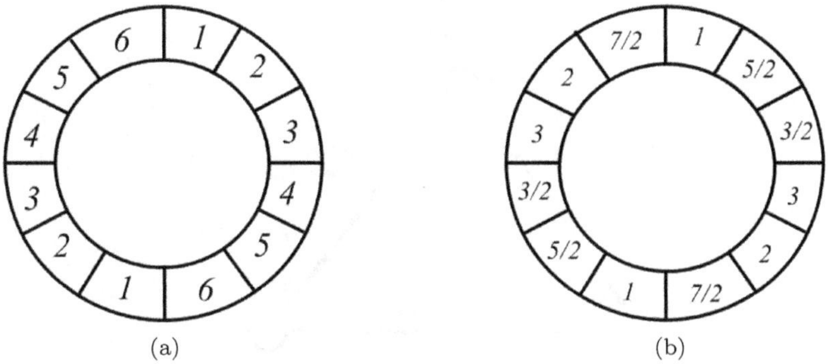

Fig. 7.1.2.1.15. Colorings of a service area of Fig. 7.1.2.1.14.

7.1.2.2. *Another new coloring problem*

We introduce another new coloring problem on graphs or networks related to the channel assignment problem. This coloring takes the degree of *cochannel interference* into consideration. In Section 7.1.1.2, if we assume the two-belt buffering system in Fig. 7.1.1.2.7, a channel A of a cell can be used in six cells which are located at two cells away from the cell at the same time. That is, a two-belt buffering system means that the buffering is sufficient for effective immunity to such interference in the worst case. However, if the channel A is used in less than six cells, the channel A could be assigned closer at the same time.

We explain this coloring using a simple example. Figure 7.1.2.2.1 illustrates an interference graph. In the graph, we need three channels A, B and C. Figure 7.1.2.2.2(a) illustrates a network called an interference network, where each edge weight represents the degree of cochannel interference. For example, if channel A is assigned to v_1 and v_2, the cochannel interference to v_3 is $1 + 0.5 = 1.5$. Now, we assume that if the cochannel interference to each cell is not greater than 1, the assignment guarantees good communication quality. This means that the threshold is one. In this case, we need only two channels (see Fig. 7.1.2.2.2(b)). Let each edge weight be 1 and the threshold be 0. Then, the coloring problem is the same as the coloring problem of graphs. Therefore, the new coloring problem is a generalization of the usual coloring problem.

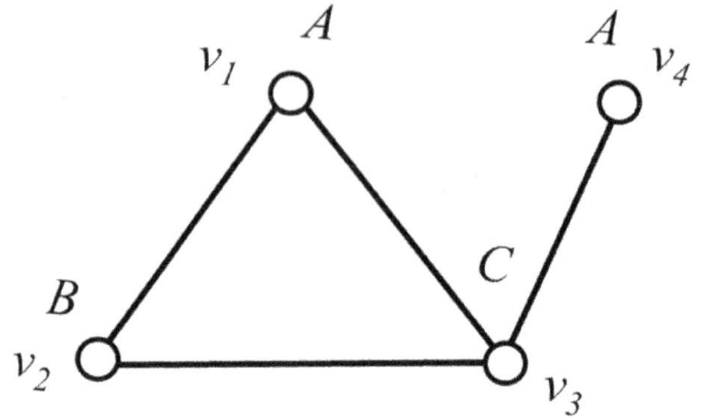

Fig. 7.1.2.2.1. A simple example of an interference graph.

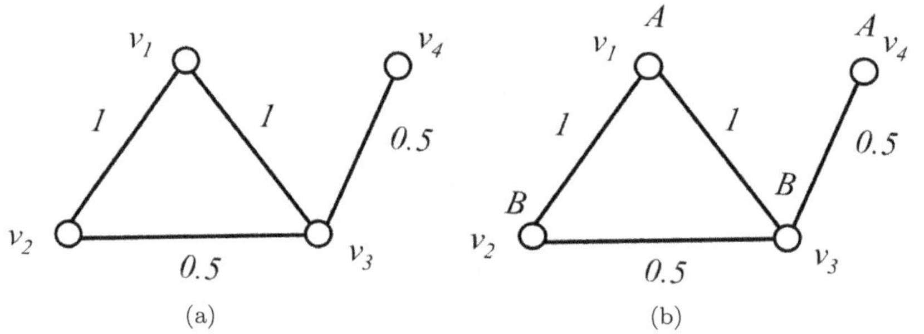

Fig. 7.1.2.2.2. An interference network and a coloring of the network.

Assuming that the ratio of all edge weights can be represented using natural numbers, we can reconstruct the interference network, that admits multiple edges, such that all edges have same positive weight. In this case, the new coloring problem is rewritten into a coloring problem of (multiple) graphs. For example, the network in Fig. 7.1.2.2.2(a) is translated into the graph in Fig. 7.1.2.2.3.

Let G be an undirected graph that admits multiple edges and no loops. The degree of vertex v of G is denoted by $d_G(v)$ and the maximum degree of G is denoted by $d_{\max}(G)$. For each vertex v of G, we assign a color. The assignment is called the *coloring with defect h* if the following conditions are satisfied. Let G_c be the induced subgraph whose vertices are assigned

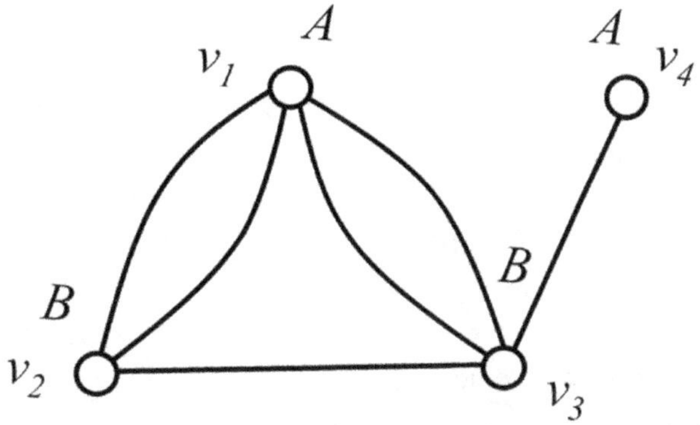

Fig. 7.1.2.2.3. An interference (multiple) graph.

the color c, (that is, subgraph induced by the vertex set assigned by c). For each G_c, $d_{\max}$ (G_c) is not greater than h.

G is *k-colorable* with defect h or simply *(k, h)-colorable* if there exists a coloring with defect h using $k'(\leq k)$ colors. The *chromatic number with defect h* is the minimum k for which G is (k, h)-colorable, denoted by $\chi_h(G)$. In the coloring with defect h, it is clear that h is threshold in the above explanation of Fig. 7.1.2.2.3.

From the above definitions, in the case that $h = 0$, the coloring problem coincides with the usual coloring problem. So, $\chi_0(G) = \chi(G)$. Therefore, given a graph G, the problem of determining whether G is $(k, 0)$-colorable or not is an NP-complete problem. Moreover, for any h, we obtain the similar result.

Theorem 7.1.2.2.3. *Given a graph G, for any nonnegative integer h_0, the problem of determining whether a graph G is (k, h_0)-colorable or not is an NP-complete problem.*

Proof. Let G be a graph. We construct a graph G' from G and we show that the problem of determining whether a graph G is k-colorable for the usual coloring or not is rewritten into the problem of determining whether a graph G' is (k, h_0)-colorable or not.

For each vertex v_i of G, we add the vertices $v_{i1}, v_{i2}, \ldots, v_{is}$, where $s = k(h_0 + 1) - 1$ and add edges (x, y) for any vertex pair x, y in the vertex set $\{v_{i1}, v_{i2}, \ldots, v_{is}\}$. Therefore, the induced subgraph of $\{v_{i1}, v_{i2}, \ldots, v_{is}\}$ becomes a clique that is denoted by K_i. Let G' be a graph such that vertices and edges are added for all vertices of G. We claim that G' is (k, h_0)-colorable if and only if G is k-colorable for the usual coloring.

Assume that G' is (k, h_0)-colorable and $f'(\cdot)$ is the coloring with defect h_0. Since the order of each clique K_i is $k(h_0 + 1)$, the number of vertices assigned a color c is correctly $h_0 + 1$ for each color c. So if $(u, v) \in E(G)$, then $f'(u) \neq f'(v)$. $f'(\cdot)$ is a usual coloring of G. Therefore, G is k-colorable for the usual coloring.

Conversely, assume that G is k-colorable for the usual coloring and $f(\cdot)$ is the coloring. If $f(v_i) = c$, we assign colors to vertices $v_{i1}, v_{i2}, \ldots, v_{is}$ of K_i as follows. We assign a color c' to $h_0 + 1$ vertices for each color $c'(\neq c)$ and assign a color c to h_0 vertices. Then $f(\cdot)$ is a coloring with defect h_0 of G'.

Therefore, G' is (k, h_0)-colorable. □

Given a graph G, for any nonnegative integer h_0, the problem of determining the chromatic number with defect h_0 is obviously an NP-hard problem from Theorem 7.1.2.3.

Next, we show an upper and a lower bound of $\chi_h(G)$ as follows [24].

Theorem 7.1.2.2.4. *Let G be a graph whose maximum degree is r and h be a nonnegative integer, then*

$$\frac{\chi(G)}{h+1} \leq \chi_h(G) \leq \left\lceil \frac{r+1}{h+1} \right\rceil, \tag{7.1.2.2.1}$$

where $\lceil x \rceil$ represents the minimum integer that is not less than x.

Example. Figure 7.1.2.2.4(a) is an interference (multiple) graph and let $h = 2$. Since $r = 5$, the upper bound is $\lceil 6/3 \rceil = 2$. An assignment using two colors is shown in Fig. 7.1.2.2.4(b). **(End of Example)**

In K_4, that is a complete graph with four vertices, since $r = 3$ and $\chi(K_4) = 4$, the equalities hold in (7.1.2.27) if $h = 1$. Therefore, the upper and lower bounds are sharp. And if $h = 0$, the upper bound is $r + 1$. This is a well-known result in graph theory.

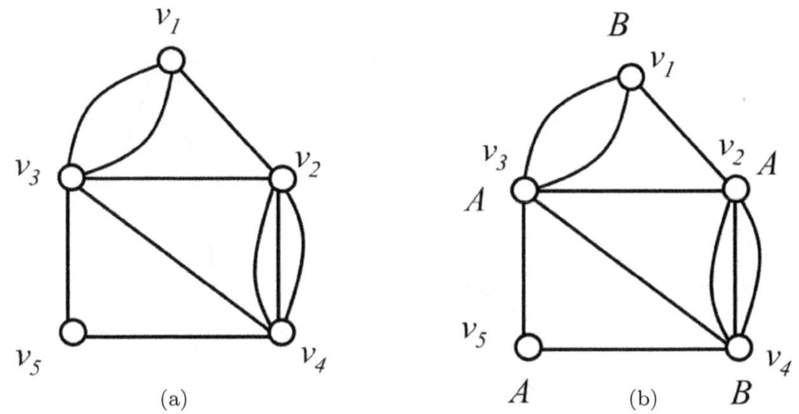

Fig. 7.1.2.2.4. An interference (multiple) graph and a coloring of the graph.

7.1.2.3. *Mixture of previous coloring concepts*

To efficiently use spectrum further, we combine the concepts of colorings in Sections 7.1.2.1 and 7.1.2.2. Roughly speaking, this coloring is a coloring taking the degree of interchannel interference, including cochannel interference, into consideration using fractions as colors. The degree of interference between two channels f_i and f_j is denoted by $d^*\ (f_i, f_j)$. Here, we define $d^*\ (f_i, f_j)$ as follows:

$$d^*(f_i, f_j) = \begin{cases} 1 - |f_i - f_j| & \text{if } |f_i - f_j| \le 1, \\ 0 & \text{otherwise.} \end{cases} \qquad (7.1.2.3.1)$$

Let G be a multiple graph and h be a fixed nonnegative integer. We consider a mapping c from a nonnegative rational number to a vertex. Let $N\ (v) = \{u \in V(G) | u \text{ is adjacent to } v\}$ and $m\ (u, v)$ be the number of edges that join u and v. If for any v,

$$\sum_{u \in N(v)} m(u, v) d^*(c(u), c(v)) \le h, \qquad (7.1.2.3.2)$$

then c called a feasible coloring with h. For a feasible coloring c with h, let us define the total bandwidth B.

$$B = \max_{u,v \in V(G)} |c(u) - c(v) + l|. \qquad (7.1.2.3.3)$$

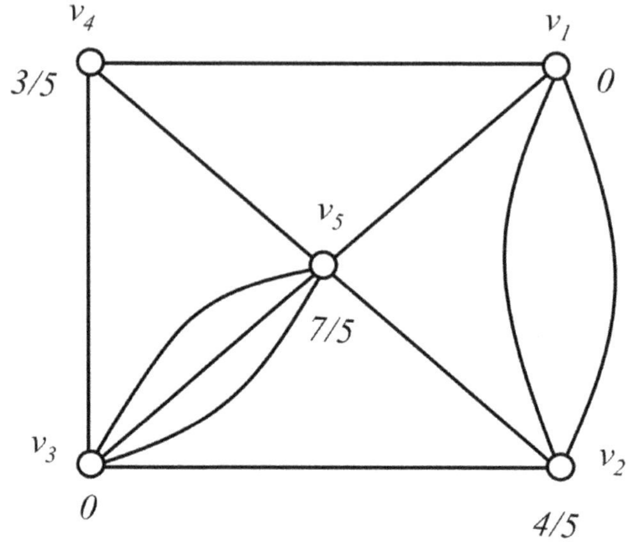

Fig. 7.1.2.3.1. A feasible coloring with 1 of a graph.

The minimum number of total bandwidths for feasible colorings with h is denoted by $\chi_h^*(G)$.

Theorem 7.1.2.3.1. *Let G be a multiple graph and h be a fixed nonnegative integer. Then,*

$$\chi_0 * (G) = \chi(G), \tag{7.1.2.3.4}$$

$$\frac{\chi(G)}{2h} \leq \chi_{h^*}(G) \leq \chi_h(G), \quad (h \neq 0). \tag{7.1.2.3.5}$$

Example. The coloring in Fig. 7.1.2.3.1 is a feasible coloring with 1. For example,

$$\begin{aligned}
&\sum_{u \in N(v_1)} m(u, v_1) d^*(c(u), c(v_1)) \\
&= 2 \cdot d^*(c(v_2), c(v_1)) + 1 \cdot d^*(c(v_4), c(v_1)) + 1 \cdot d^*(c(v_5), c(v_1)) \\
&= 2 \cdot (1 - |4/5 - 0|) + 1 \cdot (1 - |3/5 - 0|) + 1 \cdot 0 \\
&= 0.8 < 1.
\end{aligned}$$

And the total bandwidth is $(7/5) - 0 + 1 = 12/5$.

In K_4, $\chi(K_4) = 4$ and we easily show that $\chi_h^*(K_4) = 2$. In this case, the lower bound of equality in (7.1.2.3.5) holds. **(End of Example)**

7.1.3. *Rearrangement in a cellular mobile system*

7.1.3.1. *An explanation of rearrangement by an example*

In cellular mobile communication systems, which are based on mobile communication schemes, efficient use of channels is a very important issue. In a cellular system, the service area is divided into many small cells and each channel is reused in some cells, simultaneously. As mentioned in Section 7.1.1.2, channel assignment methods affect the system capacity and are classified into Fixed Channel Assignment (FCA) and Dynamic Channel Assignment (DCA). In FCA, channels are permanently allocated to each cell. Basically, all channels are available in every cell in DCA. In general, DCA gives better performance than FCA. The size of a cell becomes smaller in future cellular systems. As it becomes smaller, the variance of traffic fluctuations in each cell becomes larger. DCA is more flexible than FCA, so DCA is applicable to the traffic fluctuations. Generally, to obtain the solution of the channel assignment problem corresponds to coloring problems in a graph which is NP-complete mentioned in Section 4.4.1. So, approximate methods must be used in order to obtain the solution in a practical time when DCA is applied to the cellular system. For example, first available method, ring method and 1-clique method have been proposed [8].

In a general communication system, a channel once started using the channel continues to use the channel until the communication is completed. *Rearrangement* is a method which assigns a channel to a new call by reassigning new channels to calls already in progress. This method is applicable to various systems that adopt a common control switching network. Every channel is available in every cell in DCA. Then, channels are possible to be rearranged so that blocking probability can be made low. As a remark, when the mobile terminal crosses the cell during communication, it may be necessary to forcibly change the channel. This is also a change of channel during communication, but this is called *handover* or *handoff* (refer to Section 6.1).

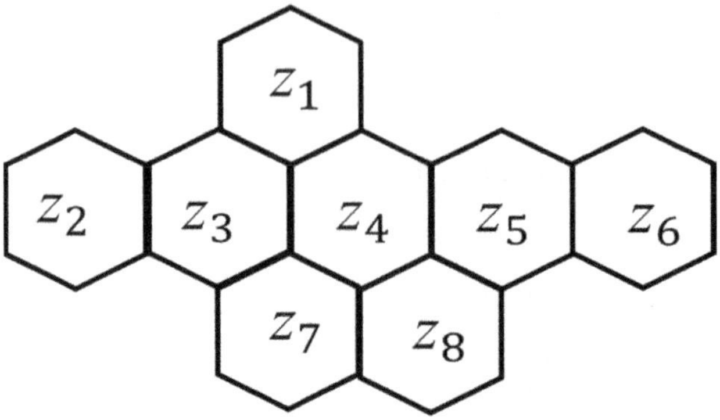

Fig. 7.1.3.1.1. A service area.

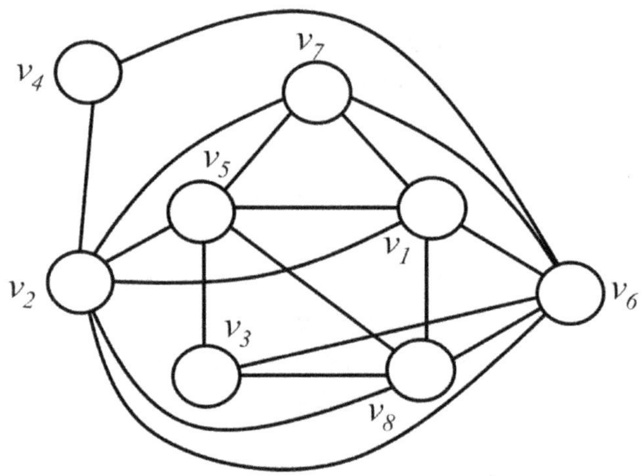

Fig. 7.1.3.1.2. The complement graph $\bar{G}$ of the interference graph G.

Let us consider a service area of Fig. 7.1.3.1.1, where the buffer cells of a cell consist of the contiguous cells to it. Figure 7.1.3.1.2 is the complement graph $\bar{G}$ of the interference graph G of Fig. 7.1.3.1.1. In $\bar{G}$,

(i) the vertex to which the channel is assigned forms a clique, and
(ii) the same channel cannot be assigned to a call in an independent set.

The system has three channels A, B and C, and let the number of calls in v_1, v_2, v_3, v_4, v_5, v_6, v_7 and v_8 be 1, 2, 1, 0, 1, 2, 1 and 1, respectively. A channel assignment is shown in Fig. 7.1.3.1.3(a) for the calls. Suppose a new call is attempted at a vertex v_4.

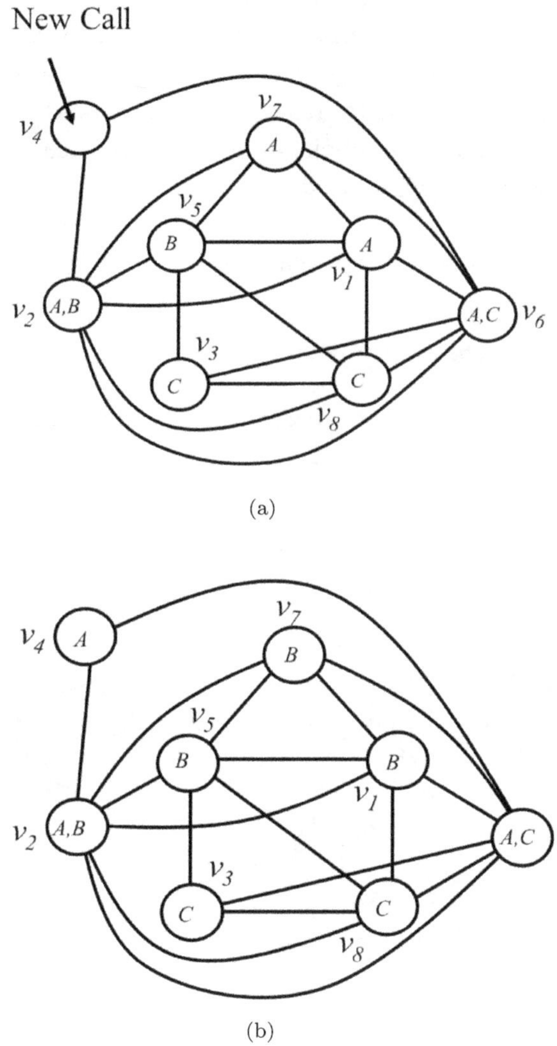

Fig. 7.1.3.1.3. A channel assignment (a) and a channel assignment (b) with rearrangement for a new call.

Using the above (i) and (ii), the clique $\{v_2, v_5\}$ using channel B can be expanded to $\{v_1, v_2, v_5, v_7\}$ which includes v_1 and v_7 in Fig. 7.1.3.1.3(a). As a result, the channel A is assigned for a new at a vertex v_4. That is, a channel A of v_1 and v_7 is rearranged to B and the channel A is assigned for a new.

In the method, rearrangement is limited to the relocation of only one channel in the buffer cells of z_4 or v_4 where the call is attempted. This method is called *first-level rearrangement* [14]. In general, the number of channels to search and the range of buffers to search are unlimited. So, there are many algorithms for rearrangement. Although the first-level rearrangement method is very simple, its high performance is shown in the computer simulation in the next section.

7.1.3.2. *A simulation result for a rearrangement*

Let us show a computer simulation result to compare FCA, DCA and DCA with rearrangement. The system is as follows.

(1) The service area consists of 61 regular hexagonal cells with a side of 1 km, as shown in Fig. 7.1.3.2.1.

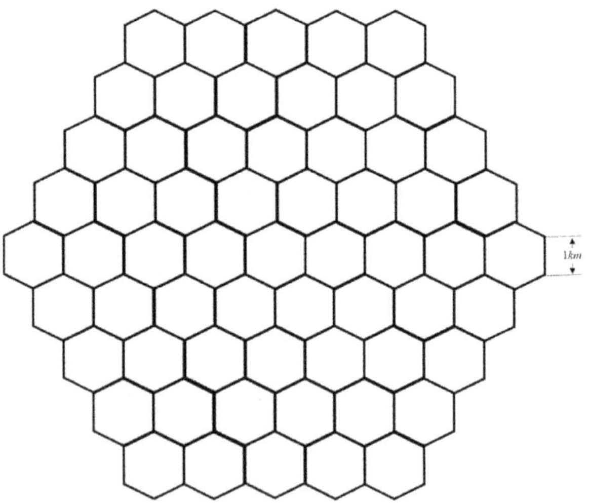

Fig. 7.1.3.2.1. A service area with 61 cells where one side of the cell is 1 km.

(2) Calls in each cell shall interfere with each other up to the next two cells, that is, a two-belt buffering system.
(3) Calls occur uniformly in each cell.
(4) Call holding times have an exponential distribution with an average of 1.5 minutes.
(5) The direction of movement at the time of call occurrence is uniform and random.
(6) The way to move is to go straight for a certain fixed time. This time is a regular distribution with an average of 1 minute and a standard deviation of 1 minute, and less than 0 hours is a cut-off distribution.
(7) The speed of going straight is 30% at 0 km/h, 20% at 15 km/h, 30 km/h, 45 km/h and 10% at 60 km/h, respectively.
(8) When communication terminal is changing the direction of movement, the average value is the direction of movement up to that point, and a normal distribution with a standard deviation of 90° is used. It is assumed that the distribution takes a course change angle of a distribution (truncated Gaussian) in which the portion less than -180° and more than 180° is cut off.

The computer simulation method is shown in the flow chart in Fig. 7.1.3.2.2. The call occurrence time is determined by a random number with a Poisson distribution. The same channel assignment method is used when a call is generated or crosses a cell. When a channel is assigned to a moving terminal, its movement is calculated to determine the time for the next cell change.

The following three types of channel assignment examined by the computer simulation.

(1) Channel assignment method 1: FCA (Fixed Channel Assignment). Since the service area is 7-colorable (see Section 7.1.1.3), the number of channels is $105 = 15$ (ch/cell) $\cdot 7$ for the entire system.
(2) Channel assignment method 2: DCA (Dynamic Channel Assignment) without rearrangement. The channel assignment uses first-available method (see Section 7.1.1.4). Number of channels is 100.

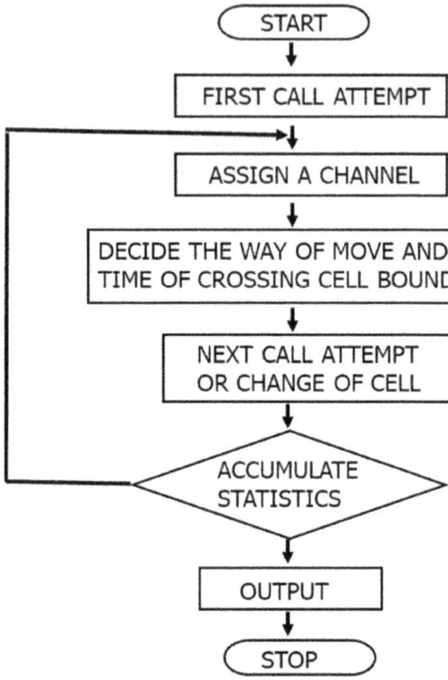

Fig. 7.1.3.2.2. The flow chart for the computer simulation.

(3) Channel assignment method 3: DCA (Dynamic Channel Assignment) with first-level rearrangement. The channel assignment uses first-available method. Number of channels is 100.

The results of the computer simulation are shown in Fig. 7.1.3.2.3.

This result is the sum of the call blocking rates at the time of call occurrence for 10,000 calls from 7.5 minutes after the start of the simulation, and the traffic carried per channel A_c is calculated by the following equation:

$$A_c = N_z \cdot A \cdot (1 - B)/N_c, \qquad (7.1.3.2.1)$$

where N_z is the number of cells (= 61), A is traffic intensity per cell (*erl*), B is blocking probability and N_c is the number of channels of the system.

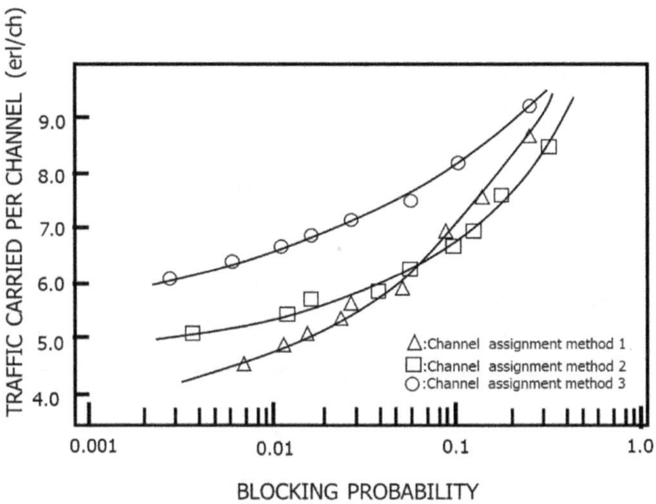

Fig. 7.1.3.2.3. The results of the computer simulation.

B aggregates only the call blocking rate at the time of the call. Calls that go out of the service area during a communication or that cannot be assigned to a channel due to changing cells are calculated as the call is completed.

In Fig. 7.1.3.2.3, looking at 1% of blocking probability the traffic carried per channel is 4.7 (*erl/ch*) for FCA, 5.4 (*erl/ch*) for DCA, and 6.7 (*erl/ch*) for DCA with first-level rearrangement. Conventionally, it is known that DCA is superior to FCA in terms of traffic characteristics (especially at low-blocking probability). This simulation result shows that the traffic characteristics can be further improved by applying the rearrangement. It should be noted that even if the simple assignment method like first-level rearrangement is applied, the degree of improvement is greater than the degree of improvement in the traffic characteristics when FCA is changed to DCA.

7.1.3.3. *Other trends regarding cellular systems*

The movement characteristics of moving objects play an important role in design and analysis in cellular systems. In the computer simulation in Section 7.1.3.2, a rough model was used. After that, a *mobile model* based

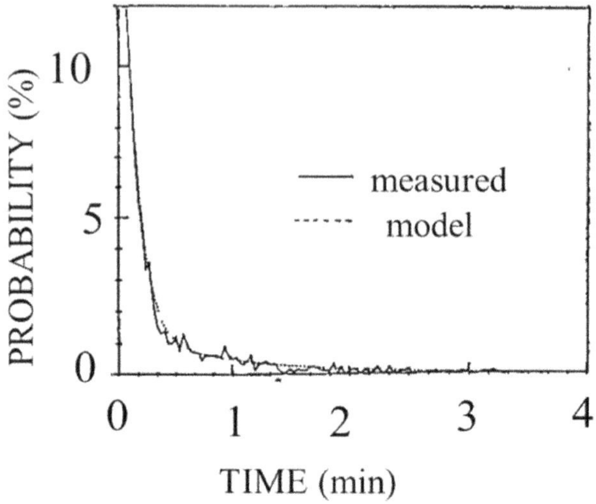

Fig. 7.1.3.3.1. The time distribution by which the subscriber moves straight.

on *actual measurement data* is proposed. An example, we show the models based on the measurements in Niigata city, Japan [17].

(1) The measured result and the model of the time distribution by which subscriber moves straight are represented in Fig. 7.1.3.3.1.
(2) The measured result and the model of the speed distribution are shown in Fig. 7.1.3.3.2.
(3) The measured result and the model of the traveling direction distribution are represented in Fig. 7.1.3.3.3.

At present, more precise modeling is carried out by grasping the location of moving terminals with GPS, collecting them by mobile communication, and processing them as big data [43]. In cellular systems, the channel assignment problem is NP-complete. So, it is not easy to solve the optimal solution. One method to obtain the solutions is to use *neural networks*. There are such reaches which applies the Hopfield neural network to the combinational optimization. The application of AI (Artificial Intelligence) to the channel assignment problems is useful.

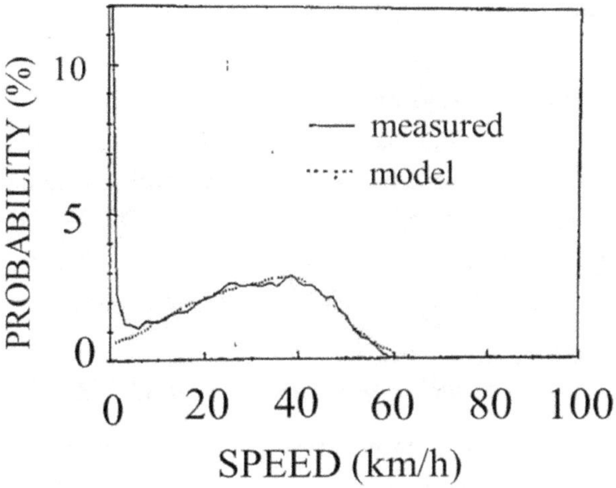

Fig. 7.1.3.3.2. The speed distribution.

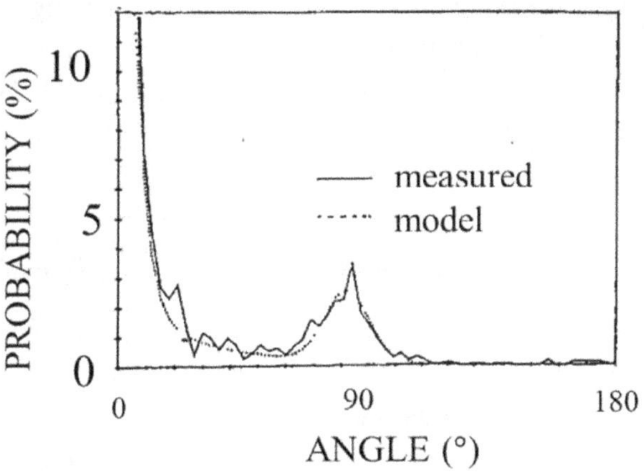

Fig. 7.1.3.3.3. The traveling direction distribution.

7.1.4. *Spatial data and mobile communications*

Spatial data (*SD*: Spatial Data) is data that represents the spatial positions and characteristics of natural and artificial features and boundaries on the earth. The types of data include a wide variety of data. The relationships that connect spatial positions and their characteristics range from direct to indirect. In other words, spatial data is data that can be directly or indirectly associated with a position on the earth. It is known that *GIS* (*Geographical Information System*) plays a very important role in utilizing spatial data. GIS is a computer information system that comprehensively processes, manages, and analyzes attribute data (spatial data) such as nature, society, and economy with information on geographical position and space, and displays the results. Let us consider the relationship between SD/GIS and mobile communication, then there are two aspects: The role of mobile communication in SD/GIS and the application of SD/GIS to mobile communication.

7.1.4.1. *The role of mobile communication in SD/GIS*

To manage spatial data, it is necessary to acquire and update the latest spatial data. In addition to satellites, remote sensing, etc., on-site actual measurement and surveys are required to acquire the latest spatial data. It is necessary to acquire and transmit 3D data (including 4D data including the time axis) while driving on the road as a mobile mapping system. In this case, high-speed and high-reliability communications are required from the interface with the sensor. In a broad sense, it also includes data collection from radar in peripheral monitoring technology in ITS (Intelligent Transport System).

7.1.4.2. *The application of SD/GIS to mobile communication*

Spaces for mobile communication (called mobile communication space, which includes urban areas where skyscrapers and trees coexist, multi-walled areas in buildings and underground spaces, etc.) are terrain shapes as well as buildings and obstacles. The use of spatial data is effective in solving the problem of *optimal area coverage* (optimal placement of antennas, optimal control of cell size) by cells (including *microcells* and *picocells*) in consideration of the placement of objects, etc.

It is becoming difficult to analyze radio wave propagation characteristics in cities due to the increasing height of structures. It is necessary to construct cells (radio cells) using microwave propagation characteristics in cities. If more accurate radio wave propagation characteristics can be calculated from detailed spatial data, it will be useful for designing mobile communication. Especially on highways, the positional relationship between vehicles changes dynamically, so the radio wave propagation characteristics also change dynamically at high speed. The problem of modeling propagation characteristics in such cases is closely related to ITS (Intelligent Transport Systems).

In mobile communication, it is important to design a system that takes into consideration the mobility of mobile terminals. It is necessary to analyze the communication traffic when the flow (movement) of people, automobiles, etc. is given as spatial data, and to optimally design the communication system using it. In the next section, a simple example will be used to show that the traffic (mobility) of mobile terminals and communication traffic are closely related.

7.1.4.3. *The mobile terminals traffic and communication traffic*

A simple example is shown to consider the relationship between the mobility of mobile terminals and communication traffic. As shown in Fig. 7.1.4.3.1, the distance traveled by a car in 3 minutes is 2 km at a speed of 40 km/h, 3 km at a speed of 60 km/h, and 4 km at a speed of 80 and 5 km at 100 km/h. If the time is 1.5 minutes (90 seconds), the distance traveled is half of these. Let us consider cells for a straight road (one-dimensional cell) with a length of 1 km as shown in Fig. 7.1.4.3.2. During a communication time of 3 minutes (or 1.5 minutes), the car will pass through multiple cells. That is, channel switching (handoff) will occur during the communication.

In the service area of Fig. 7.1.4.3.2, it is assumed that all cars are moving in one direction at a constant speed v and the cars are uniformly distributed in each cell. Suppose a call is made in a cell and a channel is assigned. In this case, the call is made when a person traveling by car receives a call, or when a person traveling by car calls. Let us find out the probability that this car will enter the next cell during a call. Consider the probability of reaching the cell boundary (handoff) while holding the channel from the point where the call occurred, referring to Figs. 6.1.4 and 6.1.6. The distance from the point where the call occurred to the cell boundary is x. Let $P\ (X \leq x)$ be the probability that the call occurred. In order for the distance from the point where the call occurred to the cell boundary to be x or less, the point where the call occurred must exist in the gray part in Fig. 7.1.4.3.3. Considering the area of the entire cell and the gray part,

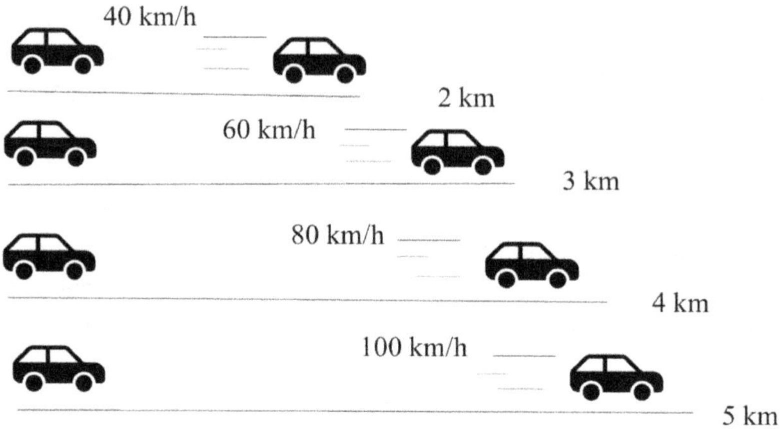

Fig. 7.1.4.3.1. Distance a car travels in 3 minutes.

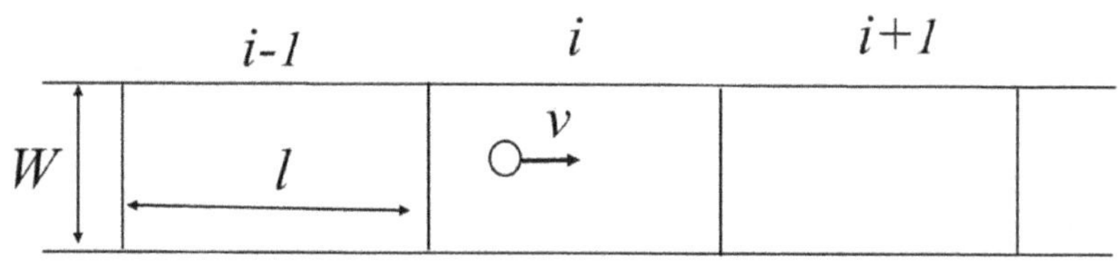

Fig. 7.1.4.3.2. Cells for a straight road (one-dimensional cell model) (see Fig. 6.1.6).

$P(X \leq x)$ is expressed as the following equation as the area ratio:

$$P(X \leq x) = \frac{xW}{lW} = \frac{x}{l} \tag{7.1.4.3.1}$$

Therefore, the probability $P(T \leq t)$ that the time that this car stays in the cell where the call occurred is t or less is as follows from the relation of $x = vt$:

$$P(T \leq t) = \frac{vt}{l}, \tag{7.1.4.3.2}$$

where $0 \leq t \leq \frac{l}{v}$. Then, the probability density function $p(t)$ is as follows:

$$p(t) = \frac{dP(T \leq t)}{dt} = \frac{v}{l}. \tag{7.1.4.3.3}$$

As in the conventional communication traffic theory, assuming that the hold time (communication time) of the call follows the exponential distribution

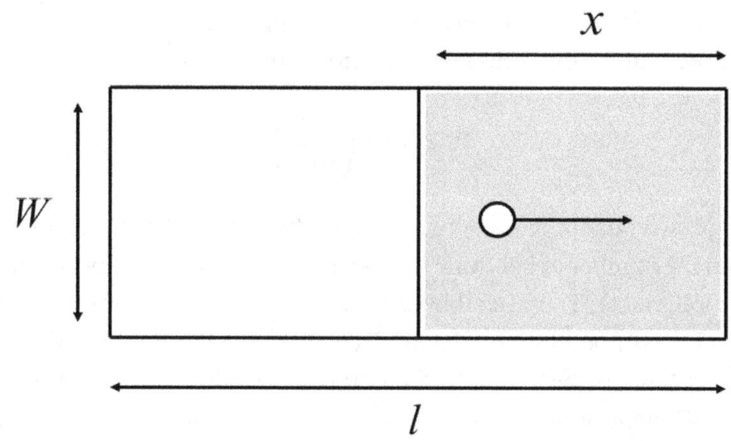

Fig. 7.1.4.3.3. Distance traveled to cell boundaries.

of the average h_o, the probability $P(H > t)$ that the hold time H becomes longer than t is obtained by the following equation:

$$P(H > t) = \exp\left(-\frac{t}{h_o}\right). \tag{7.1.4.3.4}$$

Let us assume that a car call occurs where the distance to the cell boundary is x and the time to move to the boundary is t when moving at speed v. Then, the probability that the car of the call will move to the next cell while holding the channel is the probability that the hold time of the call is longer than t. This probability is calculated by the formula (7.1.4.3.4). Therefore, the probability P_1 of moving from the cell where the call occurred to the next cell can be expressed as follows from Equations (7.1.4.3.3) and (7.1.4.3.4).

$$\begin{aligned} P_1 &= \int_0^{l/v} P(H > t)p(t)dt \\ &= \frac{vh_o}{l}\left\{1 - \exp\left(\frac{-l}{vh_o}\right)\right\}. \end{aligned} \tag{7.1.4.3.5}$$

The probability P_2 that a call that has moved from the previous cell (handoff call) moves to the next cell is different from P_1. In this case, all handoff calls are considered to have occurred on the boundary of the previous cell. Therefore, in general, P_2 has a longer hold time in that cell of the call than in the case of P_1. Using the *memorylessness of the exponential distribution*, P_2 is expressed by the following equation from (7.1.4.3.4).

$$P_2 = \exp\left(\frac{-l}{vh_o}\right). \tag{7.1.4.3.6}$$

There are two quantities closely related to these P_1 and P_2: The in-cell hold time and the number of handoffs that occur per unit time (handoff call occurrence coefficient). The in-cell hold time is the time to occupy one channel. If it can be assumed that the handoff call follows the Poisson distribution, the call volume to be added to the cell can be calculated using the occurrence coefficient of the occurrence call and the handoff call, and the in-cell hold time. Given this call volume and number of channels, the blocking probability can be calculated using the well-known Erlang B Formula (see (7.1.1.4.2)). That is, the communication traffic characteristics of this mobile communication system can be obtained.

In order to confirm the feasibility of such an analysis method, an example of comparison between the analysis result and the computer simulation is shown. Let us consider a ring-shaped one-dimensional service area ($l = 1\,km$, the number of channels is 8, the hold time is 1.5 minutes and zero-belt buffering) in Fig. 7.1.4.3.2. Figure 7.1.4.3.4 shows the result of numerical analysis of blocking probability in this service area. It also shows blocking probability obtained by computer simulation. From this result, it can be seen that the numerical analysis result and the computer simulation result are in good agreement. From Fig. 7.1.4.3.4, as the speed increases, the blocking probability decreases. The reason is that the holding time in one cell decreases due to the movement of the call (car). Therefore, as the call volume, the blocking probability and the speed increases, the rate of decrease in the blocking probability is small.

A large call volume means a high density of cars. The higher the density of the car, the slower the speed in general. Many researches have been done on this phenomenon. As a simple example, Fig. 7.1.4.3.6 shows the result of computer simulation using the Greenshields model [78] (the speed decrease is linear) as shown in Fig. 7.1.4.3.5. Figures 7.1.4.3.4 and 7.1.4.3.6 are the results of using the fixed channel assignment (FCA) method. Since

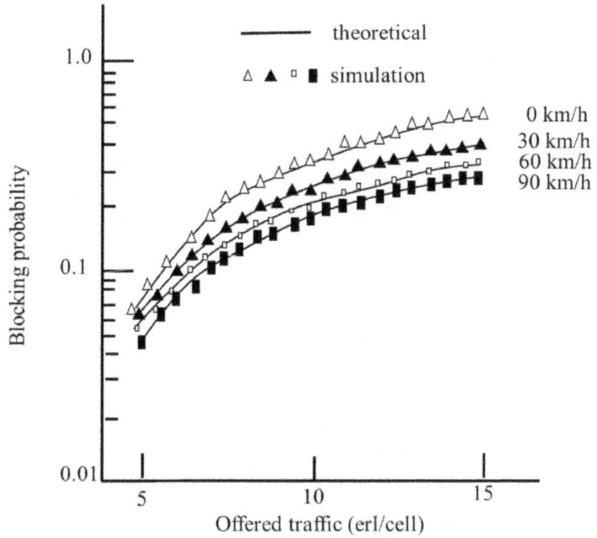

Fig. 7.1.4.3.4. Comparison between the analysis and the simulation on traffic characteristics of one-dimensional service area.

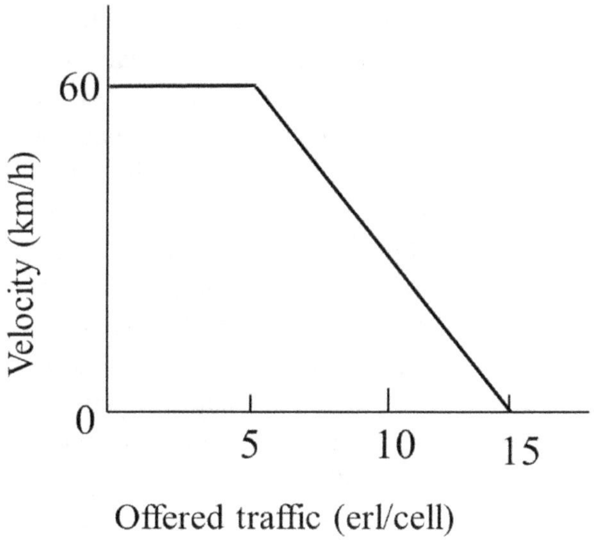

Fig. 7.1.4.3.5. A Greenshields model.

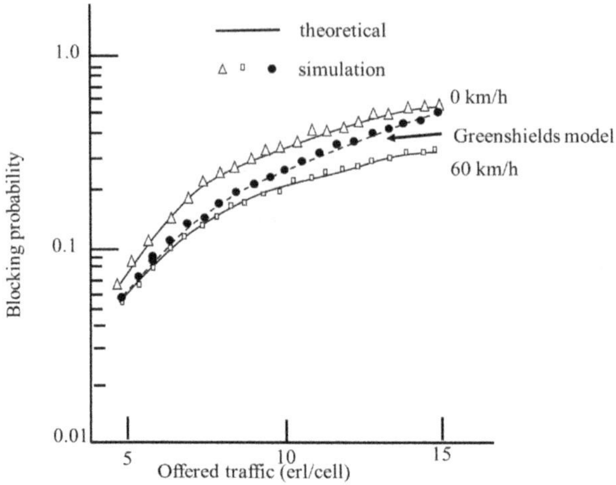

Fig. 7.1.4.3.6. Communication traffic characteristics using Greenshields model.

it is zero-belt buffering, the same result is obtained when dynamic channel assignment (DCA) is performed.

Since the above example is a simple one in which the movement of the moving terminal is constant velocity, theoretical analysis is often possible. As we can be seen from such a simple example, fixed communication in which the user does not move during a call and mobile communication in which the user moves show significantly different characteristics in communication traffic characteristics. The element that this mobile terminal moves is a unique characteristic of mobile communication. To obtain the communication traffic characteristic in mobile communication, it is necessary to obtain the cell change probability as described above. In addition, detailed spatial data such as the speed of moving objects on the road is required to obtain the cell change probability. If such spatial data is obtained, the communication traffic characteristics can be obtained, and it will be effective in the design of mobile communication.

Generally, in addition to the one-dimensional space, a two-dimensional plane, or a three-dimensional space such as inside a building is the service area for mobile communication. In the above analysis example, the call is moving in the one-dimensional service area. Let us consider analyzing communication traffic characteristics as well in a two-dimensional

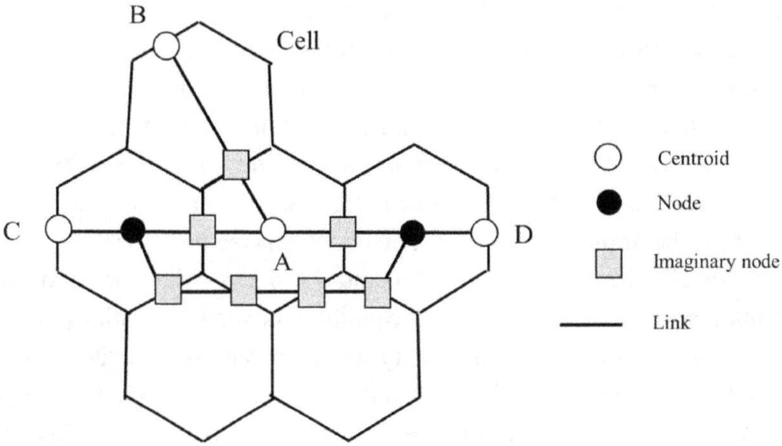

Fig. 7.1.4.3.7. The road network (graph) form two-dimensional service area.

or three-dimensional service area. For example, as shown in Fig. 7.1.4.3.7, the shape of the cell existing on the two-dimensional plane is a regular hexagon, and it is necessary to consider the entry and exit of the moving terminal from six directions for one cell. It can be expected that it will not be easy to analyze the communication traffic characteristics in the dimensional plane exactly. For this problem, the two-dimensional service area is modeled on the one-dimensional network (graph) and analyzed in a similar manner. Let me show an example. A two-dimensional system consisting of six cells as shown in Fig. 7.1.4.3.7 is approximated by a network (graph) on the road. Traffic flows on the network branch at a certain point and merge at a certain point. In the approximation method, the traffic flow is divided into several one-dimensional routes, and the above-mentioned communication traffic analysis method in the one-dimensional service area is applied to each route.

Obtain the traffic volume (spatial data) between typical points (centroids) from traffic volume surveys. From this result, the traffic volume flowing through each one-dimensional route is estimated using nodes and imaginary nodes. In Fig. 7.1.4.3.7, there are four centroids A to D. The traffic volume between any two centroids can be found from the survey results, etc. However, since it is not known what route this traffic volume will take from centroids to centroids, the traffic volume on each route will be estimated using the divided distribution method. In the divided distribution method, the traffic volume is allocated little by little to the route (shortest route) that many people will use, and the allocation is repeated until the survey result and the final result of the allocation match. At this time, since the road is crowded on the route where the traffic volume is distributed, it is considered that the speed of the moving terminal may become slow. This method can be applied to three-dimensional service area [35].

As described in Sections 7.1.4.1 and 7.1.4.2, spatial data and mobile communication are closely related. Spatial data such as spatial population density, base station, terminal density and distribution, cell size and shape are used for designing mobile communication systems. It will be indispensable data for capacity estimation, service quality estimation, etc. Also, some of these spatial data can be given only probabilistically. For example, the probability of the number of mobile stations existing in a cell. The distribution is given as a *geometric probability* (probability associated with a figure). In a multihop mobile communication system (see the next section) in which a mobile station plays the role of a relay station, the distribution of the number of relay stations will also be necessary as a geometric probability.

7.1.4.4. *Cell Structure and computational geometry*

Computational geometry is concerned with the design and analysis of algorithms for solving geometric problems. The field of computational geometry includes geographic information processing, VLSI design, pattern recognition and robotics. The *Voronoi Diagram* is one of the fundamental concepts in computational geometry with many applications.

For two points p and q, let $d\ (p, q)$ denote the Euclidean distance between p and q. For a finite set $P = \{p_1, \ldots, p_n\}$ of points in the plane, region $V\ (p_i)$ is defined by

$$V(pi) = \{p | d(p, p_i) < d(p, p_j) \quad \text{for any } j(\neq i)\}.$$

$V(p_i)$ is called the Voronoi region of p_i. The Voronoi regions $V(p_1), \ldots, V(p_n)$ make a partition of the plane, and this partition is called the Voronoi diagram for P (see Fig. 7.1.4.4.1). It is well-known that a Voronoi diagram on n points in the plane can be obtained in $O(n \log n)$

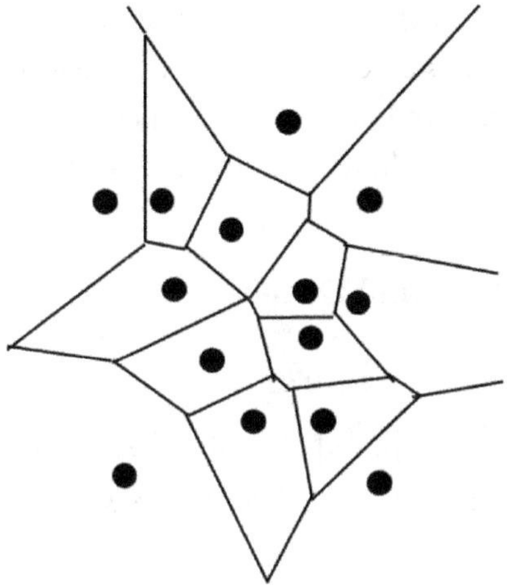

Fig. 7.1.4.4.1. A Voronoi diagram.

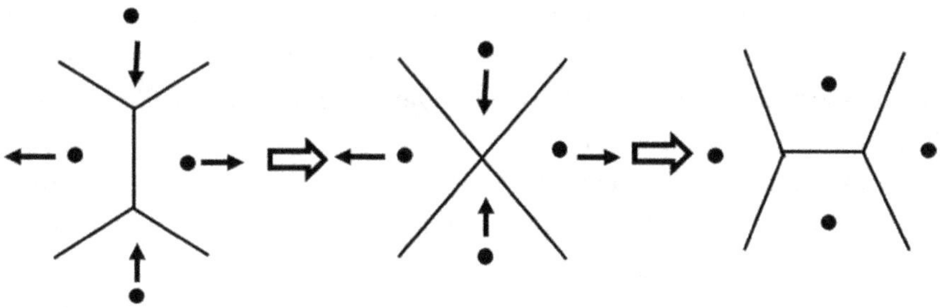

Fig. 7.1.4.4.2. Topological change of Voronoi diagram.

time and $O(n)$ space. The Voronoi diagram for moving objects has been investigated in connection with motion planning in robotics and geometric optimization problem in *computational geometry* [34]. A topological change of Voronoi diagram is shown in Fig. 7.1.4.4.2. Now if we regard each element of P as a cell site and assume that radio signal strength drops in proportion to distance, then we can regard Voronoi regions as cells. In micro (or pico) cellular systems, by miniaturization of devices, mobile units may be able to have the function of exchanges, and connect other mobile units and cell sites. In this case, not only mobile units but also cell sites move in service area. Therefore, cell structure changes dynamically (see about *dynamic spectrum sharing* in research on ITU in Section 7.1.1.1). If we know how to change cell structure, it is available for assignment of channels, channel switching and dynamic spectrum sharing. The computational geometry is applicable to cell structure of mobile communications.

7.2. Multihop Systems and Graphs

7.2.1. *Multihop wireless networks*

In cellular mobile system, mobile terminals communicate with each other via the base stations in cells. *Multihop wireless networks* consist of mobile terminals with personal communication devices. Each terminal can receive a message and then send it to another terminal. If a terminal cannot communicate with another terminal which sends data directly, some terminals relay the data. That is, each mobile terminal has the role of a base station (relay station), and it is considered that the base station moves in cellular system. Also, considering it as a network on the user side, it can be considered as a temporarily constructed network. Therefore, it is also called a *Mobile Ad Hoc Network* (MANET).

In the network, each terminal communicates with the other terminals using channels (Fig. 7.2.1.1). For example, in Fig. 7.2.1.2, terminal x_2 communicates with x_1, x_3 and x_5 directly; terminal x_2 communicates with x_6 via x_5. In this network, we assign channels to terminals. We assume that on the communication between terminals x and y, the same channel is used in the communications from x to y and from y to x. Therefore, we assign the channels to seven terminal pairs $(x_1, x_2), (x_2, x_3), \ldots, (x_7, x_8)$.

7.2.1.1. *Strong edge coloring*

For multihop wireless networks, we assign channels to communication between two terminals. This is a kind of edge coloring problem. For conventional edge coloring, we assign different colors to adjacent edges. However, in the model of conventional edge coloring, cochannel interference may occur. Figure 7.2.1.2 shows an edge coloring of the graph of Fig. 7.2.1.1. The channel A is assigned to edge (v_1, v_2) and (v_3, v_4) in Fig. 7.2.1.2. The terminal x_2 sends data in all directions when x_2 communicates with x_1

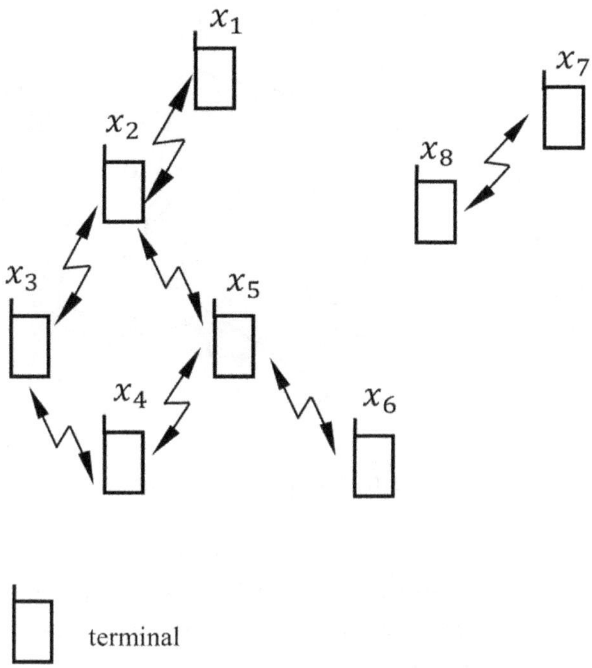

Fig. 7.2.1.1. A multihop wireless network.

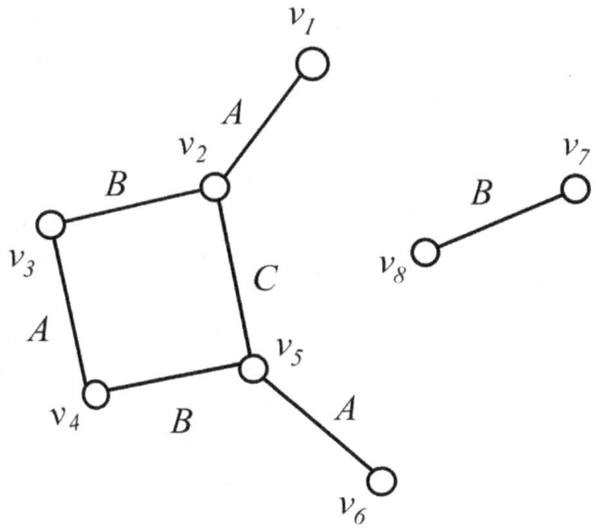

Fig. 7.2.1.2. An edge coloring of the graph of Fig. 7.2.1.1.

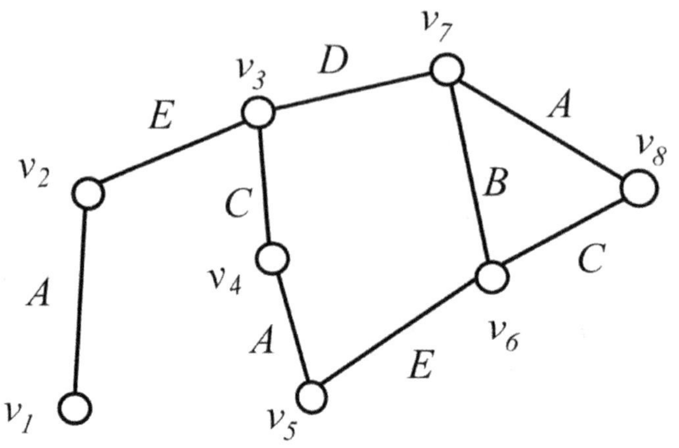

Fig. 7.2.1.1.1. A strong edge coloring.

using channel A. Since the terminal x_4 communicates with x_3 using channel A, the terminal x_3 cannot receive carrier from terminal x_4 due to the interference of channel A form x_2. Therefore, we defined the following edge coloring and called it *strong edge coloring*.

Definition. Let $G = (V, E)$ be an undirected graph such that V is the vertex set, E is the edge set. Let $N = (G, w_n)$ be an undirected network such that w_n is the edge weight function assigning a positive real number $w_n(e)$ to each edge e of N. A strong edge coloring of G is an assignment of colors to edges of G such that every two edges of distance at most two receive different colors. Two edges are of distance at most two if and only if the edges are adjacent to each other or there is an edge between the edges.

An example of a strong edge coloring of a graph is shown in Fig. 7.2.1.1.1. The attached letter on each edge in Fig. 7.2.1.1.1 represents the color. In Fig. 7.2.1.1.1, we assign five colors $A-E$. There is no strong edge coloring with four colors or less. Because, we must assign different colors to edges in the tieset $\{(v_3, v_4), (v_4, v_5), (v_5, v_6), (v_6, v_7), (v_7, v_3)\}$ with five edges. For the results of the strong edge coloring, please see [71]. Here, we represent a multihop wireless network using a graph. Each vertex represents a terminal. If two terminals communicate with each other, we join the vertices corresponding the two terminals by an edge. If no edge between two vertices, that means no interference between them, the strong edge coloring of the graph is equal to the channel assignment on the multihop wireless networks. So, it is important to find a strong edge coloring with the minimum number of colors. However, it is very hard to solve this problem.

Theorem 7.2.1.1. [71]. *The problem of finding a strong edge coloring with the minimum number of colors for general graphs is NP-complete.*

It is known that this problem can be solved in polynomial time if the graph is a tree [80].

7.2.1.2. *CIR-edge coloring*

On strong edge coloring, the degree of interference is not taken into consideration. So, let us consider a new edge coloring problem taking the degree of interference into consideration. When we assign a channel to a terminal pair, we must consider cochannel interference. In Fig. 7.2.1.2.1, terminal x_1 communicates with terminal x_1' with channel A and terminal x_2 also communicates with terminal x_2' with channel A. Now, terminal x_3 tries to communicate with terminal x_3'. In this case, terminal x_3 can use channel A or not? Here, we must consider the cochannel interference from terminals x_1 and x_2. For simplicity, we neglect the cochannel interference from terminals x_1' and x_2'. In Fig. 7.2.1.2.2, carrier C is from terminal x_3, and interference I_{x_1} and I_{x_2} are from terminals x_1 and x_2, respectively. C/I ratio (carrier-to-interference ratio, we simply call it CIR hereafter) that is not less than a desired value α is necessary for a good communication. For example, in case of FM, 30 kHz channel bandwidth and analog telephone system, it is reported that $\alpha = 18$ dB [77], while in case of a wireless *LAN* system, it is reported that $a = 18.5$ dB [72]. Here we assume that if a channel assignment in a wireless multihop network realizes the CIR that is not less than a for each terminal, good communication in the system is guaranteed. We also assume that the total cochannel interference received by a terminal is determined by the sum of each cochannel interference. Therefore, in case of Fig. 7.2.1.2.1, if

$$C/(I_{x_1} + I_{x_2}) \geq \alpha,$$

terminal x_3 can use channel A for communication with terminal x_3'.

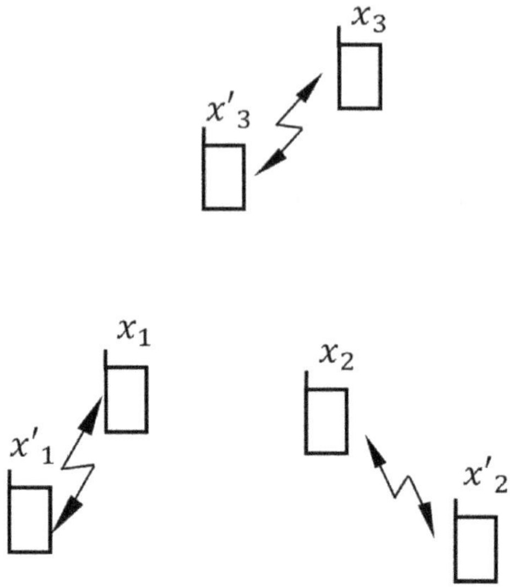

Fig. 7.2.1.2.1. Cochannel interference.

We express the channel assignment problem in different words using the terminology of graph and network theory.

Definition. Let N be an undirected network and α be a positive real number. If an edge weight function f satisfies the following conditions, f is called a *CIR-edge coloring* of N.

(1) f is a function from edges to colors.
(2) If edge e is adjacent to edge e', $f(e) \neq f(e')$.
(3) Let v be a vertex on edge $e_0 = (v, v_0)$. Let v_i be a vertex such that v_i is adjacent to v and v_i is incident to an edge e_i whose color is $f(e_0)$ ($i = 1, \ldots, k$) (see Fig. 7.2.1.2.3). Then,

$$w(e_0)/\{w(v, v_1) + w(v, v_2) + \cdots + w(v, v_k)\} \geq \alpha.$$

The minimum number of colors for all CIR-colorings of N is called the *CIR-edge coloring number* of N.

In the above definition, each vertex represents a terminal. If two terminals communicate with each other, we join the vertices corresponding to the two

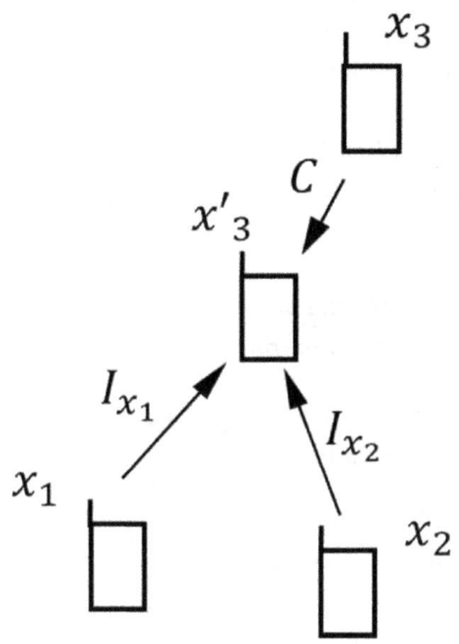

Fig. 7.2.1.2.2. Explanation of cochannel interference.

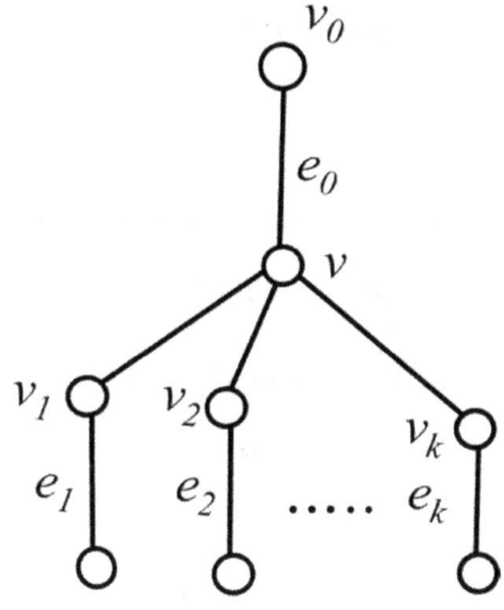

Fig. 7.2.1.2.3. Explanation of CIR-edge coloring.

terminals. The weight of an edge (u, v) represents interference in case that u and v use the same channel.

If α is large enough, a CIR-edge coloring of a network N is equal to a strong edge coloring of N. Therefore, the problem of finding the CIR-edge coloring number of N is NP-complete. Furthermore, we obtain the following result.

Theorem 7.2.1.1. *For a tree network T, the problem of finding the CIR-edge coloring number of T is NP-complete.*

Proof. Clearly, this problem belongs to class NP. Here, we consider the following NP-complete problem.
Bin Packing Problem [73].

INSTANCE: Finite set $X = \{x_1, \ldots, x_h\}$ of items, a size $s(x) \in Z^+$ for each $x \in X$, where Z^+ means the set of positive integers, a positive integer bin capacity B, and a positive integer k.

QUESTION: Is there a partition of X into distinct sets $X_1, X_2, \ldots, X_k$ such that the sum of sizes of the items in each X_i is B or less?

We construct the following tree $T = (V, E)$ from the instance of the bin packing problem.

$$V(T) = \{r, v_1, \ldots, v_k, u_1, \ldots, u_h, w_1, \ldots, w_h\},$$

$$E(T) = \{(r, v_1), \ldots, (r, v_k), (r, u_1), \ldots, (r, u_h), (u_1, w_1), \ldots, (u_h, w_h)\}.$$

Let c be the ratio of the maximum number of $s(x)$ to the minimum number of $s(x)$.

We assign the weights to edges in T as follows (see Fig. 7.2.1.2.4):

$$w(r, v_i) = (c + 1)B,$$

$$w(r, u_i) = s(x_i),$$

$$w(u_i, w_i) = a_0, \text{ where } a_0 \text{ is large enough.}$$

Let $\alpha = c + 1$. In this case, we show that there exists a partition on the bin packing problem if and only if the CIR-edge coloring number of T is at most $k + h$. Note that the CIR-edge coloring number of T is not less than $k + h$ since the degree of vertex r is $k + h$.

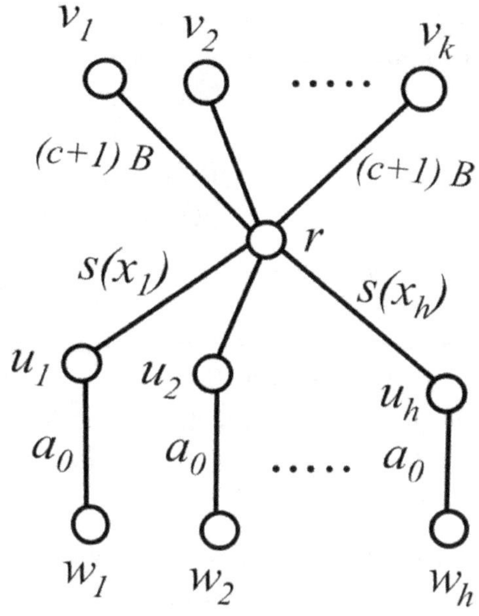

Fig. 7.2.1.2.4. A tree network T.

First, we assume there exists a partition $X_1, X_2, \ldots, X_k$ on the bin packing problem. Let $X_i = \{x_{i_1}, x_{i_2}, \ldots, x_{i_j}\}$. We assign color i to (r, v_i) and $(u_{i_1}, w_{i_1}), \ldots, (u_{i_j}, w_{i_j})$ for each i. We assign colors $k+1, \ldots, k+h$ to edges $(r, u_1), \ldots, (r, u_h)$, respectively. This assignment is the CIR-edge coloring of T with $k+h$ colors. The reason is the following.

For edge (r, v_i),

$$\begin{aligned} &w(r, v_j)/(w(r, u_{i_1}) + \cdots + w(r, u_{i_j})) \\ &\quad \geq (c+1)B/B = c + l = \alpha \quad \text{in case of } X_i \neq \emptyset, \end{aligned}$$

the color i is only assigned to (r, v_i) in case of $X_i = \emptyset$.

For edge (r, u_i), the color $k+i$ is only assigned to (r, u_i). For edge (u_i, w_i), $w\ (u_i, w_i)$ is large enough.

Conversely, let f be a CIR-edge coloring of T with $k+h$ colors. In this case, $f\ (r, u_i) \neq (u_j, w_j)$ for $i \neq j$. Because,

$$w(r, u_i)/w(r, u_j) < c < c + 1 = \alpha$$

So, we must assign different colors to (r, u_i) and (u_j, w_j). Therefore, if a color is assigned to some edges, the edges belong to $\{(r, v_1), \ldots, (r, v_k), (u_1, w_1), \ldots, (u_h, w_h)\}$. Since this coloring uses $k + h$ colors, one of the edges is (r, v_i).

Here, we assume that a color is assigned to edges (r, v_i) and $(u_{i_1}, w_{i_1}), \ldots, (u_{i_j}, w_{i_j})$. Since f is a CIR-edge coloring,

$$w(r, v_i)/(w(r, u_{i_1}) + \cdots + w(r, u_{i_j})) \geq c + 1.$$

Since $w(r, v_i) = (c + 1)B$, $(w(r, u_{i_1}) + \cdots + w(r, u_{i_j})) \leq B$. So, the sum of $s(x_{i_1}), \ldots, s(x_{i_j})$ is B or less. If the assigned color to (r, v_i) is not assigned to the edges in $\{(u_{i_1}, w_{i_1}), \ldots, (u_{i_j}, w_{i_j})\}$, let $X_i = \emptyset$. Therefore, there exists a partition of X satisfying the condition of the bin packing problem. □

The definition of the CIR-edge coloring is a special case of channel assignments on multihop wireless networks as we assign channels to all edges. In practice, although a terminal v can communicate with the other terminal u, v may not communicate with u. Therefore, the edge set is divided into communication edges and interference edges. We assign channels to communication edges only.

For example, in Fig. 7.2.1.2.5, the bold edges represent communication edges and the other edges represent interference edges. Let $\alpha = 2$. First, we consider a strong edge coloring to communication edges (see Fig. 7.2.1.2.6). In Fig. 7.2.1.2.6, a, b, c,... represent colors. So, in case of the strong edge coloring, we need six colors.

A CIR-edge coloring to communication edges is shown in Fig. 7.2.1.2.7. In Fig. 7.2.1.2.7, a, b, c, d represent colors. In this case, the number of colors is four. Therefore, the CIR-edge coloring is effective to achieve high spectral efficiency.

Next, we show the effectiveness of the CIR-edge coloring in comparison with the strong edge coloring from the practical point of view. We use the following model to explain the effectiveness of the CIR-edge coloring. We assume

(1) the 40 dB/dec mobile radio propagation path loss rule (in propagation path loss, radio signal strength drops proportionate to distance) [77].

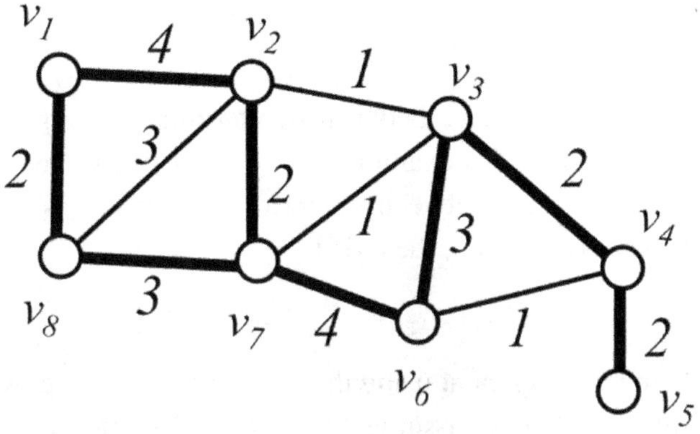

Fig. 7.2.1.2.5. A network.

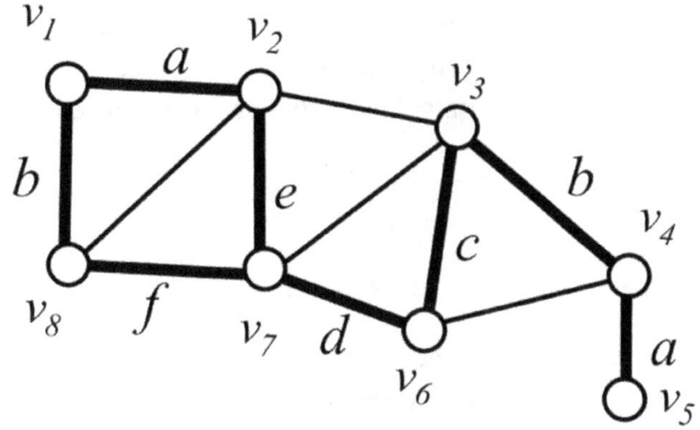

Fig. 7.2.1.2.6. A strong edge coloring network.

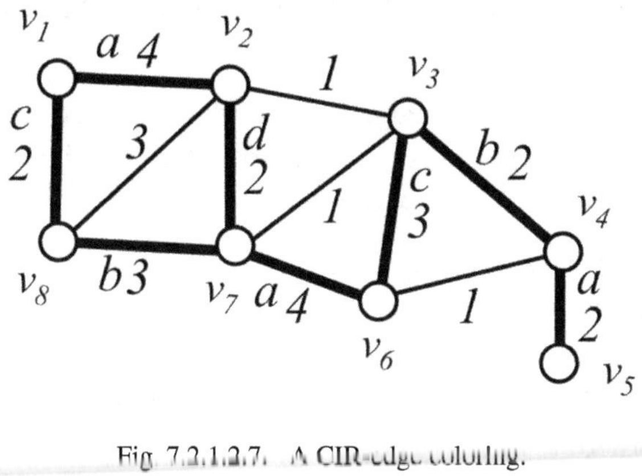

Fig. 7.2.1.2.7. A CIR-edge coloring.

(2) same transmission power for each terminal,
(3) $\alpha = 18\,\text{dB}$.

For simplicity, if the distance between two terminals is large enough, we neglect the interference. For example, in Fig. 7.2.1.2.2, we assume the distances between x_3' and x_3, between x_3' and x_1 and between x_3' and x_2 are d_1, d_2 and d_3, respectively. Then, the CIR is

$$C/(I_{x_1} + I_{x_2}) = d_1^{-4}/(d_2^{-4} + d_3^{-4}).$$

In case of channel assignment using the strong edge coloring, we construct a graph as follows. Here we assume that a terminal x_3' tries to communicate with terminal x_3 using channel A. In x_3', let C be the carrier from x_3. Let R be the distance between x_3' and x_3. We assume that if the distance between x_3' and any terminal using channel A is not less than D, x_3' can use channel A. The worst case is the following on this situation (see Fig. 7.2.1.2.8). In Fig. 7.2.1.2.8, terminals of outside use channel A. In this case,

$$C/I = R^{-4}/6D^{-4} \geq 18\,\text{dB}.$$

Then,

$$D/R \geq 4.4.$$

Let R_M be the maximum distance between terminals that can communicate with each other. We construct a graph as follows. Each vertex represents a terminal. If the distance between terminals v and u is less than 4.4 R_M, we join the vertices by an edge. If the vertex pair needs to communicate with each other, the edge is a communication edge. If not, the edge is an interference edge. Namely,

if $R_M < d < 4.4R_M$, the edge is an interference edge and
if $d < R_M$, the edge is a communication edge or an interference edge,
where d is the distance between the vertex pair.

For example, we consider eight terminals on a straight line. The adjacent terminals communicate with each other, where the distance of each adjacent terminals is one. So, communication edges form a path P_8 with eight vertices (see Fig. 7.2.1.2.9). Let $R_M = 1$. Using the above formulation, we construct a graph assigning colors using the strong edge coloring (see Fig. 7.2.1.2.10).

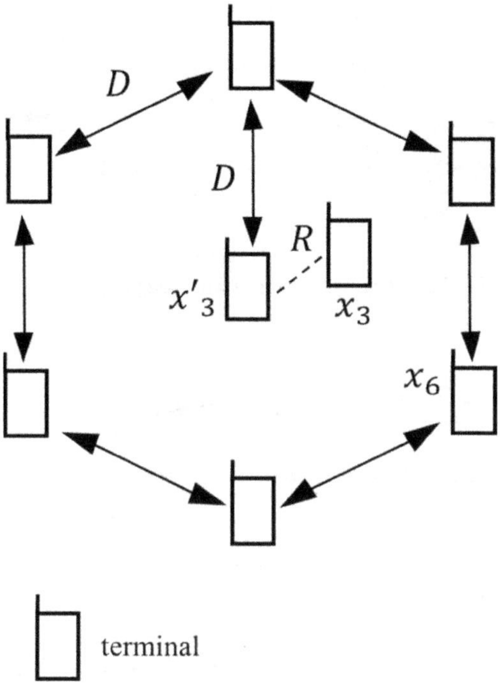

Fig. 7.2.1.2.8. The worst case on this situation.

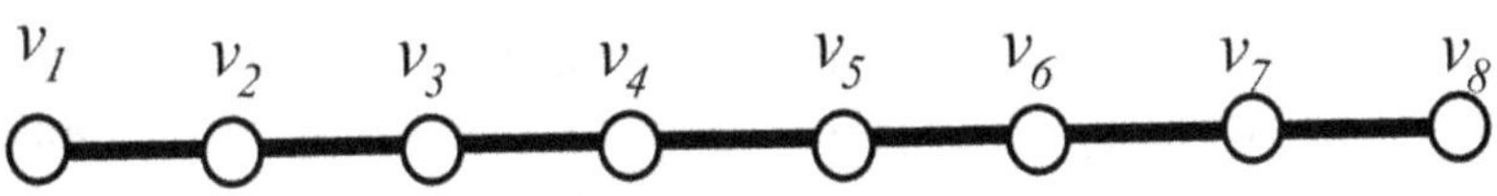

Fig. 7.2.1.2.9. A path P_8.

In this case, we need six colors. Then, we apply the CIR-edge coloring to this case. In Fig. 7.2.1.2.11, the cochannel interference in v_6 from v_1 and v_2 is 5^{-4} and 4^{-4} respectively, because the distance between v_1 and v_6 is 5 and the distance between v_2 and w is v_6. Since

$$C/I = I^{-4}/(5^{-4} + 4^{-4}) \geq 18\,\text{dB},$$

we can assign the color a to edges (v_1, v_2) and (v_6, v_7). So, there exists a coloring with only five colors. Therefore the CIR-edge coloring is effective to achieve high spectral efficiency.

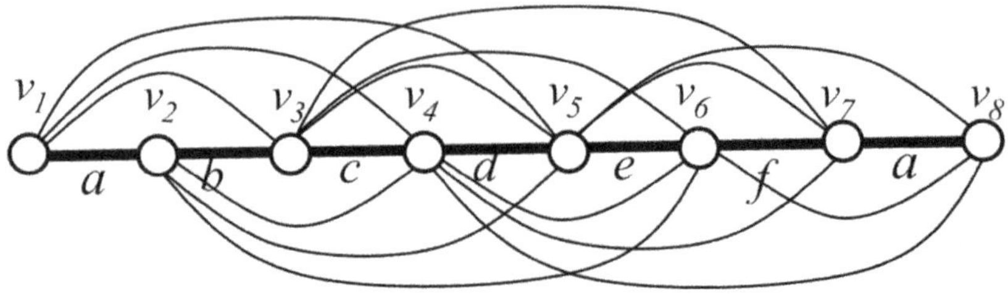

Fig. 7.2.1.2.10. A graph for a strong edge coloring.

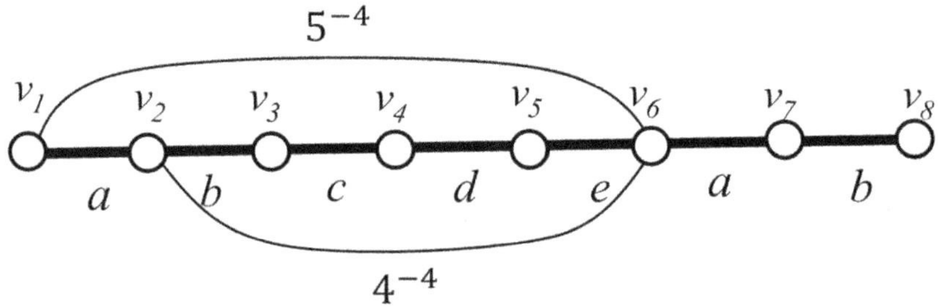

Fig. 7.2.1.2.11. Explanation of cochannel interference.

Obviously, it is easy to modify this model when the mobile radio propagation rule or the value of a are changed. Also, as described above, since the channel assignment problem in multihop wireless network has been proved to be NP-complete, it should be noted that it is an application field of approximate solution methods or neural networks (AI).

7.2.2. *Some control problems in multihop networks*

7.2.2.1. *Network coding and multihop wireless networks*

Network coding [80] is a new architecture for wireless networks, and various applications are expected with it. As a representative example, we consider the communication between vertices v_2 and v_3 in Fig. 7.2.2.1.1. In a conventional approach, four steps are needed (Fig. 7.2.2.1.2(a)). However, with the concept of network coding, we can reduce it to three steps (Fig. 7.2.2.1.2(b)) because the packet from node v_1 uses the EX-OR of the packets from v_2 and

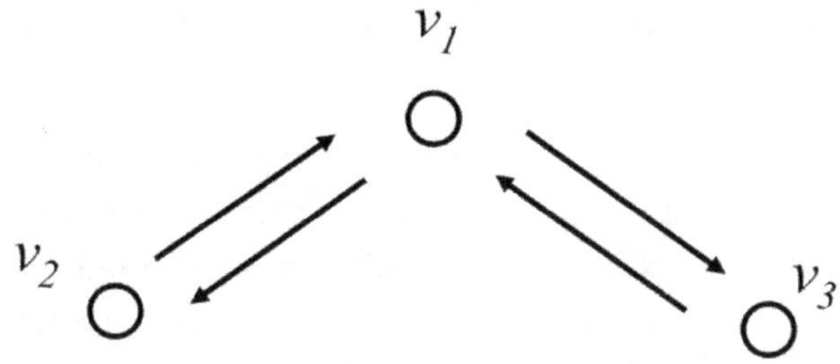

Fig. 7.2.2.1.1. Communication between vertices v_2 and v_3.

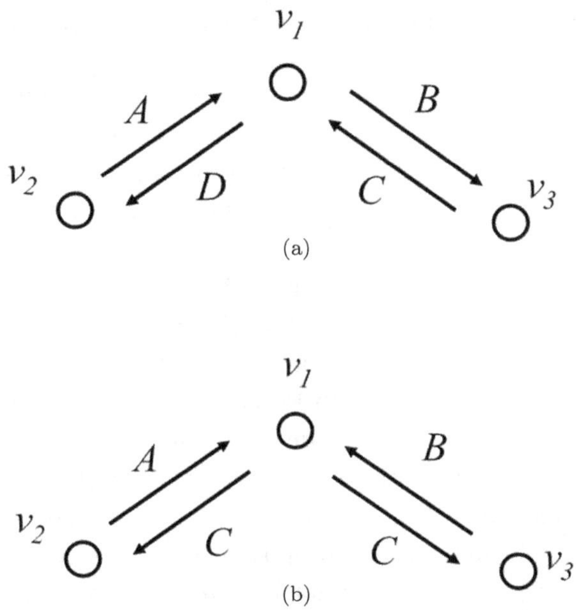

Fig. 7.2.2.1.2. Communication between vertices v_2 and v_3.

v_3, and each vertex can decode data from the received packet [81]. Efficient communications with network coding are expected, however, presently little concrete indication exists. A graph theoretical approach to this problem can be found in [82].

7.2.2.2. *Two-node-disjoint path routing in multihop wireless networks*

In a multihop wireless network, a routing protocol is executed to find at least a multihop path between a source node and a destination node. There have been many proposals for routing protocols, which are roughly classified into two kinds: reactive and proactive [92]. Reactive routing protocols only begin to make a routing table when a request for the transmission of a packet arrives at a source node. This approach reduces overhead to learn the network's topology information. Reactive routing protocols include *Dynamic Source Routing* (*DSR*) [83], *Ad hoc* On-Demand Distance Vector (AODV) [84], and so on. Proactive routing protocols prepare a routing table in advance by exchanging control messages between mobile nodes and updating the table if the network topology changes. Proactive routing protocols include Destination-Sequenced Distance-Vector (DSDV) [85].

Each of the above routing protocols constructs a multihop path between a source and a destination; however, we often encounter a situation where source and destination nodes share more than one multihop path. Furthermore, these can sometimes be disjoint paths. The introduction of disjoint paths is effective on load balancing, fault tolerance, and so on. Hence, multipath routing protocols have been proposed to construct disjoint paths in multihop networks. Split Multipath Routing (SMR) [86] is one multipath routing protocol that constructs two disjoint paths in an on-demand fashion. In SMR, the destination selects two routes: One is the shortest delay route, and the other is maximally disjoint to the first route. An efficient algorithm of finding node disjoint paths as well as an efficient algorithm of finding edge-disjoint paths with minimum total length between two nodes of a network was given in 1974 by J.W. Suurballe [87], and as a revised version of the algorithms a quick algorithm for finding shortest pairs of disjoint paths was given by J.W. Suurballe [87] and R.J. Tarjan [88] in 1984. We can apply the Suurballe–Tarjan algorithm to finding two disjoint paths in a multihop wireless network, if they exist, that minimize the total number of hops in both of the paths. One of noteworthy examples of the application of the Suurballe–Tarjan algorithm, there is a novel polynomial time algorithm that optimally solves the minimum energy 2 link-disjoint paths problem, as well as a polynomial time algorithm for the minimum energy k node-disjoint paths problem in [89].

7.3. Delay Tolerant Networks and Graphs

7.3.1. *Epidemic communications*

In a multihop wireless network, information is sent to the destination node over a multihop wireless path connected between the source and destination nodes. *Epidemic communication* and *information floating* are more flexible network technologies than multihop wireless networks. Epidemic communication provides information by spatially spreading information through direct wireless communication and the movement of mobile nodes with information. Information floating distributes information to unspecified nodes in a particular area without the chaotic spread of information by allowing direct wireless communication only in the specified area. These information distribution methods that utilize the direct exchange of information between mobile nodes and the movement of mobile nodes that carry information are being researched and developed (see [63]).

As an example, let us consider a multihop wireless network of Fig. 7.3.1.1. Figure 7.3.1.1 shows the cover area of each terminal. Each terminal moves in the service area. The information of terminal x_1 can be sent to terminal x_6, but cannot be sent to terminal x_7. Figure 7.3.1.2 shows the situation after moving where x_4 and x_6 are moving. The information sent from x_1 to x_6 by multihop and stored at x_6, is sent to x_7. This is an example of the epidemic communications. Figure 7.3.1.3(a) shows the network before moving, and Fig. 7.3.1.3(b) is the network after the move.

Epidemic communication repeatedly conveys information by distributing copies of information to surrounding mobile objects and moving them to further spread information. In other words, It spreads information spatially. Epidemic communication may lead to the disorderly diffusion of information outside the destination node and destination area, and this is a problem to be solved.

Therefore, new methods such as exploring the movement tendency of moving objects and using them to control the diffusion of information have

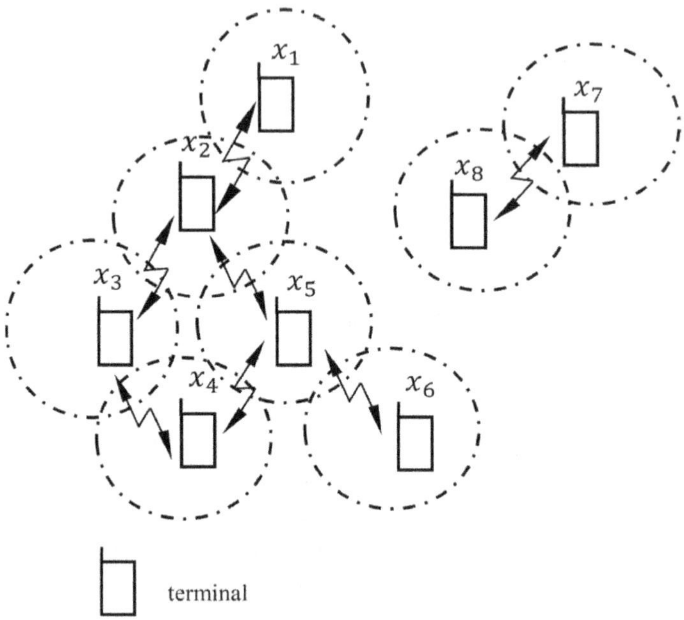

Fig. 7.3.1.1. A multihop network of Fig. 6.90 and the cover area of each terminal.

been considered. There is *information floating* as another method to directly limit the disordered diffusion. On the premise that the moving body can grasp its own position, by sending the information of the area (transmittable area) where the information may be transmitted together with the information, each moving body transmits only within the transmitting range. This is done to prevent unnecessary diffusion of information. Network problems related to information floating are not limited to the basic function, information distribution. It has been shown that information floating can not only distribute information but also accumulate spatial information. In a normal wireless sensor network, it is assumed that the information measured by the sensor is collected by a data collection device called a gateway, but information floating has a virtual data storage function without installing this gateway. It can be realized in space.

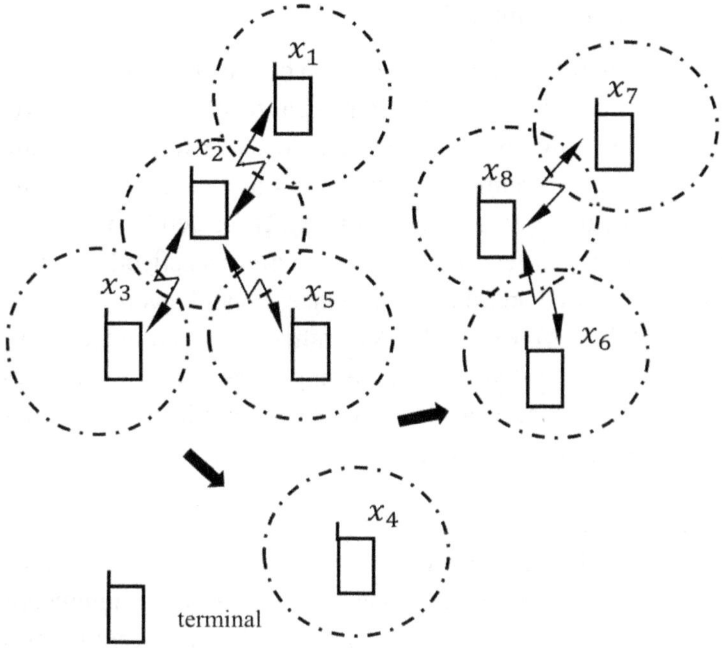

Fig. 7.3.1.2. Some terminals move and the information sent from x_1 to x_6 and stored is sent to x_7.

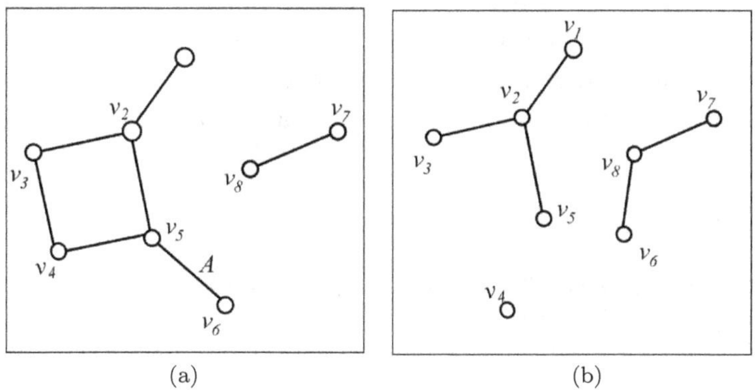

Fig. 7.3.1.3. The network (graph) models for moving terminals.

7.3.2. *Emergency communications*

Epidemic communication is considered to be effective for *emergency communication* in the event of a disaster because it can be used even when the density of moving objects is low. The idea of using the movement of a moving object to spread information in this way is an idea unique to mobile networks, as it cannot be a wired network. The essential part of epidemic communication that differs from multihop wireless networks is the idea that information is carried and then sent (hence, it can be considered as one of the *Delay Tolerant Networks* (DTN)) rather than communicating directly immediately. These systems are interesting areas related to both graph theory and probability theory, as they involve geometrical probabilities due to the involvement of movement, traffic and figures.

- **Summary:** The relationship among cellular systems and graphs is discussed. The coloring problems of graphs for channel assignment problems, generalized coloring problems in mobile communication systems and communication traffic characteristics in a cellular system are described. Note that the mobile systems include both the mobile terminals traffic and communication traffic. Multihop and epidemic communications and graphs are also described.
- **Interesting challenges:** Mobile communication is characterized by the movement of terminals. Consider a unified new model in which the graph structure changes, considering the changes in the communication environment and the movement of terminals due to the movement.

Bibliography

[1] W. C. Y. Lee, *Mobile Communications Engineering*, McGraw-Hill, 1998.

[2] M. Kuwahara, *Digital Mobile Communications*, Kagakusinbun Pub., 1992.

[3] W. C. Jakes, Jr., (Ed.), *Microwave Mobile Communications*, John Wiley & Sons, Inc., 1974.

[4] A. K. Erlang, "Solution of some problems in the theory of probabilities of significance in automatic telephone exchanges," *The Post Office Engi. J.*, vol. 10, pp. 189–197, 1917.

[5] L. Schiff, "Traffic capacity of three types of common-user mobile radio communication systems," *IEEE Trans., Comm. Tech.*, COM-18, pp. 12–21, 1970.

[6] L. Anderson "A simulation study of some dynamic channel assignment algorithms in a high capacity mobile telecommunications system," *IEEE Trans. Comm.*, vol. COM-21, no. 11, pp. 1294–1301, 1973.

[7] D. C. Cox and D. O. Reudink, "Increasing channel occupancy in large-scale mobile radio systems: Dynamic channel reassignment," *IEEE Trans. Comm.*, vol. COM-21, no. 11, pp. 1302–1306, 1973.

[8] M. Sengoku, K. Itoh and T. Matsumoto, "A dynamic frequency assignment algorithm in mobile radio communication systems," *IECE Trans.*, vol. E61, pp. 527–533, 1978.

[9] K. Appel and W. Haken, *Every Planar Map is Four Colorable*, American Mathematical Society, 1989.

[10] Deo, *Graph Theory with Applications to Engineering and Computer Science*, Prentice-Hall, 1974.

[11] G. Chartrand, L. Lesniak and P. Zhang, *Graphs & Digraphs*, Sixth Edition, Chapman and Hall/CRC, 2015.

[12] A. V. Aho, J. E. Hopcroft and J. D. Ullman, *The Design and Analysis of Computer Algorithms*, Addison-Wesley, Reading, MA, 1975.

[13] M. Sengoku, "Telephone traffic in a mobile radio communication system using dynamic frequency assignment," *IEEE Trans. Vehicular Technol.*, vol. VT-29, no. 2, pp. 270–278, 1980.

[14] M. Sengoku, M. Kurata and Y. Kajitani, "Application of rearrangement to a mobile radio communication system," *IECE Trans.*, vol. J64–B, 9, pp. 978–985, 1981.

[15] M. Sengoku, H. Kariya and T. Abe, "An approximate formulas of telephone traffic in a mobile communication system using dynamic channel assignments," *IEICE Trans.*, B J65-B, vol. no. 9, pp. 1198–1199, 1982.

[16] H. Takagi and L. Kleinrock, "Optimal transmission range for randomly distributed packet radio terminals," *IEEE Trans. Commun.*, vol. 32, no. 3, pp. 246–257, 1984.

[17] M. Sengoku, H. Izumi, A. Yoshikoshi, T. Abe, Y. Yamaguchi and T. Adachi: "Measurements of the moving of mobiles for the traffic analysis (1), (2)," *Shin'etsu Regional Conference of the IEICE*, vol. 59, no. 79. 1984, 1985.

[18] M. Sengoku, N. Minamoto, A. Ohtsuka and T. Abe, "Traffic simulations using greenshields model in a mobile communication system," Shin'etsu *Regional Conference of the IEICE*, vol. 79, 1985.

[19] M. Sengoku, W. Kudo and T. Abe, "Adaptive channel assignments in a small cell mobile radio communication system," *Proceedings of The Eleventh International Teletraffic Congress*, North-Holland, vol. 1, pp. 2311–2317, 1985.

[20] M. Sengoku, "Efficient utilization of frequency spectrum for mobile communication systems -algorithms for channel assignments-," *Proc. IEICE*, vol. 69, no. 4, pp. 350–356, 1986.

[21] A. Ohtsuka, M. Sengoku, Y. Yamaguchi and T. Abe, "Basic study on mobile flows and mobile telephone traffic," *IEICE Tech. Rep.*, vol. CAS-86-249, 1986.

[22] L. Kleinrock and J. Silvester, "Spatial reuse in multihop packet radio networks", *Proc. IEEE*, vol. 75, no. 1, pp. 156–167, 1987.

[23] M. Sengoku, H. Tamura, S. Shinoda, T. Abe and Y. Kajitani, "Graph theoretical considerations of channel offset systems in a cellular mobile system," *IEEE Trans. Vehicular Technology*, vol. 40, no. 2, pp. 405–412, 1991.

[24] H. Tamura, M. Sengoku, S. Shinoda and T. Abe, "Channel assignment problem in a cellular mobile system and a new coloring problem of networks," *IEICE Trans.*, vol. E74, no. 10, pp. 2983–2989, 1991.

[25] T. Kabasawa, T. Watanabe, M. Sengoku, Y. Yamaguchi, S. Shinoda and T. Abe, "Transient characteristics of mobile communication traffic in a band-shaped service area," *IEICE Trans.*, vol. E76-A, no. 6, pp. 961–966, 1993.

[26] K. Shimada, K. Nakano, M. Sengoku and T. Abe, "An approach to dynamic channel assignment in a cellular mobile communication system using a neural network," *IEICE Trans.*, vol. E77-A, no. 6, pp. 985–992, 1994.

[27] M. Sengoku, H. Tamura, S. Shinoda and T. Abe, "Development in graph-and/or network-theoretic research of cellular mobile communication channel assignment problems," *IEICE Trans.*, vol. E77-A, no. 7, pp. 1117–1124, 1994.

[28] K. Nakano, M. Sengoku, S. Shinoda and T. Abe, "Clique packing in cellular mobile communication systems," *IEEE 1995 Inte. Conf. Universal Personal Communications (ICUPC '95) Records*, pp. 71–75, 1995.

[29] K. Nakano, N. Karasawa, M. Sengoku, S. Shinoda and T. Abe, "Characteristics of dynamic channel assignment in cellular systems with reuse partitioning," *IEICE Trans.*, vol. E79-A, No. 7, pp. 983–989, 1996.

[30] H. Tamura, M. Sengoku and S. Shinoda, "New applications of computational geometry and graph theory to cellular mobile communications," *Proc. IEEE Asia Pacific Conference on Circuits and Systems (APCCAS '96)*, pp. 199–202, Nov. 1996.

[31] G. Montenegro, M. Sengoku, Y. Yamaguchi and T. Abe, "Time-dependent analysis of mobile communication traffic in a ring-shaped service area with nonuniform vehicle distribution", *IEEE Trans. Vehicular Technol.*, vol. 41, no. 3, pp. 243–254, 1992.

[32] T. S. Kim, M. Y. Chung, D. K. Sung and M. Sengoku: "Mobility Modeling and Traffic Analysis in Three-Dimensional Indoor Environments," *IEEE Trans. Vehicular Technol.*, vol. 47, no. 2, pp. 546–557, 1998.

[33] K. Imai, "Voronoi diagram for moving points and its application," *Trans. Japan Society Indust. Appli. Math.*, vol. 1, no. 2, pp. 127–134, 1991.

[34] F. P. Preparata and M. I. Shamos, *Computational Geometry*, Springer-Verlag, 1985.

[35] K. Nakano, K. Saita, M. Sengoku, Y. Yamada and S. Shinoda, "Mobile communication traffic analysis on a road systems model," *Performance and Management of Complex Communication Networks, Published by International Federation for Information Processing (IFIP)*, Chapman & Hall, pp. 1–20, 1998.

[36] S. Watanabe, T. Sato, M. Sengoku and T. Abe, "Two dimensional equalization scheme of orthogonal coding multi-carrier CDMA," *IEICE Trans.*, vol. E81-A, no. 6, pp. 1079–1088, 1998.

[37] H. Tamura, M. Sengoku, K. Nakano and S. Shinoda, "Graph theoretic or computational geometric research of cellular mobile communications," *Proc. 1999 IEEE Inte. Symp. Circuits and Systems (ISCAS '99)*, (May 1999).

[38] K. Nakano, M. Sengoku and S. Shinoda, "Characteristics of multi-hop radio networks," *2000 IEEE 51st Vehicular Technol. Confe. (VTC2000-Spring) Proc.*, vol. 2, pp. 1328–1332, 2000.

[39] N. Shinagawa, T. Kobayashi, K. Nakano and M. Sengoku, "Evaluation of teletraffic in cellular communication systems using multi-connections for soft handoff," *IEICE Trans.*, vol. E83-A, no. 7, pp. 1318–1327, July 2000.

[40] K. Nakano, M. Sengoku and S. Shinoda, "Fundamental characteristics of multi-hop wireless communication networks", *Telecommunication Systems, Baltzer Science Publishers*, vol. 15, nos. 1, 2, pp. 79–91, 2000.

[41] K. Mase, M. Sengoku and S. Shinoda, "A perspective to next-generation ad hoc networks – a proposal for an open community network-," *IEICE Trans.*, vol. E84-A, no. 1, pp. 98–106, 2001.

[42] H. S. Lee, N. Karasawa, K. Nakano and M. Sengoku, "Clique packing approximation for analysis of teletraffic characteristics of dynamic channel assignment considering mobility," *IEICE Trans.*, vol. E84-A, no. 7, pp. 1651–1659, 2001.

[43] H. S. Lee, K. Mase, T. Adachi, T. Oosawa, K, Nakano, M. Sengoku, H. Hidaka, N. Shinagawa and T. Kobayashi, "Pedestrian tracking using GPS, pedometer and magnetic compass", *IEICE Trans.*, vol. J84-B, no.12, pp. 2254–2263, 2001.

[44] K. Mase, Y. Wada, N. Mori, K. Nakano and M. Sengoku, "Flooding Schemes for Clustered Ad Hoc Networks", *IEICE Trans.*, vol. E85-B, no. 3, pp. 605–613, 2002.

[45] A. Tanaka, K. Nakano, M. Sengoku and S. Shinoda, "Analysis of communication traffic characteristics of a two-hop wireless network," *IEICE Trans.*, vol. E85-A, no. 7, pp. 1436–1444, 2002.

[46] K. Nakano, Y. Shirai, M. Sengoku and S. Shinoda, "On connectivity and mobility in mobile multi-hop wireless networks," *IEEE Vehicular Technol. Conf.* (*VTC2003-Spring*) *Proc.*, 2003.

[47] H. Tamura, K. Watanabe, M. Sengoku and S. Shinoda, "A channel assignment problem in multihop wireless networks and graph theory," *J. Circuits. Syst. Comput.*, vol. 13, no. 2, pp. 375–385, 2004.

[48] F. Tasaki, F. Uta, H. Tamura, M. Sengoku and S. Shinoda, "Effect of a new channel assignment strategy on multihop wireless networks," *IEICE Trans.*, vol. E87-B, no. 5, pp. 1095–1103, 2004.

[49] K. Nakano, R. Panta, S. Narita, M. Sengoku, N. Umeda and S. Shinoda, "Multi-hop networking with relaying capacity for cellular mobile systems," *IEEE Vehicular Technol. Confe.* (*VTC2004-Spring*) *Proce.*, 2004.

[50] R. Panta, K. Nakano, M. Sengoku, N. Umeda and S. Shinoda, "Analytic estimation of cell extension with finite relaying capacity of relay nodes," *Proc. 2004 47th IEEE Midwest Symp. Circuits and Systems* (*MWSCAS 2004*), 2004.

[51] S. Maruyama, K. Nakano, K. Meguro, M. Sengoku and S. Shinoda, "On location of relay facilities to improve connectivity of multi-hop wireless networks," *Proc. 2004 Joint Conf. 10th Asia-Pacific Conf. Communications and 5th Int. Symp. Multi-Dimensional Mobile Communications* (*APCC/MDMC '04*), 2004.

[52] K. Nakano, R. Panta, M. Sengoku and S. Shinoda, "On performance of a charging/rewarding scheme in mobile ad-hoc networks," *IEEE Int. Symp. Circuits and Systems* (*ISCAS 2005*), 2005.

[53] J. P. Park, K. Nakano, A. Kamakura, Y. Kakizaki, M. Sengoku and S. Shinoda, "On spreading of computer viruses in mobile ad hoc networks," *Systems Modeling and Simulation, Theory and Applications, Asia Simulation Conference 2006*, Springer, 2006.

[54] K. Nakano, K. Miyakita, M. Sengoku and S. Shinoda, "Analysis and relative evaluation of connectivity of a mobile multi-hop network," *IEICE Trans.*, vol. E91-B, no. 6, pp. 1874–1885, 2008.

[55] K. Miyakita, K. Nakano, M. Sengoku and S. Shinoda, "Theoretical analysis of route expected transmission count in multi-hop wireless networks," *IEICE Trans.*, vol. E91-B, no. 8, pp. 2533–2544, 2008.

[56] K. Miyakita, K. Nakano, Y. Morioka, M. Sengoku and S. Shinoda, "Characterization of minimum route ETX in multi-hop wireless networks," *IEICE Trans.*, vol. E92-B, no. 3, pp. 745–754, 2009.

[57] K. Miyakita, K. Nakano, M. Sengoku and S. Shinoda, "Characterization of Minimum Route MTM in One-dimensional Multi-hop Wireless Networks", *IEICE Trans.* vol. E92-A, no. 9, pp. 2227–2235, 2009.

[58] K. Nakano, K. Miyakita, A. Otsuka, M. Sengoku and S. Shinoda, "Analysis of mean waiting time for delivery of a message in mobile multi-hop networks," *IEICE Trans.* vol. E92-A, no. 9, pp. 2236–2247, 2009.

[59] M. Tsuge, M. Tokunaga, K. Nakano and M. Sengoku, "On estimation of link travel time in floating car systems," *Inte. J. Intelligent Transportation Systems Research*, vol. 8, no. 3, pp. 175–187 (Oct. 2010).

[60] A. Urabe, K. Miyakita, K. Nakano, M. Sengoku and S. Shinoda, "On relation between limitation of transmission and spread of information in epidemic transmission," *IEICE Trans.*, vol. J94-B, no. 2, pp. 122–136, 2011.

[61] K. Nakano, K. Miyakita, M. Sengoku and S. Shinoda, "Modeling of a charging network for electric vehicles," *First Int. Workshop on Smart Grid Modeling and Simulation Organized in the 2nd IEEE Int. Conf. Smart Grid Communications*, pp. 19–24, 2011.

[62] Y. P. Kim, K. Nakano, K. Miyakita, M. Sengoku and Y. J. Park, "A routing protocol for considering the time variant mobility model in delay tolerant network," *IEICE Trans.*, vol. E95-D, no. 2, pp. 451–461, 2012.

[63] K. Nakano, "Epidemic communication, information floating and safety/security" *IEICE Fundamental Review*, vol. 10–4, pp. 282–292, 2016.

[64] M. Takano, A. Kashiwabara, H. Tamura, K. Nakano and M. Sengoku, "A consideration of method of information sources moving with covering problem for distribution of regional information," *J. Advanced Simulation Science Eng.* (*JASSE*), vol. 11, no. 2, pp. 15–22, 2019.

[65] S. Matsumoto, H. Tamura, K. Nakano and M. Sengoku, "On a coloring to matching of graphs and channel assignment on multi-hop wireless networks," *IEICE Trans.*, J102-A, no. 6, pp. 194–200, 2019.

[66] H. Tamura, S. Matsumoto, K. Nakano and M. Sengoku, "On graph coloring with relation to channel assignment in wireless networks," *IEICE Trans.*, vol. J102-A, no.7, pp. 214–223, 2019.

[67] ITU-R, "Monte Carlo simulation methodology for the use in sharing and compatibility studies between different radio services or systems," *Report ITU-R* SM. 2028–2, 2017.

[68] Federal Communications Commissions, "Wireless telecommunications bureau and office of engineering and technology approve five spectrum access system administrators to begin initial commercial deployments in the 3.5 GHz band," 2019.

[69] H. Harada, "Future prospects of dynamic spectrum sharing technology," *Proc. IEICE*, vol. 104, no. 12, pp. 1245–1250, 2021.

[70] R. J. Faudree, A. Gyarfas, R. H. Schelp and Zs. Tuza, "The chromatic index of graphs," *Ars Combinatoria.*, vol. 29B, pp. 205–211, 1990.

[71] M. Mahdian, "On the computational complexity of strong edge coloring," *Discrete Appl. Math.*, vol. 118, pp. 239–248, 2002.

[72] K. Suda, T. Furuno, T. Taga and T. Tanaka, "Radio link design for high-speed wireless LAN systems," *IEICE Trans.*, vol. J84-B (11), pp. 2123–2132, 1999.

[73] M. R. Garay and D. S. Johnson, *Computers and Intractability*, Freeman, 1979.

[74] R. I. Wilkinson, "Theories for toll traffic engineering in the U.S.A.," *Bell Syst. Tech. J.*, vol. 35, pp. 421–514, 1956.

[75] M. Akiyama, *Modern Communication Switching Engineering*, Denkishoin Pub., 1972.

[76] P. Raymond, "Performance analysis of cellular networks," *IEEE Trans. Commun.* vol. 39, no. 12, pp. 1787–1793, 1991.

[77] W. C. Y. Lee, "New cellular schemes for spectral efficiency, " *IEEE Trans. Veh. Technol.*, vol. VT-36, pp. 188–192, 1987.

[78] Ed. N. Gartner, C. J. Messer and A. K. Rathi, *Traffic Flow Theory*, Organized by the Committee on Traffic Flow Theory and Characteristics (AHB45), 2001.

[79] K. Cameron, "Induced matchings, *Discrete Appl. Math.*, 24, pp. 97–102. 1989.

[80] R. Ahlswede, S. Li and R. Yeung, "Network information flow," *IEEE Trans. Inf. Theory*, vol. 46, no. 4, pp. 1404–1216, 2000.

[81] S. Katti, H. Rahul, W. Hu, D. Katabi, M. Medard and J. Crowcroft, " XOR's in the air: Practical wireless network coding," *Proc. ACM SIGCOMM 2006*, pp. 243–254, 2006.

[82] H. Tamura, Y. Shimizu, M. Sengoku and S. Shinoda, "The shortest step algorithm with network coding on multihop wireless cyclic networks," *Int. Technical Conf. Circuits/Systems, Computers and Communications* (*ITC-CSCC2009*), pp. 58–61, 2009.

[83] D. B. Johnson, "Routing in ad hoc networks of mobile hosts," *Proc. Workshop on Mobile Computing Systems and Applications*, pp. 158–163, 1994.

[84] C. E. Perkins and E. M. Royer. "*Ad hoc* on demand distance vector routing," *Proc. IEEE Workshop on Mobile Computing Systems and Applications*, pp. 90–100, 1999.

[85] C. E. Perkins and P. Bhagwat, "Highly dynamic destination- sequenced distance-vector routing (DSDV) for mobile computers," *ACM SIGCOMM Computer Communication Review*, vol. 24, no. 4, pp. 234–244, 1994.

[86] S. L. Lee and M. Gerla, "Split multipath routing with maximally disjoint paths in ad hoc networks," *IEEE International Conf. Commun.*, no. 1, pp. 3201–3205, 2001.

[87] J. W. Suurballe, "Disjoint paths in a network," *Networks*, vol. 4, pp. 125–145, 1974.

[88] J. W. Suurballe and R. E. Tarjan, "A quick method for finding shortest pairs of disjoint paths," *Networks*, vol. 14, pp. 325–336, 1984.

[89] A. Srinivas and E. Modiano, "Finding minimum energy disjoint paths in wireless ad-hoc networks," *Wireless Networks*, vol. 11, pp. 401–417, 2005.

[90] H. A. Shah, K. S. Kwak, M. Sengoku and S. Shinoda, "Reliable cooperative spectrum sensing through multi-bit quantization with presence of multiple primary users in cognitive radio networks," *Int. Technical Conf. Circuits/Systems, Computers and Communications* (*ITC-CSCC 2019*), pp. 14–16, 2019.

[91] H. Tamura, K. Nakano and M. Sengoku, "On a Grundy coloring divided into two steps in graphs," *Int. Technical Conf. Circuits/Systems, Computers and Communications* (*ITC-CSCC 2020*), pp. 145–148, 2020.
[92] C. E. Perkins, *Ad Hoc Networking*, Addison-Wesley, 2001.
[93] M. Sengoku, "Big data, open data and open science," *IEICE Tech. Rep.*, ICTSSL-2016-75, 2017.
[94] H. Tamura, K. Nakano and M. Sengoku, *Network Engineering*, IEICE, Corona Pub., 2020.

Index

D

CPSIA information can be obtained
at www.ICGtesting.com
Printed in the USA
JSHW020800090323
38455JS00006B/70

9 789811 255298